Introduction to Bioinformatics

SECOND EDITION

Arthur M. Lesk
The Pennsylvania State University

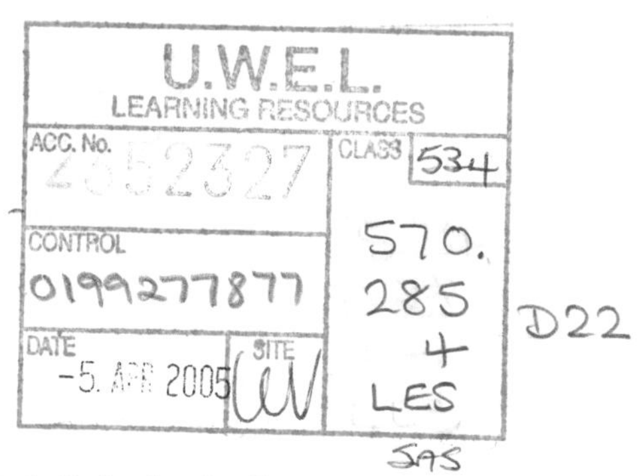

In nature's infinite book of secrecy
A little I can read.
– *Antony and Cleopatra*

OXFORD
UNIVERSITY PRESS

OXFORD
UNIVERSITY PRESS

Great Clarendon Street, Oxford OX2 6DP

Oxford University Press is a department of the University of Oxford.
It furthers the University's objective of excellence in research, scholarship,
and education by publishing worldwide in

Oxford New York

Auckland Cape Town Dar es Salaam Hong Kong Karachi
Kuala Lumpur Madrid Melbourne Mexico City Nairobi
New Delhi Shanghai Taipei Toronto

With offices in

Argentina Austria Brazil Chile Czech Republic France Greece
Guatemala Hungary Italy Japan South Korea Poland Portugal
Singapore Switzerland Thailand Turkey Ukraine Vietnam

Published in the United States
by Oxford University Press Inc., New York

First published 2005

British Library Cataloguing in Publication Data

Data available

Library of Congress Cataloging in Publication Data

Data available

ISBN 0 19 9277877 (hbk)

10 9 8 7 6 5 4 3 2 1

Typeset by Newgen Imaging Systems (P) Ltd., Chennai, India
Printed in Great Britain
on acid-free paper by Ashford Colour Press Limited, Gosport, Hampshire

Dedicated to Eda, with whom I have merged my genes.

Preface

On 26 June 2000, the sciences of biology and medicine changed forever. Prime Minister of the United Kingdom, Tony Blair, and President of the United States, Bill Clinton, held a joint press conference, linked via satellite, to announce the completion of the draft of the Human Genome. *The New York Times* ran a banner headline: 'Genetic Code of Human Life is Cracked by Scientists'. The sequence of three billion bases was the culmination of over a decade of work, during which the goal was always clearly in sight and the only questions were how fast the technology could progress and how generously the funding would flow. The Box shows some of the landmarks along the way.

Next to the politicians stood the scientists. John Sulston, Director of The Sanger Centre in the UK, had been a key player since the beginning of high-throughput sequencing methods. He had grown with the project from the earliest 'one man and a dog' stages to the current international consortium. In the US, appearing with President Clinton were Francis Collins, director of the US National Human Genome Research Institute, representing the US publicly-funded efforts; and J. Craig Venter, President and Chief Scientific Officer of Celera Genomics Corporation, representing the commercial sector. It is difficult to introduce these two without thinking, 'In this corner . . . and in this corner . . . ' Although never actually coming to blows, there was certainly intense competition, in the later stages a race.

The race was more than an effort to finish first and receive scientific credit for priority. Indeed, it was a race after which the contestants would be tested not for whether they had taken drugs, but whether they and others could discover them. Clinical applications were a prime motive for support of the human genome project. Once the courts had held that gene sequences were patentable—with enormous potential payoffs for drugs based on them—the commercial sector rushed to submit patent applications on sets of sequences that they determined, and the academic groups rushed to place each bit of sequence that *they* determined into the public domain to *prevent* Celera—or anyone else—from applying for patents.

The academic groups lined up against Celera were a collaborating group of laboratories primarily but not exclusively in the UK and USA. These included The Sanger Centre in England, Washington University in St. Louis, Missouri, the Whitehead Institute at the Massachusetts Institute of Technology in Cambridge, Massachusetts, Baylor College of Medicine in Houston, Texas, the Joint Genome Institute at Lawrence Livermore National Laboratory in Livermore, California, and the RIKEN Genomic Sciences Center, now in Yokahama, Japan.

Both sides could dip into deep pockets. Celera had its original venture capitalists; its current parent company, PE Corporation; and, after going public, anyone who cared to take a flutter. The Sanger Centre was supported by the UK Medical

Research Council and The Wellcome Trust. The US academic labs were supported by the US National Institutes of Health and Department of Energy.

On 26 June 2000 the contestants agreed to declare the race a tie, or at least a carefully out-of-focus photo finish.

Landmarks in the Human Genome Project

1953	Watson–Crick structure of DNA published.
1975	F. Sanger, and independently A. Maxam and W. Gilbert, develop methods for sequencing DNA.
1977	Bacteriophage ΦX-174 sequenced: first 'complete genome'.
1980	US Supreme Court holds that genetically-modified bacteria are patentable. This decision was the original basis for patenting of genes.
1981	Human mitochondrial DNA sequenced: 16 569 base pairs.
1984	Epstein-Barr virus genome sequenced: 172 281 base pairs.
1990	International Human Genome Project launched—target horizon 15 years.
1991	J. Craig Venter and colleagues identify active genes via Expressed Sequence Tags—sequences of initial portions of DNA complementary to messenger RNA.
1992	Complete low resolution linkage map of the human genome.
1992	Beginning of the *Caenorhabditis elegans* sequencing project.
1992	Wellcome Trust and United Kingdom Medical Research Council establish The Sanger Centre for large-scale genomic sequencing, directed by J. Sulston.
1992	J. Craig Venter forms The Institute for Genome Research (TIGR), associated with plans to exploit sequencing commercially through gene identification and drug discovery.
1995	First complete sequence of a bacterial genome, *Haemophilus influenzae*, by TIGR.
1996	High-resolution map of human genome—markers spaced by ~600 000 base pairs.
1996	Completion of yeast genome, first eukaryotic genome sequence.
May 1998	Celera claims to be able to finish human genome by 2001. Wellcome responds by increasing funding to Sanger Centre.
1998	*Caenorhabditis elegans* genome sequence published.
1 September 1999	*Drosophila melanogaster* genome sequence announced, by Celera Genomics; released Spring 2000.
1999	Human Genome Project states goal: working draft of human genome by 2001 (90% of genes sequenced to >95% accuracy).
1 December 1999	Sequence of first complete human chromosome published.
26 June 2000	Joint announcement of complete draft sequence of human genome.
2003	Fiftieth anniversary of discovery of the structure of DNA. Completion of high-quality human genome sequence by public consortium.

The human genome is only one of the many complete genome sequences known. Taken together, genome sequences from organisms distributed widely among the branches of the tree of life give us a sense, only hinted at before, of the very great unity *in detail* of all life on Earth. They have changed our perceptions, much as the first pictures of the Earth from space engendered a unified view of our planet.

The sequencing of the human genome ranks with the Manhattan project that produced atomic weapons during the Second World War, and the space program that sent people to the Moon, as one of the great bursts of technological achievement of the last century. These projects share a grounding in fundamental science, and large-scale and expensive engineering development and support. For biology, neither the attitudes nor the budgets will ever be the same. Soon a 'one man and a dog project' will refer only to an afternoon's undergraduate practical experiment in sequencing and comparison of two mammalian genomes.

The human genome is fundamentally about information, and computers were essential both for the determination of the sequence and for the applications to biology and medicine that are already flowing from it. Computing contributed not only the raw capacity for processing and storage of data, but also the mathematically-sophisticated methods required to achieve the results. The marriage of biology and computer science has created a new field called bioinformatics.

Today bioinformatics is an applied science. We use computer programs to make inferences from the data archives of modern molecular biology, to make connections among them, and to derive useful and interesting predictions.

This book is aimed at students and practising scientists who need to know how to access the data archives of genomes and proteins, the tools that have been developed to work with these archives, and the kinds of questions that these data and tools can answer. In fact, there are a lot of sources of this information. Sites treating topics in bioinformatics are sprawled out all over the Web. The challenge is to select an essential core of this material and to describe it clearly and coherently, at an introductory level.

It is assumed that the reader already has some knowledge of modern molecular biology, and some facility at using a computer. The purpose of this book is to build on and develop this background. It is suitable as a textbook for advanced undergraduates or beginning postgraduate students. Many worked-out examples are integrated into the text, and references to useful web sites and recommended reading are provided.

Problems test and consolidate understanding, provide opportunities to practise skills, and explore additional subjects. Three types of problems appear at the ends of chapters. Exercises are short and straightforward applications of material in the text. Answers to exercises appear at the end of the book. Problems also require no information not contained in the text, but require lengthier answers or in some cases calculations. The third category, 'Web–lems', require access to the World Wide Web. Weblems are designed to give readers practice with the tools required for further study and research in the field.

What has made it possible to try to write such a book now is the extent to which the World Wide Web has made easily accessible both the archives themselves and the programs that deal with them. In the past, it was necessary to install programs and data on one's own system, and run calculations locally. Of course this meant that everything was dependent on the facilities available. Now it is possible to channel all the work through an interface to the Web. The web site linked with this book will ease the transition (see inside front cover). To ensure that readers will be able freely to pursue discussions in the book onto the Web, descriptions of and references to commercial software have been avoided, although many commercial packages are of very high quality.

A serious problem with the Web is its volatility. Sites come and go, leaving trails of dead links in their wake. There are so many sites that it is necessary to try to find a few gateways that are stable—not only continuing to exist but also kept up to date in both their contents and links. I have suggested some such sites, but many others are just as good. The problem is not to create a long list of useful sites—this has been done many times, and is relatively easy—but to create a short one—this is much harder!

Computing is introduced in this book based on the widely available language PERL. Examples of simple PERL programs appear in the context of biological problems. Many simple PERL tasks are assigned as exercises or problems at the ends of the chapters.

My goal is that readers of this book will emerge with:

- An appreciation of the nature of the very large amount of detailed information about ourselves and other species that has become available.
- A sense of the range of applications of bioinformatics to molecular biology, clinical medicine, pharmacology, biotechnology, forensic science, anthropology and other disciplines.
- A useful knowledge of the techniques by which, through the World Wide Web, we gain access to the data and the methods for their analysis.
- An appreciation of the role of computers and computer science in the investigations and applications of the data.
- Confidence in the reader's basic skills in information retrieval, and calculations with the data, and in the ability to extend these skills by self-directed 'field work' on the Web.
- A sense of optimism that the data and methods of bioinformatics will create profound advances in our understanding of life, and improvements in the health of humans and other living things.

Where might the reader turn next? This book is designed as a companion volume—in current parlance, a 'prequel'—to *Introduction to Protein Architecture: The Structural Biology of Proteins* (Oxford University Press, 2000). *Introduction to Protein Science: Architecture, Function, and Genomics* (Oxford University Press, 2004) attempts a broader synthesis of that field. Of course those titles are recommended. Other books on sequence analysis range from those oriented towards biology to others in the field of computer science. The goal is that each reader will come to recognize his or her own interests, and be equipped to follow them up.

Preface to Second Edition

Bioinformatics has grown since the first edition of this book appeared.

The most striking change has been a refocus on integration; that is, on trying to see life processes as unified systems. I wrote at the end of *Introduction to Protein Science: Architecture, Function and Genomics*, 'During the last century, molecular biologists have been taking living things apart. Our task now is to put them back together.' We have had large amounts of data. Now we are trying to see how they interrelate. At the heart of life processes are complicated patterns of interaction among the components, in space and in time. To understand these patterns, the field has moved towards combining information into networks, and trying to understand their structures and dynamics.

Supporting this venture are the growing streams of data. The human genome, available in draft form when the first edition appeared, is now complete. It is joined by the complete genomes of 20 archaea, 194 bacteria, over 36 eukarya, and many other organelle and viral sequences. These genomes illuminate one another. One story that they tell is about unsuspected underlying unities of all living things, despite the obvious and profound differences in morphology and lifestyle.

Genomic sequences are supplemented by other data streams, notably the proteome. Patterns of gene expression, and networks of regulatory interactions show how cells and organisms implement the information in the DNA. The potential for the life of an organism is contained in its genome, but it would be impossible to deduce a biography from it. Genomes are not formulas or scripts. It is in the proteins, and their interactions with themselves and with DNA, that we must seek the set of activities, contingent on and responsive to the environment. Proteomics is giving us the information we need to see how the system works.

Research and applications require that the data be available in useful form. It is not enough to make the data public. The information must be subjected to quality control, annotation, and a logical structure must be imposed on it to make information retrieval possible. For this we are indebted to the institutions that archive, curate, organize and distribute the data. A recent trend has seen mergers of these groups into collaborative projects spanning the continents. In accord with the need to integrate the study of different types of data, we are moving in the direction of a single biological data repository. Individual scientists will be able to define 'virtual databanks' tailoring access to the information to suit particular needs and interests.

A gratifying consequence of academic bioinformatics is its contributions to applications in medicine, agriculture, and technology. A better understanding of life processes empowers us to deal with them when they go wrong.

Acknowledgements

I am grateful to many colleagues for discussions and advice during the preparation of this book, to the universities of Uppsala, Umeå, Rome 'Tor Vergata' and Cambridge for the opportunity to try out this material, and to the Institut des Haute Études Scientifiques, Bures-sur-Yvette, for hospitality during the writing of part of this book.

I thank D. J. Abraham, S. Aparicio, M. M. Babu, T. Baglin, D. Baker, A. Bench, G. Blatch, M. Brand, A. Brazma, A. Buckle, C. Cantor, R. W. Carrell, C. Chothia, A. Coulson, D. Crowther, T. Dafforn, R. Foley, A. Friday, M. B. Gerstein, T. Gibson, J. Irving, B. Jorden, J. Karn, K. Karplus, P. Klappa, E. V. Koonin, M. Krichevsky, P. Lawrence, A. Lister, E. L. Lesk, M. E. Lesk, V. E. Lesk, V. I. Lesk, M. Levitt, D. Liberles, O. Lichtarge, L. Lo Conte, D. A. Lomas, K. Lutfy, G. Magklaras, J. Magré, A. D. MacKerell, Jr., M. McFall-Ngai, J. McInerney, P. Miller, C. Mitchell, J. Moult, E. Nacheva, C. Notredame, H. Parfrey, D. Parkinson, A. Pastore, D. Penny, J. Pettitt, C. A. Praul, F. W. Roberts, G. D. Rose, P. B. Rosenthal, B. Rost, E. J. Simon, J. Sulston, M. Segal, O. Skovegaard, E. L. Sonnhammer, R. Srinivasan, R. Staden, I. Tickle, A. Tramontano, A. A. Travers, A. R. Venkitaraman, P. L. Welcsh, G. Vriend, J. C. Whisstock, M. Widersten, A. S. Wilkins, S. H. White, and E. B. Ziff for advice and critical reading.

I thank the staff of Oxford University Press for their skills and patience in producing this book.

A. M. L.
Cambridge, December 2004

Contents

Plan of the book *xix*

1 Introduction *1*

Life in space and time *3*

Evolution is the change over time in the world of living things *4*

Dogmas: central and peripheral *6*

Observables and data archives *9*

Information flow in bioinformatics *12*

Curation, annotation, and quality control *13*

The World Wide Web *14*

Electronic publication *15*

Computers and computer science *16*

Programming *17*

Biological classification and nomenclature *21*

Use of sequences to determine phylogenetic relationships *24*

Use of SINES and LINES to derive phylogenetic relationships *30*

Searching for similar sequences in databases: PSI-BLAST *32*

Introduction to protein structure *40*

The hierarchical nature of protein architecture *41*

Classification of protein structures *44*

Protein structure prediction and engineering *51*

Critical Assessment of Structure Prediction (CASP) *52*

Protein engineering *52*

Proteomics *52*

DNA microarrays *53*

Mass spectrometry *54*

Systems biology *54*

Clinical implications *55*

The future *57*

Recommended reading 57

Exercises, Problems, and Weblems 59

2 Genome organization and evolution *67*

Genomes and proteomes *68*

Genes *69*

Proteomes *71*

Eavesdropping on the transmission of genetic information *72*
Mappings between the maps *77*
High-resolution maps *78*

Picking out genes in genomes *80*

Genomes of prokaryotes *81*
The genome of the bacterium *Escherichia coli* *82*
The genome of the archaeon *Methanococcus jannaschii* *85*
The genome of one of the simplest organisms: *Mycoplasma genitalium* *86*

Genomes of eukaryotes *87*
The genome of *Saccharomyces cerevisiae* (baker's yeast) *89*
The genome of *Caenorhabditis elegans* *93*
The genome of *Drosophila melanogaster* *94*
The genome of *Arabidopsis thaliana* *95*

The genome of *Homo sapiens* (the human genome) *96*
Protein coding genes *97*
Repeat sequences *99*
RNA *100*

Single-nucleotide polymorphisms (SNPs) *101*

Genetic diversity in anthropology *102*
Genetic diversity and personal identification *103*
Genetic analysis of cattle domestication *104*

Evolution of genomes *104*
Please pass the genes: horizontal gene transfer *108*
Comparative genomics of eukaryotes *109*

Recommended reading *111*

Exercises, Problems, and Weblems *112*

3 Archives and information retrieval *117*

Introduction *118*
Database indexing and specification of search terms *118*
Follow-up questions *120*
Analysis of retrieved data *121*

The archives *121*
Nucleic acid sequence databases *122*
Genome databases *124*
Protein sequence databases *124*
Databases of structures *128*
Specialized, or 'boutique' databases *135*
Expression and proteomics databases *136*
Databases of metabolic pathways *138*
Bibliographic databases *139*
Surveys of molecular biology databases and servers *139*

Gateways to archives *140*
Access to databases in molecular biology *141*

ENTREZ *141*
The Sequence Retrieval System (SRS) *148*
The Protein Identification Resource (PIR) *149*
ExPASy—Expert Protein Analysis System *150*
Ensembl *151*

Where do we go from here? *152*

Recommended reading 152

Exercises, Problems, and Weblems 153

4 Alignments and phylogenetic trees *157*

Introduction to sequence alignment *158*

The dotplot *160*

Dotplots and sequence alignments *165*

Measures of sequence similarity *171*
Scoring schemes *171*

Computing the alignment of two sequences *175*
Variations and generalizations *175*
Approximate methods for quick screening of databases *176*

The dynamic programming algorithm for optimal pairwise sequence alignment *176*

Significance of alignments *182*

Multiple sequence alignment *186*

Applications of multiple sequence alignments to database searching *188*
Profiles *189*
PSI-BLAST *191*
Hidden Markov Models *193*

Phylogeny *198*

Phylogenetic trees *203*
Clustering methods *205*
Cladistic methods *206*
The problem of varying rates of evolution *207*
Computational considerations *208*

Recommended reading 209

Exercises, Problems, and Weblems 210

5 Protein structure and drug discovery *219*

Introduction *220*

Protein stability and folding *223*
The Sasisekharan-Ramakrishnan-Ramachandran plot describes allowed mainchain conformations *223*
The sidechains *225*
Protein stability and denaturation *225*
Protein folding *228*

Applications of hydrophobicity *229*

Superposition of structures, and structural alignments *233*

DALI (Distance-matrix ALIgnment) *235*

Evolution of protein structures *236*
Classifications of protein structures *238*
SCOP *239*
Protein structure prediction and modelling *240*
Critical Assessment of Structure Prediction (CASP) *242*
Secondary structure prediction *244*
Homology modelling *250*
Fold recognition *252*
Conformational energy calculations and molecular dynamics *255*
ROSETTA *259*
LINUS *259*
Assignment of protein structures to genomes *263*
Prediction of protein function *265*
Divergence of function: orthologues and paralogues *266*
Drug discovery and development *269*
The lead compound *271*
Bioinformatics in drug discovery and development *273*
Recommended reading 284
Exercises, Problems, and Weblems 285

6 Proteomics and systems biology *291*

DNA microarrays *293*
Analysis of microarray data *295*
Mass spectrometry *301*
Identification of components of a complex mixture *301*
Protein sequencing by mass spectrometry *304*
Genome sequence analysis by mass spectrometry *306*
Systems biology *311*
Networks and graphs *313*
Network structure and dynamics *318*
Protein complexes and aggregates *320*
Properties of protein-protein complexes *321*
Protein interaction networks *324*
Regulatory networks *329*
Structures of regulatory networks *330*
Structural biology of regulatory networks *336*
Recommended reading 339
Exercises, Problems, and Weblems 339

Conclusions *345*
Answers to Exercises *347*
Glossary *353*
Index *357*
Colour plates

Plan of the book

- Chapter 1 sets the stage and introduces all of the major players: DNA and protein sequences and structures, genomes and proteomes, databases and information retrieval, the World Wide Web, computer programming. Before developing individual topics in detail it is important to see the framework of their interactions.
- Chapter 2 presents the nature of individual genomes, including the Human Genome, and the relationships among them, from the biological point of view.
- Chapter 3 imparts basic skills in using the Web in bioinformatics. It describes archival databanks, and leads the reader through sample sessions involving information retrieval from some of the major archival databases in molecular biology.
- Chapter 4 treats the analysis of relationships among sequences—alignments and phylogenetic trees. These methods underlie some of the major computational challenges of bioinformatics: detecting distant relatives, understanding relationships among genomes of different organisms, and tracing the course of evolution at the species and molecular levels.
- Chapter 5 moves into three dimensions, treating protein structure and folding. Sequence and structure must be seen as full partners, with bioinformatics developing methods for moving back and forth between them as fluently as possible. Understanding protein structures in detail is essential for determining their mechanisms of action, and for clinical and pharmacological applications.
- Chapter 6 treats proteomics and systems biology, including new high-throughput sources of information about the expression and distribution of proteins in cells, and attempts to synthesize the information to reveal patterns of organization.

CHAPTER 1

Introduction

Chapter contents

Life in space and time *3*

Evolution is the change over time in the world of living things *4*

Dogmas: central and peripheral *6*

Observables and data archives *9*

Information flow in bioinformatics *12*

Curation, annotation, and quality control *13*

The World Wide Web *14*

Electronic publication *15*

Computers and computer science *16*

Programming *17*

Biological classification and nomenclature *21*

Use of sequences to determine phylogenetic relationships *24*

Use of SINES and LINES to derive phylogenetic relationships *30*

Searching for similar sequences in databases: PSI-BLAST *32*

Introduction to protein structure *40*

The hierarchical nature of protein architecture *41*

Classification of protein structures *44*

Protein structure prediction and engineering *51*

Critical Assessment of Structure Prediction (CASP) *52*

Protein engineering *52*

Proteomics *52*

DNA microarrays *53*

Mass spectrometry *54*

Systems biology *54*

Clinical implications *55*

The future *57*

Recommended reading 57

Exercises, Problems, and Weblems 59

Biology has traditionally been an observational rather than a deductive science. Although recent developments have not altered this basic orientation, the nature of the data has radically changed. It is arguable that until recently all biological observations were fundamentally anecdotal—admittedly with varying degrees of precision, some very high indeed. However, in the last generation the data have become not only much more quantitative and precise, but, in the case of nucleotide and amino acid sequences, they have become *discrete*. It is possible to determine the genome sequence of an individual organism or clone not only completely, but in principle *exactly*. Experimental error can never be avoided entirely, but for modern genomic sequencing it is extremely low.

Not that this has converted biology into a deductive science. Life does obey principles of physics and chemistry, but for now life is too complex, and too dependent on historical contingency, for us to deduce its detailed properties from basic principles. Whether this impoverishes or enriches the subject is a matter of opinion.

A second obvious property of the data of bioinformatics is their *very very large amount*. Currently the nucleotide sequence databanks contain 80000×10^6 bases (abbreviated 80000 Mbp). If we use the approximate size of the human genome—3×10^9 letters—as a unit, this amounts to 26 HUman Genome Equivalents (or 26 huges, an apt name). For a comprehensible standard of comparison, 1 *huge* is comparable to the number of characters appearing in 6 complete years of issues of *The New York Times*. The database of macromolecular structures contains 30 000 entries, the full three-dimensional coordinates of proteins, of average length ~400 residues. Not only are the individual databanks large, but their sizes are increasing at a very high rate. Figure 1.1 shows the growth over the past decade of GenBank (archiving nucleic acid sequences) and the Protein Data Bank (archiving macromolecular structures). It would be precarious to extrapolate.

This quality and quantity of data have encouraged scientists to aim at commensurately ambitious goals:

- To have it said that they 'saw life clearly and saw it whole'. That is, to understand integrated aspects of the biology of organisms, viewed as coherent complex systems.
- To interrelate sequence, three-dimensional structure, expression patterns, interactions, and function of individual proteins, nucleic acids and protein-nucleic acid complexes.
- To integrate the data on different aspects of the life of a cell or organism into a 'systems' description of the structure and dynamics.
- To use data on contemporary organisms as a basis for travel backward and forward in time—back to deduce events in evolutionary history, forward to greater deliberate scientific modification of biological systems.
- To support applications to medicine, agriculture and technology.

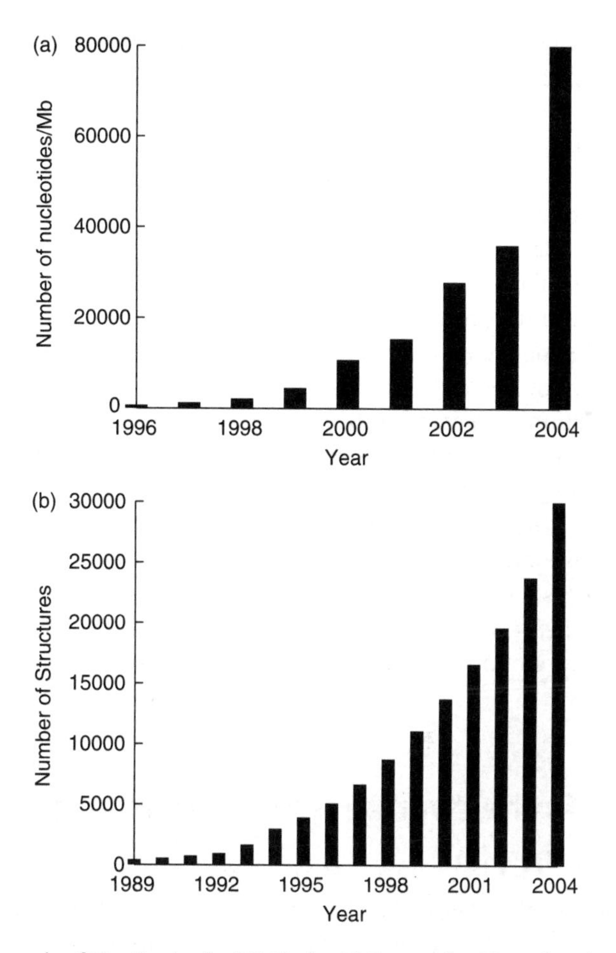

Fig. 1.1 (a) Growth of GenBank, the US National Center for Biotechnology Information genetic sequence archival databank. (b) Growth of Protein Data Bank, archive of three-dimensional biological macromolecular structures.

Life in space and time

It is difficult to define life, and it may be necessary to modify its definition—or to live, uncomfortably, with the old one—as computers grow in power and the silicon-life interface grows more intimate. For now, try this: A biological organism is a naturally-occurring, self-reproducing device that effects controlled manipulations of matter, energy and information.

From the most distant perspective, life on Earth is a complex self-perpetuating system distributed in space and time. It is of the greatest significance that it is largely composed of *discrete* individual organisms, each with a finite lifetime and in most cases with unique features.

Spatially, starting far away and zooming in progressively, one can distinguish, within the biosphere, local **ecosystems**, stable until their environmental conditions change or they are invaded. Each species within an ecosystem is composed of **organisms** carrying out individual if not independent activities. Organisms are composed of **cells**. Every cell is an intimate localized ecosystem, not isolated from its environment but interacting with it in specific and controlled ways. Eukaryotic cells contain a complex internal structure of their own, including nuclei and other subcellular organelles, and a cytoskeleton. And finally we come down to the level of molecules.

Life is extended not only in space but in time. We see today a snapshot of one stage in a history of life that extends back in time for at least 3.5 billion years. The theory of natural selection has been extremely successful in rationalizing the process of life's development. However, historical accident plays too dominant a role in determining the course of events to allow much detailed prediction. Nor does fossil DNA afford substantial access to any historical record at the molecular level. Instead, we must try to read the past in contemporary genomes. US Supreme Court Justice Felix Frankfurter once wrote that '... the American constitution is not just a document, it is a historical stream'. This is also true of genomes, which contain records of their own evolution.

Evolution is the change over time in the world of living things

The processes of evolution change distributions of genotypes and phenotypes in successive generations. The **genotype** is an organism's genetic information, the sequence of its genome. All other observable features of an organism—macroscopic and biochemical—comprise its **phenotype**. The genotype is inherited from a parent or parents, subject to modification by mutation or lateral transfer of genetic material. The phenotype depends on the genotype, which controls the development of the organism under the influence of its environment.

The asymmetry between genotype and phenotype is the engine of evolution:

- Changes in genotypes are inheritable. Effects on the phenotype of the environment or lifestyle—for instance, better nutrition leading to larger body size, or debilitating effects of disease or injury—are not directly inheritable.
- During the development of any organism, genotype constrains phenotype. Phenotype does not influence genotype.
- Many genotypes can create the same phenotype:
  - Many mutations in genes coding for proteins leave amino acid sequences unchanged, or make modifications with no effect on function.
  - **Alleles** are different forms (sequences) of the same gene. Any organism that contains two or more copies of a gene can repeat the same allele

In mammals ~20% of loci are heterozygous.

(homozygosity) or contain different alleles (heterozygosity). Homozygotes and heterozygotes have different genotypes, but if one allele is dominant, and a single gene has exclusive control over a trait, homozygotes and heterozygotes may have the same phenotype.

At what levels does evolution operate? Most life consists of discrete organisms. A **population** is a group of similar organisms that interact: populations of sexually-reproducing organisms interbreed; individuals in all populations compete for resources. Evolution alters the composition and distribution of the gene pools and phenotypes in populations.

What is the mechanism of evolution? Within a population, individuals with a variety of genotypes arise, displaying a corresponding variety of phenotypes. Although evolution has no *direct* leverage on genotype, individuals with different phenotypes show differential success at reproduction. As a result, the new generation may have an altered distribution of genotypes and phenotypes. **Natural selection**—enhanced reproduction by 'fitter' individuals—is the most important mechanism of evolution. Another is **genetic drift**, the random change in allelic frequencies, not in response to selection. Genetic drift is especially important in small, isolated populations.

Mechanisms that produce genetic variety create the *potential* for evolution:

- **Mutations**, such as point substitutions, insertions and deletions, and transpositions. Rates of generation of point mutations are estimated to be about 10^{-12} – 10^{-10} per base pair per generation. (This is *not* the same as the rate of allelic replacement in a population. Mutations only propose candidates for evolutionary change.)
- **Recombination** can bring different loci together, or split them apart. Recombination within a gene can create a new allele. Recombination between genes can alter the relationship between genes and regulatory elements.
- **Gene duplication**, followed by divergence.
- **Gene flow**, from mixing of populations, or from lateral gene transfer between species.

Evolution can increase or decrease the variety in gene pools. If a novel mutation confers selective advantage only in the homozygous state, the gene may spread throughout a population. Adoption of the allele by all members of a population may *decrease* the variety in the gene pool. If a gene arises that confers selective advantage only in the heterozygous state, the pool may move towards greater variety. Some mutations create recessive alleles that are deleterious only in the homozygous state. These are hard to remove from a population, especially if heterozygotes have some compensating advantage, as in sickle-cell anaemia, in which heterozygotes have enhanced resistance to malaria.

Microevolution refers to relatively small changes in a few genes, leading in most cases to relatively small changes in phenotypes. Microevolution affects the individuals *within* a population. Modern techniques allow us to follow microevolution at the molecular level, through measurements of genome sequences and protein expression patterns. **Macroevolution** refers to larger-scale changes in populations as a whole, including formation of new species. The fossil record provides a (partial) history of macroevolution, using geological methods to date events. Comparative anatomy and physiology, and embryology, provide additional clues.

Observations of micro- and macroevolution illuminate each other. Genome sequences help in the classification of species. The fossil record permits dating of past events that have had consequences on the molecular scale that we observe now. A major challenge to modern biology is to understand how large-scale events such as development of new species can occur as a result of microevolutionary events.

Dogmas: central and peripheral

The information archive in each organism—the blueprint of potential development and activity—is the genetic material, DNA, or, in some viruses, RNA. DNA molecules are long, *linear*, chain molecules containing a message in a four-letter alphabet (see Box). Even for micro-organisms the message is long, typically 10^6 characters. Implicit in the structure of the DNA are mechanisms for self-replication and for translation of genes into proteins. The double-helix, and its internal self-complementarity providing for accurate replication, are well known (see Plate I). Near-perfect replication is essential for stability of inheritance; but some imperfect replication, or mechanism for import of foreign genetic material, is also essential, else evolution could not take place in asexual organisms.

The strands in the double-helix are antiparallel; directions along each strand are named 3′ and 5′ (for positions in the deoxyribose ring). In translation to protein, the DNA sequence is always read in the 5′ → 3′ direction.

The implementation of genetic information occurs, initially, through the synthesis of RNA and proteins. Proteins are the molecules responsible for much of the structure and activities of organisms. Our hair, muscle, digestive enzymes and antibodies are all proteins. Both nucleic acids and proteins are long, linear chain molecules. The genetic 'code' is in fact a cipher: Successive triplets of letters from the DNA sequence specify successive amino acids; stretches of DNA sequences encipher amino acid sequences of proteins. Typically, proteins are 200–400 amino acids long, requiring 600–1200 letters of expressed DNA message to specify them. Synthesis of RNA molecules, for instance the RNA components of the ribosome, are also directed by DNA sequences.

Sydney Brenner distinguished 'junk' from 'garbage': garbage you throw away, junk you keep around.

However, not all DNA is expressed as proteins or structural RNA. Most genes in higher organisms contain internal untranslated regions, or introns. Some regions of the DNA sequence are devoted to control mechanisms, and a substantial amount of the genomes of higher organisms appears to be 'junk'. (Which may mean merely that we do not yet understand its function.)

The four naturally-occurring nucleotides in DNA (RNA)

a adenine g guanine c cytosine t thymine (u uracil)

The twenty naturally-occurring amino acids in proteins

Nonpolar amino acids

G	glycine	A	alanine	P	proline	V	valine
I	isoleucine	L	leucine	F	phenylalanine	M	methionine

Polar amino acids

S	serine	C	cysteine	T	threonine	N	asparagine
Q	glutamine	H	histidine	Y	tyrosine	W	tryptophan

Charged amino acids

D aspartic acid E glutamic acid K lysine R arginine

Other classifications of amino acids can also be useful. For instance, histidine, phenylalanine, tyrosine, and tryptophan are aromatic, and are observed to play special structural roles in membrane proteins.

Amino acid names are frequently abbreviated to their first three letters, for instance Gly for glycine; except for isoleucine, asparagine, glutamine and tryptophan, which are abbreviated to Ile, Asn, Gln and Trp, respectively. The rare amino acid selenocysteine has the three-letter abbreviation Sec and the one-letter code U.

It is conventional to write nucleotides in lower case and amino acids in upper case. Thus atg = adenine-thymine-guanine and ATG = alanine-threonine-glycine.

In DNA the molecules comprising the alphabet are chemically similar, and the structure of DNA is, to a first approximation, uniform (although some DNA-protein interactions distort the DNA structure). Proteins, and structural RNAs, in contrast, show a great variety of three-dimensional conformations. These are necessary to support their very diverse functional roles.

Alternative genetic codes appear in organelles—chloroplasts and mitochondria—and in some species.

The amino acid sequence of a protein dictates its three-dimensional structure. For each natural amino acid sequence, there is a unique stable *native state* that under proper conditions is adopted spontaneously. If a purified protein is heated, or otherwise brought to conditions far from the normal physiological environment, it will 'unfold' to a disordered and biologically-inactive structure. (This is why our bodies contain mechanisms to maintain nearly-constant internal conditions.) When normal conditions are restored, protein molecules will generally readopt the native structure, indistinguishable from the original state.

The functions of proteins depend on their adopting the native three-dimensional structure. For example, the native structure of an enzyme may have a cavity on its

The standard genetic code

ttt	Phe	tct	Ser	tat	Tyr	tgt	Cys
ttc	Phe	tcc	Ser	tac	Tyr	tgc	Cys
tta	Leu	tca	Ser	taa	STOP	tga	STOP
ttg	Leu	tcg	Ser	tag	STOP	tgg	Trp
ctt	Leu	cct	Pro	cat	His	cgt	Arg
ctc	Leu	ccc	Pro	cac	His	cgc	Arg
cta	Leu	cca	Pro	caa	Gln	cga	Arg
ctg	Leu	ccg	Pro	cag	Gln	cgg	Arg
att	Ile	act	Thr	aat	Asn	agt	Ser
atc	Ile	acc	Thr	aac	Asn	agc	Ser
ata	Ile	aca	Thr	aaa	Lys	aga	Arg
atg	Met	acg	Thr	aag	Lys	agg	Arg
gtt	Val	gct	Ala	gat	Asp	ggt	Gly
gtc	Val	gcc	Ala	gac	Asp	ggc	Gly
gta	Val	gca	Ala	gaa	Glu	gga	Gly
gtg	Val	gcg	Ala	gag	Glu	ggg	Gly

From one dimension to three

The spontaneous folding of proteins to form their native states is the point at which Nature makes the giant leap from the one-dimensional world of gene and protein sequences to the three-dimensional world we inhabit. There is a paradox: The translation of DNA sequences to amino acid sequences is very simple to describe logically; it is specified by the genetic code. The folding of the polypeptide chain into a precise three-dimensional structure is very difficult to describe logically. However, translation requires the immensely complicated machinery of the ribosome, tRNAs and associated molecules; but protein folding occurs spontaneously.

surface that binds a small molecule, and juxtaposes it to catalytic residues. Many regulatory mechanisms depend on the binding of proteins to other proteins or to DNA. We thus have the paradigm:

- DNA sequence determines protein sequence
- Protein sequence determines protein structure
- Protein structure determines protein function

- Regulatory mechanisms, including but not limited to control of expression patterns, deliver the right amount of the right function to the right place at the right time

Much of the organized activity of bioinformatics has been focussed on the analysis of the data related to these processes.

So far, this paradigm does not include levels higher than the molecular of structure and organization, including, for example, such questions as how tissues become specialized during development or, more generally, how environmental effects exert control over genetic events. In some cases of simple feedback loops, it is understood at the molecular level how increasing the amount of a reactant causes an increase in the production of an enzyme that catalyzes its transformation. More complex are the programs of development during the lifetime of an organism. These fascinating problems about the information flow and control within an organism have now come within the scope of mainstream bioinformatics. The topic of **systems biology** focusses on the integration and control of the activities of cells and organisms.

Observables and data archives

A databank includes an archive of information, a logical organization or 'structure' of that information, and tools to gain access to it. Databanks in molecular biology cover nucleic acid and protein sequences, macromolecular structures and functions, expression patterns and networks of metabolic pathways and control cascades. They include:

- Archival databanks of biological information:
  - DNA and protein sequences, including annotation
  - Variations, such as compilations of haplotypes
  - Nucleic acid and protein structures, including annotation
  - Databanks focussed on organisms, including genome databases
  - Databanks of protein expression patterns
  - Databanks of metabolic pathways
  - Databanks of interaction patterns and regulatory networks
- Derived databanks: These contain information collected from the archival databanks, and from the analysis of their contents. For instance:
  - sequence motifs (characteristic 'signature patterns' of families of proteins)
  - mutations and variants in DNA and protein sequences
  - classifications or relationships (connections and common features of entries from archives; for instance a databank of sets of protein sequence families, or a hierarchical classification of protein folding patterns)

- Bibliographic databanks
- Databanks of web sites:
  - databanks of databanks containing biological information
  - links between databanks

Web resources: Nucleic acid and protein sequences

The archive of nucleic acid sequences is maintained by a triple partnership: *GenBank*, based at the US National Center for Biotechnology Information, in Bethesda, Maryland, USA; *The EMBL Nucleotide Sequence Database*, based at the European Bioinformatics Institute, in Hinxton, UK; and *The Center for Information Biology and DNA DataBank of Japan*, at the National Institute of Genetics in Mishima, Japan. The three sites exchange incoming submissions daily, to ensure common coverage. However, the format, annotation and embedded links differ among the corresponding entries released by the different databases.

The archive of amino acid sequences of proteins, now determinined almost exclusively from translation of gene sequences, is maintained by the *United Protein Database (UniProt)*, a merger of the databases *SWISS-PROT*, *The Protein Identification Resource (PIR)* and *Translated EMBL (TrEMBL)*.

Associated with the archives are tools for selection and retrieval of sequences. The *Sequence Retrieval System (SRS)*, a product of Lion Bioscience AG, is available freely to academic users through the European Bioinformatics Institute and numerous mirror sites. The US National Center for Biotechnology Information offers *ENTREZ*. Both allow parallel searches in multiple data archives.

Many full-genome sequencing projects maintain databases focussed on individual species. Notable are the *ENSEMBL* (Sanger Centre, Hinxton, UK) and University of California at Santa Cruz browsers, for the human and other genomes.

Many derived databanks assemble families of proteins or subunits based on the similarities of their sequences. An 'umbrella' database, Interpro, integrates the contents, features, and annotation of several individual databases of protein families, domains, and functional sites; and contains links to others, including the Gene Ontology Consortium™ functional classification. Interpro intends to assimilate additional databases. ('Resistance is futile.')

The mechanism of access to a databank is the set of tools for answering questions such as:

- 'Does the databank contain the information I require?' (Example: In which databanks can I find amino acid sequences of alcohol dehydrogenases?)
- 'How can I assemble selected information from the databank in a useful form?' (Example: How can I compile a list of globin sequences, or even better, a table of aligned globin sequences?)

- Indices of databanks are useful in asking 'Where can I find some specific piece of information?' (Example: What databanks contain the amino acid sequence of porcupine trypsin?) Of course if I know and can specify exactly what I want, the problem is relatively straightforward.

A databank without effective modes of access is merely a data graveyard. How to achieve effective access is an issue of database design that ideally should remain hidden from users. It has become clear that effective access cannot be provided by bolting a query system onto an unstructured archive. Instead, the logical organization of the storage of the information must be designed with the access in mind—what kinds of questions users will want to ask—and the structure of the archive must mesh smoothly with the information-retrieval software.

A variety of database queries can arise in bioinformatics. These include:

(1) Given a sequence, or fragment of a sequence, find sequences in the database that are similar to it. This is a central problem in bioinformatics. We share such string-matching problems with many fields of computer science. For instance, word processing and editing programs support string-search functions.

(2) Given a protein structure, or fragment, find protein structures in the database that are similar to it. This is the generalization of the string matching problem to three dimensions.

(3) Given a sequence of a protein of unknown structure, find *structures* in the database that adopt similar three-dimensional structures. One is tempted to cheat—to look in the sequence data banks for proteins with sequences similar to the probe sequence: For if two proteins have sufficiently similar sequences, they will have similar structures. However, the converse is not true, and one can hope to create more powerful search techniques that will find proteins of similar structure even though their sequences have diverged beyond the point where they can be recognized as similar by sequence comparison.

(4) Given a protein structure, find *sequences* in the databank that correspond to similar structures. Again, one can cheat by using the structure to probe a structure databank, but this can offer only limited success because there are so many more sequences known than structures. It is therefore desirable to have a method that can pick out the structure from the sequence.

(1) and (2) are solved problems; such searches are carried out thousands of times a day. (3) and (4) are active fields of research.

Tasks of greater subtlety arise when one wishes to study relationships between information contained in separate databanks. This requires links that facilitate simultaneous access to several databanks. Here is an example: 'For which proteins of known structure involved in diseases of purine biosynthesis in humans, are there related proteins in yeast?' We are setting conditions on: known structure, specified function, detection of relatedness, correlation with disease, specified species. Today, the quality of a database depends not only on the information it contains, but on the effectiveness of its links to other sources of information. The growing importance of simultaneous access to databanks has led to research in databank interactivity—how can databanks 'talk to one another' without too great a sacrifice of the freedom of each one to structure its own data in ways appropriate to the individual features of the material it contains.

A problem that has not yet arisen in molecular biology is control of updates to the archives. A database of airline reservations must prevent many agents from selling the same seat to different travellers. In bioinformatics, users can read or extract information from archival databanks, or submit material for processing by the staff of an archive, but not add or alter entries directly. This situation may change. From a practical point of view, the amount of data being generated is increasing so rapidly as to swamp the ability of archive projects to assimilate it. There is already movement towards greater involvement of scientists at the bench in preparing data for the archive.

Although there are good arguments for unique control over the archives, there is no need to limit the modes of access to them—colloquially, the design of 'front ends'. Specialized user communities may extract subsets of the data, or recombine data from different sources, and provide specialized avenues of access. Such 'boutique' databases depend on the primary archives as the source of the information they contain, but redesign the organization and presentation. Indeed, different derived databases can slice and dice the same information in different ways. A reasonable extrapolation suggests the concept of specialized 'virtual databases' (an idea first proposed in 1981), grounded in the archives but providing individual scope and function, tailored to the needs of individual research groups or even individual scientists.

Information flow in bioinformatics

Data enter the bioinformatics establishment when a scientist deposits an experimental result in an appropriate archive. The archive curates and annotates the data, to create an entry of proper contents and format. The entry is added to the public release of the archive. Note that the division of the archive into entries is determined by the provenance of the data rather than the biological unit or context; that is, an entry corresponds to one coherent set of experimental measurements; often corresponding to one published article.

Other information-retrieval projects, either associated with an archive or independent, may integrate the newly-released entry into their individual systems. They may select or reorganize the data structure, and provide tools for analysis.

Reorganization of the data may involve:

- Simply integrating the new entry into a general or specialized search engine.
- Extracting useful subsets of the data. Examples include (1) identification of genes in a connected DNA sequence, such as a bacterial genome or a eukaryotic chromosome; and (2) the selection of a nonredundant set of protein sequences, both to shorten searches and to reduce statistical bias.
- Deriving new types of information from the original data. A simple example: release of a protein-coding gene by a DNA sequence archive will be followed by the appearance of its amino acid sequence translation in databases of protein sequences.

- Recombining data in different ways. Many projects group sequences or structures of families of homologous proteins, or proteins that share function. Examples include the MEROPS protease database and the Protein Kinase Resource. (Archives tend to keep related entries separate to preserve clarity of provenance.)
- Reannotating the data, including provision of different constellations of links. The integration may be horizontal or vertical. That is, links may indicate relationships to other entries of the same type (for instance, correspondences within a genome among homologous genes or among genes associated with the same metabolic pathway). Or, links may adduce a variety of information about a gene or protein (for instance, links between a gene and the clinical consequences of mutations).

Many sites serve as gateways between the archives and computational tools available for data analysis. Information retrieval permits selection and extraction of data to provide the ingredients of a research project. Many bioinformatics resources not only offer information retrieval, but facilitate the 'downstream' processing of the entries selected. A typical example would be to retrieve the sequences of a set of homologous genes, and then to align them. The goal is to provide smooth integration of all the data-processing steps required for a research project, by intimate links among the tools for data storage, retrieval and analysis.

There is a very strong trend towards merging and integration of data resources in bioinformatics. Only national or commercial rivalries would seem to stand in the way of an extrapolation that there will soon be a single worldwide database. Because of the danger that the result will prove unwieldy, the unification of the archives will be accompanied by a fragmentation of the routes of access.

Curation, annotation, and quality control

The scientific and medical communities are dependent on the quality of databanks. Indices of quality, even if they do not permit correction of mistakes, may help us avoid arriving at wrong conclusions.

Databank entries comprise raw experimental results, and supplementary information, or annotations. Each of these has its own sources of error.

The most important determinant of the quality of the data themselves is the state of the art of the experiments. Older data were limited by older techniques; for instance, amino acid sequences of proteins were first determined by peptide sequencing, but are now translated from DNA sequences (except for partial sequencing by mass spectrometry; see Chapter 6). One consequence of the data explosion is that most sequence data are new data, governed by current technology, which in most cases does quite a good job.

Annotations include information about the source of the data and the methods used to determine them. They identify the investigators responsible, and cite relevant publications. They provide links to related information in other databanks. In sequence databanks, annotations include **feature tables**: lists of segments of the sequences that have biological significance—for instance, regions of a DNA sequence that code for proteins. These appear in computer-parsable formats, and their contents may be restricted to a controlled vocabulary. Note that agreement between databases, on a controlled vocabulary and the definitions of the terms, is essential for information-retrieval operations involving interactions between multiple databases, such as distributed queries.

Formerly, a typical DNA sequence entry was produced by a single research group, investigating a gene and its products in a coherent way. Annotations were grounded in locally-measured experimental data and written by specialists. In contrast, full-genome sequencing projects offer no experimental confirmation of the expression of most putative genes, nor characterization of their products. Curators at databanks base their annotations on the analysis of the sequences by computer programs.

Annotation is the weakest component of the genomics enterprise. Automation of annotation is possible only to a limited extent; getting it right remains labour-intensive, and allocated resources are inadequate. But the importance of proper annotation cannot be underestimated. P. Bork has commented that errors in gene assignments vitiate the high quality of the sequence data themselves.

Growth of genomic data will permit improvement in the quality of annotation, as statistical methods increase in accuracy. This will allow improved *reannotation* of entries. The improvement of annotations will be a good thing. It implies however the disturbing concomitant that annotation will be in flux. The problem is aggravated by the proliferation of web sites, with increasingly dense networks of links. They provide useful avenues for applications. But the Web is also a vector of contagion, propagating errors in raw data, in immature data subsequently corrected but the corrections not passed on, and in variant annotations.

The only possible solution is a distributed and dynamic error-correction and annotation process. Distributed, in that databank staff will have neither the time nor the expertise for the job; specialists will have to act as curators. Dynamic, in that progress in automation of annotation and error identification/correction will permit reannotation of databanks. We will have to give up the safe idea of a stable databank composed of entries that are correct when first distributed, and that stay fixed. Databanks will become a seething broth of information, growing in size, and maturing—we must hope—in quality.

The World Wide Web

All readers will have used the World Wide Web, for reference material, for news, for access to databases in molecular biology, for checking out personal information about individuals—friends or colleagues or celebrities—or just for browsing. The Web is a means of interpersonal and intercomputer contact over networks. It provides a complete global village, containing the equivalent of library, post-office, shops, and schools.

The Web can be thought of as a giant worldwide multimedia bulletin board. It contains text, images, movies, and sounds. Virtually anything that can be stored on a computer can be made available and accessed via the Web. An interesting example is a site treating the poetry of Walt Whitman (www.whitmanarchive.org). The highest level page contains a table of contents. The site contains printed text of different poems. You can compare different editions. You can access critical analysis of the poems. You can see versions of some poems in manuscripts. There is even a link to an audio file, containing a recording of Whitman himself reading part of a poem.

The links embedded in a web site can be internal or external. Internal links may take you to other portions of the text of a current document, or to associated images, movies, or sounds. External links may allow you to move *down* to more specialized documents, *up* to more general ones (perhaps providing background to technical material), *sideways* to parallel documents (other papers on the same subject), or *over* to directories that show what other relevant material is available.

The main thing to do, to get started using the Web effectively, is to find useful entry points. Once a session is launched, links will take you where you want to go. Among the most important sites are **search engines**, such as Google, that index the entire Web and permit retrieval by keywords. You can enter one or more terms, such as 'phosphorylase', 'allosteric change', 'crystal structure', and the search program will return a list of links to sites on the Web that contain these terms.

Once you have completed a successful session, when you next log in the intersession memory facilities of the browsers allow you to pick up cleanly where you left off. During any session, when you find yourself viewing a document to which you will want to return, you can save the link in a file of **bookmarks** or **favourites**. In a subsequent session you can return to any site on this list directly, not needing to follow the trail of links that led you there in the first place.

A personal **home page** is a short autobiographical sketch (with links, of course). Your colleagues will have their own home pages which typically include name, institutional affiliation, addresses for paper and electronic mail, telephone and fax numbers, a list of publications and current research interests. It is not uncommon for home pages to include personal information, such as hobbies, pictures of the individual with his or her spouse and children, and even with the family dog!

Nor is the Web solely a one-way street. Many Web documents include forms in which you can enter information, and launch a program that returns results within your session. Search engines are common examples. Many calculations in bioinformatics are now launched via such web servers. If the calculations are lengthy the results may not be returned within the session, but sent by e-mail.

Electronic publication

We are in an era of a transition to paper-free publishing. More and more publications are appearing on the Web. A scientific journal may post only its table of contents, a table of contents together with abstracts of articles, or even full articles. Many institutional publications—newsletters and technical reports—appear on the Web. Many other magazines and newspapers are showing up as well. You might want to try http://www.nytimes.com. Many printed publications now contain references to Web links containing supplementary material that never appears on paper. [As the book was going to press, Google announced collaborations with academic libraries to place entire collections of books on-line. This will create new avenues to the recovery and delivery of information.]

Computers and computer science

Bioinformatics would not be possible without advances in computing hardware and software. Fast and high-capacity storage media are essential even to maintain the archives. Information retrieval and analysis require programs; some fairly straightforward and others extremely sophisticated. Distribution of the information requires the facilities of computer networks and the World Wide Web.

Computer science is a young and flourishing field with the goal of making most effective use of information technology hardware. Certain areas of theoretical computer science impinge most directly on bioinformatics. Let us consider them with reference to a specific biological problem: 'Retrieve from a database all sequences similar to a probe sequence.' A good solution of this problem would appeal to computer science for:

- **Analysis of algorithms** *An algorithm is a complete and precise specification of a method for solving a problem.* For the retrieval of similar sequences, we need to measure the similarity of the probe sequence to every sequence in the database. It is possible to do much better than the naive approach of checking every pair of positions in every possible juxtaposition, a method that even without allowing gaps would require a time proportional to the product of the number of characters in the probe sequence times the number of characters in the database. A speciality in computer science known colloquially as 'stringology' focusses on developing efficient methods for this type of problem, and analysing their effective performance.
- **Data structures and information retrieval** How can we organize our data for efficient response to queries? For instance, are there ways to index or otherwise 'preprocess' the data to make our sequence-similarity searches more efficient? How can we provide interfaces that will assist the user in framing and executing queries?
- **Software engineering** Hardly ever anymore does anyone write programs in the native language of computers. Programmers work in higher-level languages, such as C, C++, PERL (Practical Extraction and Report Language), JAVA or even FORTRAN. The choice of programming language depends on the nature of the algorithm and associated data structure, and the expected use of the program. Of course most complicated software used in bioinformatics is now written by specialists. Which brings up the question of how much programming expertise a bioinformatician needs.

Programming

Programming is to computer science what bricklaying is to architecture. Both are creative; one is an art and the other a craft.

Many students of bioinformatics ask whether it is essential to learn to write complicated computer programs. My advice (not agreed upon by everyone in the field) is: 'Don't. Unless you want to specialize in it.' To work in bioinformatics, you will need to develop expertise in using tools available on the Web. Learning how to create and maintain a web site is essential. And of course you will need facility in the use of the operating system of your computer. Some skill in writing simple scripts in a language like PERL provides an essential extension to the basic facilities of the operating system.

On the other hand, the size of the data archives, and the growing sophistication of the questions we wish to address, demand respect. Truly creative programming in the field is best left to specialists, well-trained in computer science. Nor does *using* programs, via highly-polished (not to say flashy) Web interfaces, provide any indication of the nature of the activity involved in writing and debugging programs. Bismarck once said: 'Those who love sausages or the law should not watch either being made.' Perhaps computer programs should be added to his list.

I recommend learning some basic skills with PERL, or with one of the related languages Python or Ruby. These languages provide very powerful tools. They make it very easy to carry out many very useful simple tasks, and are available on most computer systems.

How should you learn enough PERL to be useful in bioinformatics? Many institutions run courses. Learning from colleagues is fine, depending on the ratio of your aptitude to their patience. Books are available. A very useful approach is to find lessons on the Web—ask a search engine for 'PERL tutorial' and you will turn up many useful sites that will lead you by the hand through the basics.

And of course use it as much as you can. This book will not teach you PERL, but it will provide opportunities to practise what you learn elsewhere. Should your programming ambitions go beyond simple tasks, check out the Bioperl Project, a source of freely available PERL programs and components in the field of bioinformatics (see: http://bio.perl.org/).

Examples of *simple* PERL programs appear in this book. The strength of PERL at character-string handling makes it suitable for sequence analysis tasks in biology. Here is a very simple PERL program to translate a nucleotide sequence into an amino acid sequence according to the standard genetic code. The first line, `#!/usr/bin/perl`, is a signal to the UNIX (or LINUX) operating system that what follows is a PERL program. Within the program, all text commencing with a #, through to the end of the line on which it appears, is merely comment. The line `__END__` signals that the program is finished and what follows is the input data. (All material that the reader might find useful to have in computer-readable form, including all programs, appear in the web site associated with this book: http://www.oup.co.uk/booksites/biosciences/bioinf.)

PERL Example 1.1 Translation of a DNA sequence to an amino acid sequence using the standard genetic code

```
#!/usr/bin/perl
#translate.pl -- translate nucleic acid sequence to protein sequence
#                according to standard genetic code

#   set up table of standard genetic code

%standardgeneticcode = (
 "ttt"=> "Phe",   "tct"=> "Ser", "tat"=> "Tyr",   "tgt"=> "Cys",
 "ttc"=> "Phe",   "tcc"=> "Ser", "tac"=> "Tyr",   "tgc"=> "Cys",
 "tta"=> "Leu",   "tca"=> "Ser", "taa"=> "TER",   "tga"=> "TER",
 "ttg"=> "Leu",   "tcg"=> "Ser", "tag"=> "TER",   "tgg"=> "Trp",
 "ctt"=> "Leu",   "cct"=> "Pro", "cat"=> "His",   "cgt"=> "Arg",
 "ctc"=> "Leu",   "ccc"=> "Pro", "cac"=> "His",   "cgc"=> "Arg",
 "cta"=> "Leu",   "cca"=> "Pro", "caa"=> "Gln",   "cga"=> "Arg",
 "ctg"=> "Leu",   "ccg"=> "Pro", "cag"=> "Gln",   "cgg"=> "Arg",
 "att"=> "Ile",   "act"=> "Thr", "aat"=> "Asn",   "agt"=> "Ser",
 "atc"=> "Ile",   "acc"=> "Thr", "aac"=> "Asn",   "agc"=> "Ser",
 "ata"=> "Ile",   "aca"=> "Thr", "aaa"=> "Lys",   "aga"=> "Arg",
 "atg"=> "Met",   "acg"=> "Thr", "aag"=> "Lys",   "agg"=> "Arg",
 "gtt"=> "Val",   "gct"=> "Ala", "gat"=> "Asp",   "ggt"=> "Gly",
 "gtc"=> "Val",   "gcc"=> "Ala", "gac"=> "Asp",   "ggc"=> "Gly",
 "gta"=> "Val",   "gca"=> "Ala", "gaa"=> "Glu",   "gga"=> "Gly",
 "gtg"=> "Val",   "gcg"=> "Ala", "gag"=> "Glu",   "ggg"=> "Gly"
);

#   process input data

while ($line = <DATA>) {                              # read in line of input
    print "$line";                                    # transcribe to output
    chop();                                           # remove end-of-line character
    @triplets = unpack("a3" x (length($line)/3), $line); # pull out successive triplets
    foreach $codon (@triplets) {                      # loop over triplets
        print "$standardgeneticcode{$codon}";         # print out translation of each
    }                                                 # end loop on triplets
    print "\n\n";                                     # skip line on output
}                                                     # end loop on input lines

#   what follows is input data

__END__
atgcatccctttaat
tctgtctga
```

Running this program on the given input data produces the output:

```
atgcatccctttaat
MetHisProPheAsn

tctgtctga
SerValTER
```

Even this simple program displays several features of the PERL language. The file contains background data (the genetic code translation table), statements that tell the computer to do something, and the input data (appearing after the __END__ line). Comments summarize sections of the program and describe the effect of each statement.

The program is structured as blocks enclosed in curly brackets: `{...}`, which are useful in controlling the flow of execution. Within blocks, individual statements (each ending in a `;`) are executed in order of appearance. The outer block is a **loop**:

```
while ($line = <DATA>) {
        ...
}
```

Here <DATA> refers to the lines of input data (appearing after the __END__). The block is executed once for each line of input; that is, `while` there is any line of input remaining.

Three types of data structures appear in the program. The line of input data, referred to as `$line`, is a simple **character string**. It is split into an **array** or vector of triplets. An array stores several items in a linear order, and individual items of data can be retrieved from their positions in the array. For ease of looking up the amino acid coded for by any triplet, the genetic code is stored as an **associative array**. An associative array, or hash table, is a generalization of a simple or sequential array. Whereas the elements of a simple array are indexed by consecutive integers, the elements of an associative array are indexed by *any* character strings, in this case the 64 triplets. We utilize the input triplets *in order of their appearance* in the nucleotide sequence, but we need to access the elements of the genetic code table *in an arbitrary order* as dictated by the succession of triplets. A simple array or vector of character strings is appropriate for processing successive triplets, and the associative array is appropriate for looking up the amino acids that correspond to them.

PERL Example 1.2 Assembly of overlapping fragments

Here is another PERL program, that illustrates additional aspects of the language.* This program reassembles the sentence:

All the world's a stage,
And all the men and women merely players;
They have their exits and their entrances,
And one man in his time plays many parts.

after it has been chopped into random overlapping fragments (\n in the fragments represents end-of-line in the original):

the men and women merely players;\n
one man in his time
All the world's
their entrances,\nand one man
stage,\nAnd all the men and women
They have their exits and their entrances,\n
world's a stage,\nAnd all
their entrances,\nand one man
in his time plays many parts.
merely players;\nThey have

This kind of calculation is important in assembling DNA sequences from overlapping fragments. (See Problems 1.5 and 1.6.)

* This section may be skipped on a first reading.

→

PERL Example 1.2 *(continued)*

```
#!/usr/bin/perl
#assemble.pl -- assemble overlapping fragments of strings
#  input of fragments
while ($line = <DATA>) {                    #   read in fragments, 1 per line
   chop($line);                             #   remove trailing carriage return
   push(@fragments,$line);                  #   copy each fragment into array
}
#  now array  @fragments  contains fragments
#  we need two relationships between fragments:
#  (1) which fragment shares no prefix with suffix of another fragment
#       * This tells us which fragment comes first
#  (2) which fragment shares longest suffix with a prefix of another
#       * This tells us which fragment  follows  any fragment
#  First set array of prefixes to the default value   "noprefixfound".
#       Later, change this default value when a prefix is found.
#       The one fragment that retains the default value must be come first.
#  Then loop over pairs of fragments to determine maximal overlap.
#       This determines successor of each fragment
#       Note in passing that if a fragment has a successor then the
#           successor must have a prefix
foreach $i (@fragments) {                   #   initially set  prefix of each fragment
    $prefix{$i} = "noprefixfound";          #        to  "noprefixfound"
    }                                       #   this will be overwritten when a prefix is found
#  for each pair, find longest overlap of suffix of one with prefix of the other
#        This tells us which fragment FOLLOWS any fragment
foreach $i (@fragments) {                   #   loop over fragments
    $longestsuffix = "";                    #   initialize longest suffix to null
    foreach $j (@fragments) {               #   loop over fragment pairs
        unless ($i eq $j) {                 #   don't check fragment against itself
            $combine = $i . "XXX" . $j;     #   concatenate fragments, with fence XXX
            $combine =~ /([\S ]{2,})XXX\1/;    #   check for repeated sequence
            if (length($1) > length($longestsuffix)) {    # keep longest overlap
                $longestsuffix = $1;        #   retain longest suffix
                $successor{$i} = $j;        #   record that $j follows $i
            }
        }
    }
    $prefix{$successor{$i}} = "found";      #  if $j follows $i then $j must have a prefix
}

foreach (@fragments) {                      #  find fragment that has no prefix; that's the start
    if ($prefix{$_} eq "noprefixfound") {$outstring = $_;}
}
$test = $outstring;                         #   start with fragment without prefix
while ($successor{$test}) {                 #   append fragments in order
     $test = $successor{$test};             #   choose next fragment
     $outstring = $outstring . "XXX" . $test;  # append to string
     $outstring =~ s/([\S ]+)XXX\1/\1/;     #   remove overlapping segment
}
$outstring =~ s/\\n/\n/g;                   #   change signal \n to real carriage return
print "$outstring\n";                       #   print final result
__END__
the men and women merely players;\n
one man in his time
All the world's
their entrances,\nand one man
stage,\nAnd all the men and women
They have their exits and their entrances,\n
world's a stage,\nAnd all
their entrances,\nand one man
in his time plays many parts.
merely players;\nThey have
```

Biological classification and nomenclature

Back to the eighteenth century, when academic life at least was in some respects simpler.

Biological nomenclature is based on the idea that living things are divided into units called species—groups of similar organisms with a common gene pool. (Why living things should be 'quantized' into *discrete* species is a very complicated question.) Linnaeus, a Swedish naturalist, classified living things according to a hierarchy: Kingdom, Phylum, Class, Order, Family, Genus and Species (see Box). Modern taxonomists have added additional levels. For identification it generally suffices to specify the **binomial**: Genus and Species; for instance *Homo sapiens* for human or *Drosophila melanogaster* for fruit fly. Each binomial uniquely specifies a species that may also be known by one or more common names; for instance, *Bos taurus* = cow. Of course, most species have no common names.

Classifications of human and fruit fly

	Human	Fruit fly
Kingdom	Animalia	Animalia
Phylum	Chordata	Arthropoda
Class	Mammalia	Insecta
Order	Primata	Diptera
Family	Hominidae	Drosophilidae
Genus	*Homo*	*Drosophila*
Species	*sapiens*	*melanogaster*

Originally the Linnaean system was only a classification based on observed similarities. With the discovery of evolution it emerged that the system largely reflects biological ancestry. The question of which similarities truly reflect common ancestry must now be faced. Characteristics derived from a common ancestor are called **homologous**; for instance an eagle's wing and a human's arm. Other apparently similar characteristics may have arisen independently by **convergent evolution**; for instance, an eagle's wing and a bee's wing: The most recent common ancestor of eagles and bees did not have wings. Conversely, truly homologous characters may have diverged to become very dissimilar in structure and function. The bones of the human middle ear are homologous to bones in the jaws of primitive fishes; our eustachian tubes are homologues of gill slits. In most cases experts can distinguish true homologies from similarities resulting from convergent evolution.

Sequence analysis gives the most unambiguous evidence for the relationships among species. The system works well for higher organisms, for which sequence analysis and the classical tools of comparative anatomy, palaeontology and embryology usually give a consistent picture. Classification of micro-organisms is more difficult, partly because it is less obvious how to select the features on which to classify them and partly because a large amount of lateral gene transfer threatens to overturn the picture entirely.

Ribosomal RNAs turned out to have the essential feature of being present in all organisms, with the right degree of divergence. (Too much or too little divergence and relationships become invisible.)

On the basis of 15S ribosomal RNAs, C. Woese divided living things most fundamentally into three Domains (a level *above* Kingdom in the hierarchy): Bacteria, Archaea and Eukarya (see Fig. 1.2). Bacteria and archaea are prokaryotes; their cells do not contain nuclei. Bacteria include the typical micro-organisms responsible for many infectious diseases, and, of course, *Escherichia coli*, the mainstay of molecular biology. Archaea comprise extreme thermophiles and halophiles, sulphate reducers and methanogens. We ourselves are Eukarya—organisms containing cells with nuclei, including yeast and all multicellular organisms.

A census of the species with sequenced genomes reveals emphasis on bacteria, because of their clinical importance, and for the relative ease of sequencing genomes of prokaryotes. However, fundamentally we may have more to learn about ourselves from archaea than from bacteria. For despite the obvious differences in lifestyle, and the absence of a nucleus, archaea are in some ways more closely related on a molecular level to eukarya than to bacteria. It is also likely that the archaea are the closest living organisms to the root of the tree of life.

Figure 1.2 shows the deepest levels of the tree of life. The Eukarya branch includes animals, plants, and fungi. At the ends of the Eukarya branch are the metazoa (multicellular organisms) (Fig. 1.3). We and our closest relatives are deuterostomes (Fig. 1.4).

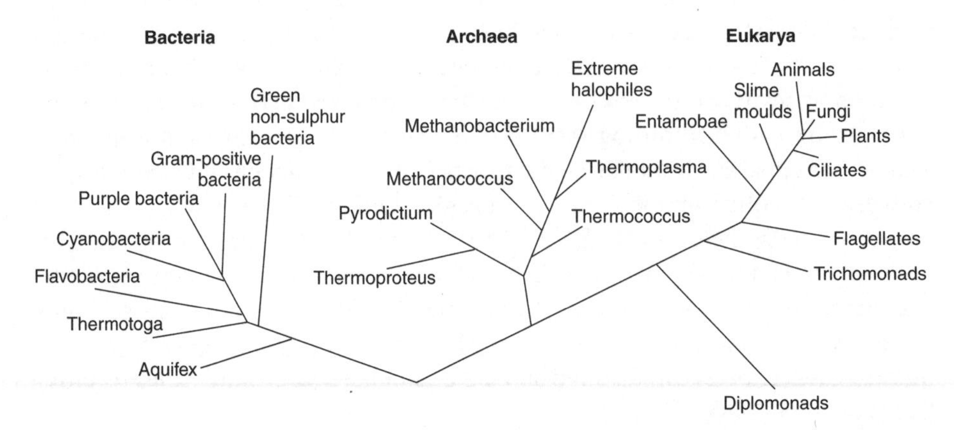

Fig. 1.2 Major divisions of living things, derived by C. Woese on the basis of 15S RNA sequences.

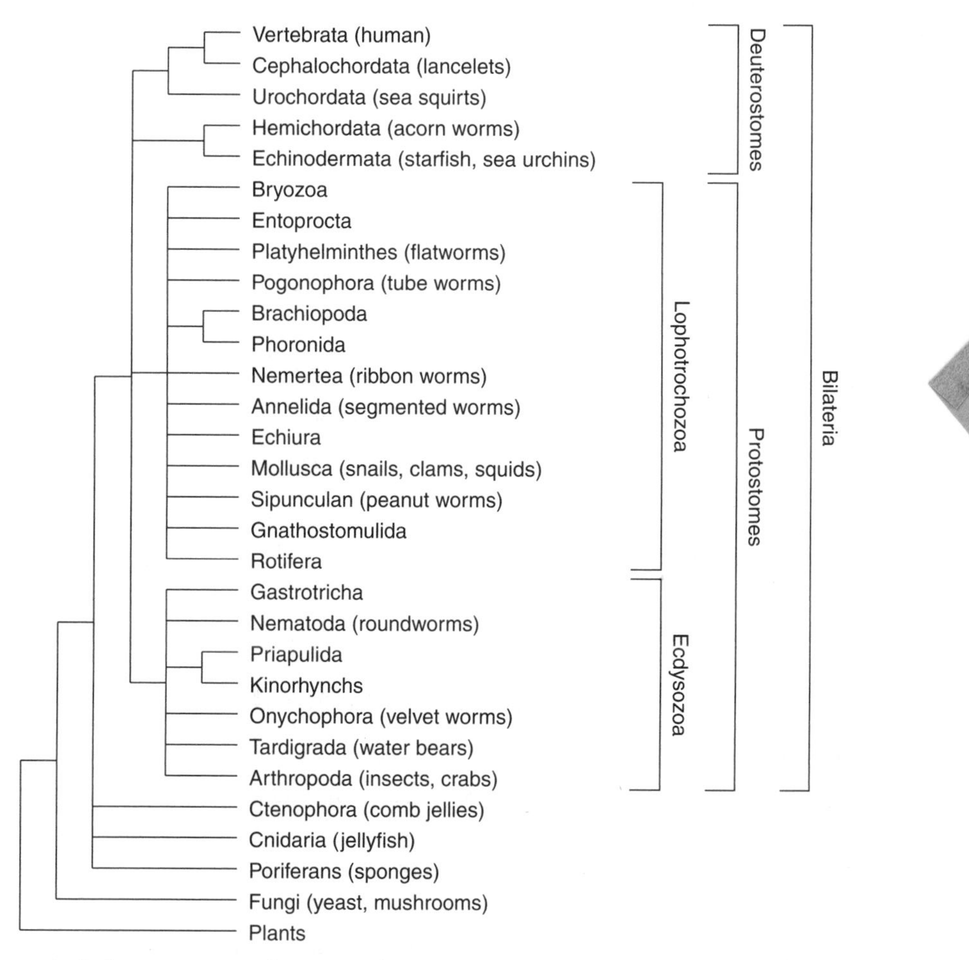

Fig. 1.3 Phylogenetic tree of metazoa (multicellular animals). Bilaterians include all animals that share a left-right symmetry of body plan. Protostomes and deuterostomes are two major lineages that separated at an early stage of evolution, estimated at 670 million years ago. They show very different patterns of embryological development, including different early cleavage patterns, opposite orientations of the mature gut with respect to the earliest invagination of the blastula, and the origin of the skeleton from mesoderm (deuterostomes) or ectoderm (protostomes). Protostomes comprise two subgroups distinguished on the basis of 18S RNA (from the small ribosomal subunit) and HOX gene sequences. Morphologically, Ecdysozoa have a moulting cuticle—a hard outer layer of organic material. Lophotrochozoa have soft bodies. (Based on Adouette, A., Balavoine, G., Lartillot, N., Lespinet, O., Prud'homme, B. & de Rosa, R. (2000), The new animal phylogeny: Reliability and implications, *Proceedings of the National Academy of Sciences USA*, **97**, 4453–4456.)

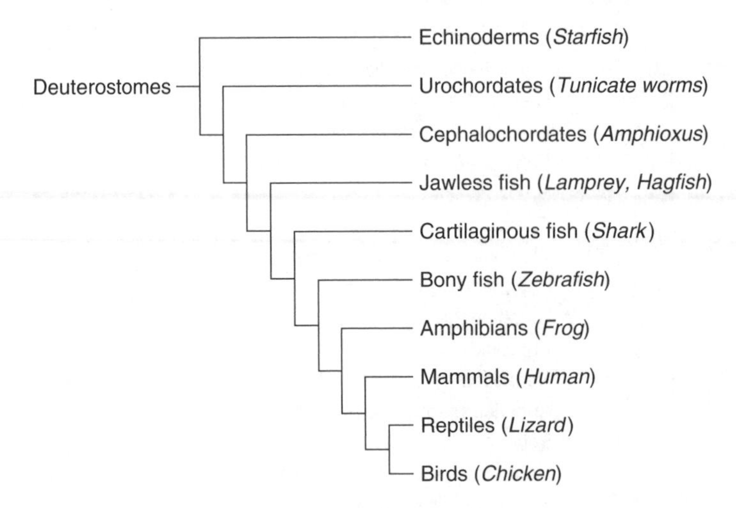

Fig. 1.4 Phylogenetic tree of vertebrates and our closest relatives. Chordates, including vertebrates, and echinoderms are all deuterostomes.

Use of sequences to determine phylogenetic relationships

Previous sections introduced sequence databanks and biological relationships. Here are examples of applications of retrieval of sequences from databanks, and of sequence comparisons to analysis of biological relationships.

Case Study 1.1 Retrieve the amino acid sequence of horse pancreatic ribonuclease

Use the ExPASy server at the Swiss Institute for Bioinformatics: The URL is: http://www.expasy.org/cgi-bin/sprot-search-ful. Type in the keywords `horse pancreatic ribonuclease` and press the ENTER key. Select `RNP_HORSE` and then `FASTA format` (see Box: FASTA format). This will produce the following (the first line has been truncated):

```
>sp|P00674|RNP_HORSE RIBONUCLEASE PANCREATIC (EC 3.1.27.5) (RNASE 1)...
KESPAMKFERQHMDSGSTSSSNPTYCNQMMKRRNMTQGWCKPVNTFVHEP
LADVQAICLQKNITCKNGQSNCYQSSSSMHITDCRLTSGSKYPNCAYQTS
QKERHIIVACEGNPYVPVHFDASVEVST
```

which can be cut and pasted into other programs.

For example, we could retrieve several sequences and align them (see Box: Sequence Alignment). Analysis of patterns of similarity among aligned sequences are useful properties in assessing closeness of relationships.

FASTA format

A very common format for sequence data is derived from conventions of FASTA, a program for FAST Alignment by W. R. Pearson. Many programs use FASTA format for reading sequences, or for reporting results.

A sequence in FASTA format:

- Begins with a single-line description. A > must appear in the first column. The rest of the title line is arbitrary but should be informative.
- Subsequent lines contain the sequence, one character per residue.
- Use one-letter codes for nucleotides or amino acids specified by the International Union of Biochemistry and International Union of Pure and Applied Chemistry (IUB/IUPAC).

  See http://www.chem.qmw.ac.uk/iupac/misc/naabb.html
  and http://www.chem.qmw.ac.uk/iupac/AminoAcid/

  Use Sec and U as the three-letter and one-letter codes for selenocysteine: http://www.chem.qmw.ac.uk/iubmb/newsletter/1999/item3.html
- Lines can have different lengths; that is, 'ragged right' margins.
- Most programs will accept lower case letters as amino acid codes.

An example of FASTA format: Bovine glutathione peroxidase

```
>gi|121664|sp|P00435|GSHC_BOVIN GLUTATHIONE PEROXIDASE
MCAAQRSAAALAAAAPRTVYAFSARPLAGGEPFNLSSLRGKVLLIENVASLUGTTVRDYTQMNDLQRRLG
PRGLVVLGFPCNQFGHQENAKNEEILNCLKYVRPGGGFEPNFMLFEKCEVNGEKAHPLFAFLREVLPTPS
DDATALMTDPKFITWSPVCRNDVSWNFEKFLVGPDGVPVRRYSRRFLTIDIEPDIETLLSQGASA
```

The title line contains the following fields:

> is obligatory in column 1.

gi|121664 is the *geninfo number*, an identifier assigned by the US National Center for Biotechnology Information (NCBI) to every sequence in its ENTREZ databank. The NCBI collects sequences from a variety of sources, including primary archival data collections and patent applications. Its gi numbers provide a common and consistent 'umbrella' identifier, superimposed on different conventions of source databases. When a source database updates an entry, the NCBI creates a new entry with a new gi number if the changes affect the sequence, but updates and retains its entry if the changes affect only non-sequence information, such as a literature citation.

sp|P00435 indicates that the source database was SWISS-PROT, and that the accession number of the entry in SWISS-PROT was P00435.

GSHC_BOVIN GLUTATHIONE PEROXIDASE is the SWISS-PROT identifier of sequence and species, (GSHC_BOVIN), followed by the name of the molecule.

Sequence Alignment

Sequence alignment is the assignment of residue-residue correspondences. We may wish to find:

- a *Global match*: align all of one sequence with all of the other.

```
And.--so,.from.hour.to.hour,.we.ripe.and.ripe
||||    |||||||||||||||||||||||||   ||||||
And.then,.from.hour.to.hour,.we.rot-.and.rot-
```

  This illustrates mismatches, insertions and deletions.

- a *Local match*: find a region in one sequence that matches a region of the other.

```
  My.care.is.loss.of.care,.by.old.care.done,
    |||||||||    |||||||||||||   |||||| ||
Your.care.is.gain.of.care,.by.new.care.won
```

  For local matching, overhangs at the ends are not treated as gaps. In addition to mismatches, seen in this example, insertions and deletions within the matched region are also possible.

- a *Motif match*: find matches of a short sequence in one or more regions internal to a long one.

  A perfect match:

```
    match
    |||||
The match is made; she seals it with a curtsy.
```

  One can allow msmatching characters:

```
          match
          |||||
for the watch to babble and to talk is most tolerable
```

```
or:  match                                    match
       |||                                    ||  |
  And witch the world with noble horsemanship.
```

  or insertions and/or deletions:

```
                mat--ch       match
                ||    |       ||  |
     Fear not, Macbeth; no manthat's born of woman
     Shall e'er have power upon thee.
```

- a *Multiple alignment*: a mutual alignment of many sequences.

```
no.sooner.---met.---------but.they.-look'd
no.sooner.look'd.---------but.they.-lo-v'd
no.sooner.lo-v'd.---------but.they.-sigh'd
no.sooner.sigh'd.---------but.they.--asked.one.
  another.the.reason
no.sooner.knew.the.reason.but.they.-------------
  sought.the.remedy
no.sooner.                .but.they.
```

The last line shows characters conserved in all sequences in the alignment.

See chapter 4 for an extended discussion of alignment.

Case Study 1.2

Determine, from the sequences of pancreatic ribonuclease from horse (*Equus caballus*), minke whale (*Balaenoptera acutorostrata*) and red kangaroo (*Macropus rufus*), which two of these species are most closely related.

Knowing that horse and whale are placental mammals and kangaroo is a marsupial, we expect horse and whale to be the closest pair. Retrieving the three sequences as in the previous example and pasting the following:

```
>RNP_HORSE
KESPAMKFERQHMDSGSTSSSNPTYCNQMMKRRNMTQGWCKPVNTFVHEP
LADVQAICLQKNITCKNGQSNCYQSSSSMHITDCRLTSGSKYPNCAYQTS
QKERHIIVACEGNPYVPVHFDASVEVST
>RNP_BALAC
RESPAMKFQRQHMDSGNSPGNNPNYCNQMMMRRKMTQGRCKPVNTFVHES
LEDVKAVCSQKNVLCKNGRTNCYESNSTMHITDCRQTGSSKYPNCAYKTS
QKEKHIIVACEGNPYVPVHFDNSV
>RNP_MACRU
ETPAEKFQRQHMDTEHSTASSSNYCNLMMKARDMTSGRCKPLNTFIHEPK
SVVDAVCHQENVTCKNGRTNCYKSNSRLSITNCRQTGASKYPNCQYETSN
LNKQIIVACEGQYVPVHFDAYV
```

into the multiple sequence alignment program CLUSTAL-W
http://www.ebi.ac.uk/clustalw/
(or alternatively, T-coffee: http://www.ch.embnet.org/software/TCoffee.html) produces the following:

```
CLUSTAL W (1.8) multiple sequence alignment

RNP_HORSE    KESPAMKFERQHMDSGSTSSSNPTYCNQMMKRRNMTQGWCKPVNTFVHEPLADVQAICLQ 60
RNP_BALAC    RESPAMKFQRQHMDSGNSPGNNPNYCNQMMMRRKMTQGRCKPVNTFVHESLEDVKAVCSQ 60
RNP_MACRU    -ETPAEKFQRQHMDTEHSTASSSNYCNLMMKARDMTSGRCKPLNTFIHEPKSVVDAVCHQ 59
              *:** **:*****:  :......*** **  *.**.* ***:***:**.   *.*:* *

RNP_HORSE    KNITCKNGQSNCYQSSSSMHITDCRLTSGSKYPNCAYQTSQKERHIIVACEGNPYVPVHF 120
RNP_BALAC    KNVLCKNGRTNCYESNSTMHITDCRQTGSSKYPNCAYKTSQKEKHIIVACEGNPYVPVHF 120
RNP_MACRU    ENVTCKNGRTNCYKSNSRLSITNCRQTGASKYPNCQYETSNLNKQIIVACEG-QYVPVHF 118
             :*: ****::***:*.* : **:** *..****** *:**: :::*******  ******

RNP_HORSE    DASVEVST 128
RNP_BALAC    DNSV---- 124
RNP_MACRU    DAYV---- 122
             *  *
```

In this table, a * under the sequences indicates a position that is conserved (the same in all sequences), and : and . indicate positions at which all sequences contain residues of very similar physicochemical character (:), or somewhat similar physicochemical character (.).

Large patches of the sequences are identical. There are numerous substitutions but only one internal deletion. By comparing the sequences *in pairs*, the numbers of identical residues shared among pairs in this alignment (not the same as counting *s) is:

Number of identical residues in aligned Ribonuclease A sequences (out of a total of 122–128 residues)			
Horse	and	Minke whale	95
Minke whale	and	Red kangaroo	82
Horse	and	Red kangaroo	75

→

Horse and whale share the most identical residues. The result appears significant, and therefore confirms our expectations. *Warning: Or is the logic really the other way round?*

Case Study 1.3

Let's try a harder one:

The two living genera of elephant are represented by the African elephant (*Loxodonta africana*) and the Indian (*Elephas maximus*). It has been possible to sequence the mitochondrial cytochrome *b* from a specimen of the Siberian woolly mammoth (*Mammuthus primigenius*) preserved in the Arctic permafrost. To which modern elephant is this mammoth more closely related?

Retrieving the sequences and running CLUSTAL-W:

```
Indian Elephant   MTHTRKFHPLFKIINKSFIDLPTPSNISTWWNFGSLLGACLITQILTGLFLAMHYTPDTM 60
Siberian Mammoth  MTHIRKSHPLLKILNKSFIDLPTPSNISTWWNFGSLLGACLITQILTGLFLAMHYTPDTM 60
African Elephant  MTHIRKSHPLLKIINKSFIDLPTPSNISTWWNFGSLLGACLITQILTGLFLAMHYTPDTM 60

                  *** ** ***:**:**********************************************

Indian Elephant   TAFSSMSHICRDVNYGWIIRQLHSNGASIFFLCLYTHIGRNIYYGSYLYSETWNTGIMLL 120
Siberian Mammoth  TAFSSMSHICRDVNYGWIIRQLHSNGASIFFLCLYTHIGRNIYYGSYLYSETWNTGIMLL 120
African Elephant  TAFSSMSHICRDVNYGWIIRQLHSNGASIFFLCLYTHIGRNIYYGSYLYSETWNTGIMLL 120
                  ************************************************************

Indian Elephant   LITMATAFMGYVLPWGQMSFWGATVITNLFSAIPYIGTNLVEWIWGGFSVDKATLNRFFA 180
Siberian Mammoth  LITMATAFMGYVLPWGQMSFWGATVITNLFSAIPYIGTDLVEWIWGGFSVDKATLNRFFA 180
African Elephant  LITMATAFMGYVLPWGQMSFWGATVITNLFSAIPYIGTNLVEWIWGGFSVDKATLNRFFA 180
                  **************************************:*********************

Indian Elephant   FHFILPFTMVALAGVHLTFLHETGSNNPLGLTSDSDKIPFHPYYTIKDFLGLLILILLLL 240
Siberian Mammoth  LHFILPFTMIALAGVHLTFLHETGSNNPLGLTSDSDKIPFHPYYTIKDFLGLLILILFLL 240
African Elephant  LHFILPFTMIALAGVHLTFLHETGSNNPLGLTSDSDKIPFHPYYTIKDFLGLLILILLLL 240
                  :********:***********************************************:**

Indian Elephant   LLALLSPDMLGDPDNYMPADPLNTPLHIKPEWYFLFAYAILRSVPNKLGGVLALFLSILI 300
Siberian Mammoth  LLALLSPDMLGDPDNYMPADPLNTPLHIKPEWYFLFAYAILRSVPNKLGGVLALLLSILI 300
African Elephant  LLALLSPDMLGDPDNYMPADPLNTPLHIKPEWYFLFAYAILRSVPNKLGGVLALLLSILI 300
                  ******************************************************:*****

Indian Elephant   LGLMPLLHTSKHRSMMLRPLSQVLFWTLTMDLLTLTWIGSQPVEHPYIIIGQMASILYFS 360
Siberian Mammoth  LGIMPLLHTSKHRSMMLRPLSQVLFWTLATDLLMLTWIGSQPVEYPYIIIGQMASILYFS 360
African Elephant  LGLMPLLHTSKHRSMMLRPLSQVLFWTLTMDLLTLTWIGSQPVEYPYIIIGQMASILYFS 360
                  **:*************************: *** **********:***************

Indian Elephant   IILAFLPIAGMIENYLIK 378
Siberian Mammoth  IILAFLPIAGMIENYLIK 378
African Elephant  IILAFLPIAGVIENYLIK 378
                  **********:*******
```

The mammoth and African elephant sequences have 4 mismatches, and the mammoth and Indian elephant sequences have 10 mismatches. It appears that mammoth is more closely related to African elephants. However, this result is less satisfying than the previous one. There are fewer differences. Are they significant? (In this case it is harder to decide whether the differences are significant because we have no preconceived idea of what the answer should be.)

This example raises a number of questions:

(1) We 'know' that African and Indian elephants and mammoths must be close relatives—just look at them. But could we tell *from these sequences alone* that they are from closely-related species?

(2) Given that the differences are few, do they represent true natural selection or merely random noise or drift? We need sensitive statistical criteria for judging the significance of the similarities and differences. It would be useful to look at the genes themselves, and check the ratio of the number of nonsynonymous substitutions to the number of synonymous substitutions. (A synonomous substitution is a change in a nucleic acid sequence that does not alter the protein sequence that it encodes. See pages 7–8.) A high ratio of nonsynonymous to synonymous substitutions suggests divergence under selective pressure.

As background to such questions, let us emphasize the distinction between **similarity** and **homology**. *Similarity* is the observation or measurement of resemblance and difference, independent of the source of the resemblance. *Homology* means, specifically, that the sequences and the organisms in which they occur are descended from a common ancestor, with the implication that the similarities are shared ancestral characteristics. Similarity of sequences (or of macroscopic biological characters) is observable in data collectable *now*, and involves no historical hypotheses. In contrast, assertions of homology are statements of historical events that are almost always unobservable. Homology must be an *inference* from observations of similarity. Only in a few special cases is homology directly observable; for instance in pedigrees of families showing unusual phenotypes such as the Hapsburg lip, or in laboratory populations, or in clinical studies that follow the course of viral infections at the sequence level in individual patients.

The assertion that the cytochromes *b* from African and Indian elephants and mammoths are homologous *means* that there was a common ancestor, presumably containing a unique cytochrome *b*, that by alternative mutations gave rise to the proteins of mammoths and modern elephants. Does the very high degree of similarity of the sequences justify the conclusion that they are homologous; or are there other explanations?

- It might be that a functional cytochrome *b requires* so many conserved residues that cytochromes *b* from all animals are as similar to one another as the elephant and mammoth proteins are. We can test this by looking at cytochrome *b* sequences from other species. The result is that cytochromes *b* from other animals differ substantially from those of elephants and mammoths.

- A second possibility is that there are special physiological requirements for a cytochrome *b* to function well in an animal with the size and form of an elephant, that the three cytochrome *b* sequences started out from independent ancestors, and that common selective pressures forced them to become similar. (Remember that we are asking what can be deduced from cytochrome *b* sequences alone.)

- The mammoth may be more closely related to the Indian elephant, but since the time of the last common ancestor the cytochrome *b* sequence of the Indian elephant has evolved faster than that of the African elephant or the mammoth, accumulating more mutations.

- Still a fourth possible hypothesis is that all common ancestors of elephants and mammoths had very dissimilar cytochromes *b*, but that living elephants and mammoths gained a common gene by transfer from an unrelated organism via a virus.

Suppose, however, we conclude that the similarity of the elephant and mammoth sequences is taken to be high enough to imply homology, what then about the ribonuclease sequences in the previous example? Are the *larger* differences among the pancreatic ribonucleases of horse, whale and kangaroo evidence that they are *not* homologues?

How can we answer these questions? Specialists have undertaken careful calibrations of sequence similarities and divergences, among many proteins from many species for which the taxonomic relationships have been worked out by classical methods. In the example of pancreatic ribonucleases, the reasoning from similarity to homology is justified. The question of whether mammoths are closer to African or Indian elephants was decided only recently, in favour of African elephants. Analysis of sequence similarities in genomes and proteins is now sufficiently well-established that it is considered the most reliable method for establishing phylogenetic relationships, even though sometimes the results may not be significant, while in other cases they even give incorrect answers. There are a lot of data available, effective tools for retrieving what is necessary to bring to bear on a specific question, and powerful analytic tools. None of this replaces the need for thoughtful scientific judgement.

Use of SINES and LINES to derive phylogenetic relationships

Major problems with inferring phylogenies from comparisons of gene and protein sequences are (1) the wide range of variation of similarity, which may dip below statistical significance, and (2) the effects of different rates of evolution along different branches of the evolutionary tree. In many cases, even if sequence similarities confidently establish relationships, it may be impossible to decide the *order* in which sets of taxa have split. The phylogeneticist's dream—features that have 'all-or-none' character, and the appearance of which is irreversible so that the order of branching events can be decided—is in some cases afforded by certain non-coding sequences in genomes.

SINES and LINES (Short and Long Interspersed Nuclear ElementS) are repetitive non-coding sequences that form large fractions of eukaryotic genomes—at least 30% of human chromosomal DNA, and over 50% of some higher plant genomes. Typically, SINES are ~70–500 base pairs long, and up to 10^6 copies may appear. LINES may be up to 7000 base pairs long, and up to 10^5 copies may appear. SINES enter the genome by reverse transcription of RNA. Most SINES contain a 5′ region homologous to tRNA, a central region unrelated to tRNA, and a 3′ AT-rich region.

Features of SINES that make them useful for phylogenetic studies include:

- A SINE is either present or absent. Presence of a SINE at any particular position is a property that entails no complicated and variable measure of similarity.
- SINES are inserted at random in the non-coding portion of a genome. Therefore appearance of similar SINES at the same locus in two species implies that

the species share a common ancestor in which the insertion event occurred. No analogue of convergent evolution muddies this picture, because there is no selection for the *site* of insertion.

- SINE insertion appears to be irreversible: no mechanism for *loss* of SINES is known, other than rare large-scale deletions that include the SINE. Therefore if two species share a SINE at a common locus, *absence* of this SINE in a third species implies that the first two species must be more closely related to each other than either is to the third.
- Not only do SINES show relationships, they imply which species came first. The last common ancestor of species containing a common SINE must have come *after* the last common ancestor linking these species and another that lacks this SINE.

N. Okada and colleagues applied SINE sequences to questions of phylogeny.

Whales, like Australians, are mammals that have adopted an aquatic lifestyle. But what—in the case of the whales—are their closest land-based relatives? Classical palaeontology linked the order *Cetacea*—comprising whales, dolphins and porpoises—with the order *Artiodactyla*—even-toed ungulates (including

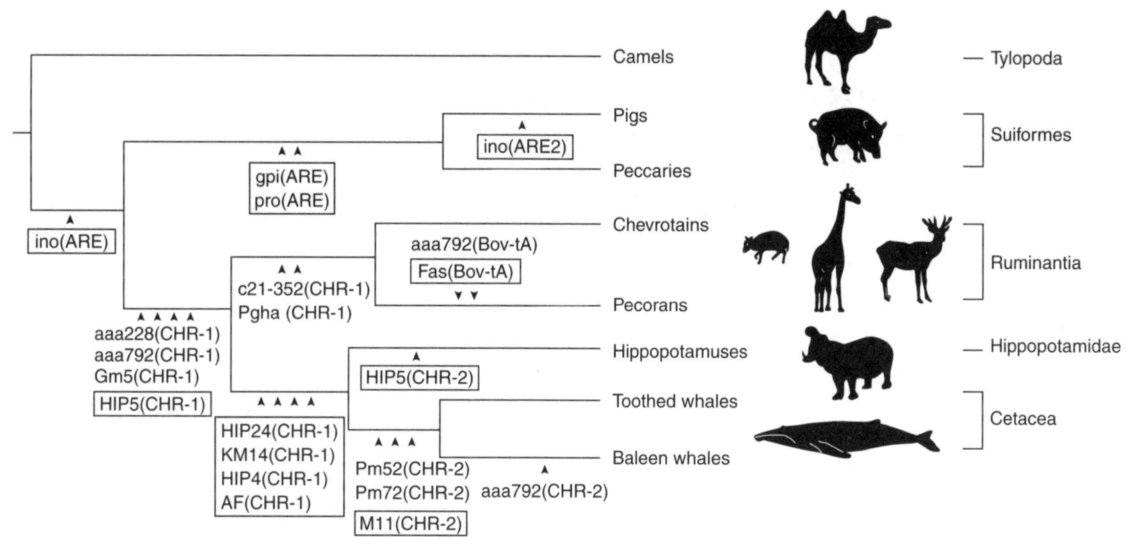

Fig. 1.5 Phylogenetic relationships among cetaceans and other artiodactyl subgroups, derived from analysis of SINE sequences. Small arrowheads mark insertion events. Each arrowhead indicates the presence of a particular SINE or LINE at a specific locus in all species to the *right* of the arrowhead. Lower-case letters identify loci, upper-case letters identify sequence patterns. For instance, the ARE2 pattern appears only in pigs, at the ino locus. The ARE pattern appears twice in the pig genome, at loci gpi and pro, and in the peccary genome at the same loci. The ARE insertion occurred in a species that was ancestral to pigs and peccaries but to no other species in the diagram. This implies that pigs and peccaries are more closely related to each other than to any of the other animals studied. (From Nikaido, M., Rooney, A. P. & Okada, N. (1999), Phylogenetic relationships among cetartiodactyls based on insertions of short and long interpersed elements: hippopotamuses are the closest extant relatives of whales, *Proceedings of the National Academy of Sciences USA*, **96**, 10261–10266. (Copyright 1999, National Academy of Sciences, USA. Reproduced by permission.))

cattle). Cetaceans were thought to have diverged before the common ancestor of the three extant Artiodactyl suborders: *Suiformes* (pigs), *Tylopoda* (including camels and llamas), and *Ruminantia* (including deer, cattle, goats, sheep, antelopes, giraffes, etc.). To place cetaceans properly among these groups, several studies were carried out with DNA sequences. Comparisons of mitochondrial DNA, and genes for pancreatic ribonuclease, γ-fibrinogen, and other proteins, suggested that the closest relatives of the whales are hippopotamuses, and that cetaceans and hippopotamuses form a separate group within the artiodactyls, most closely related to the *Ruminantia* (see Weblem 1.7).

Analysis of SINES confirms this relationship. Several SINES are common to *Ruminantia*, hippopotamuses and cetaceans. Four SINES appear in hippopotamuses and cetaceans only. These observations imply the phylogenetic tree shown in Fig. 1.5, in which the SINE insertion events are marked.

Recently-discovered fossils of land-based ancestors of whales confirm the link between whales and artiodactyls. This is a good example of the complementarity between molecular and palaeontological methods: DNA sequence analysis can specify relationships among living species quite precisely, but fossils reveal relationships among their *extinct* ancestors.

Searching for similar sequences in databases: PSI-BLAST

A common theme of the examples we have treated is the search of a database for items similar to a probe. For instance, if you are studying a novel genome, or if you identify within the human genome a gene responsible for some disease, you will wish to determine whether related genes appear in other species. The ideal method is both **sensitive**—that is, it picks up even very distant relationships—and **selective**—that is, all the relationships that it reports are true.

Sensitivity and selectivity

Database search methods involve a trade-off between *sensitivity* and *selectivity*. Does the method find all or most of the examples that are actually present, or does it miss a large fraction? Conversely, how many of the 'hits' that it reports are incorrect? Suppose a database contains 1000 globin sequences. Suppose a search of this database for globins reported 900 results, 700 of which were really globin sequences and 200 of which were not. This result would be said to have 300 false negatives (misses) and 200 false positives. Lowering a tolerance threshold will increase the numbers of *both* false negatives and false positives. Often one is willing to work with low thresholds to make sure of not missing anything that might be important; but this requires detailed examination of the results to eliminate the resulting false positives.

A powerful tool for searching sequence databases with a probe sequence is PSI-BLAST, from the US National Center for Biotechnology Information (NCBI). PSI-BLAST stands for 'Position Sensitive Iterated–Basic Linear Alignment Sequence Tool'. A previous program, BLAST, worked by identifying local regions of similarity without gaps and then piecing them together. The PSI in PSI-BLAST refers to enhancements that identify patterns within the sequences at preliminary stages of the database search, and then progressively refine them. Recognition of conserved patterns can sharpen both the selectivity and sensitivity of the search. PSI-BLAST involves a repetitive (or iterative) process, as the emergent pattern becomes better defined in successive stages of the search.

Case Study 1.4 Homologues of the human PAX-6 gene

PAX-6 genes control eye development in a widely divergent set of species (see Box). The human PAX-6 gene encodes the protein appearing in SWISS-PROT entry P26367. To run PSI-BLAST, go to the following URL: http://www.ncbi.nlm.nih.gov/blast/index.shtml and select PHI- and PSI-BLAST, under Protein. Enter the sequence, and use the default options for selections of the database to search and the similarity matrix used.

The program returns a list of entries similar to the probe, sorted in decreasing order of statistical significance. (Extracts from the response are shown in the Box, Results of PSI-BLAST search for human PAX-6 protein, page 35. Only a few lines are shown, merely to illustrate the format.) A typical line, from farther down the list, appears as follows:

```
pir||I45557  eyeless, long form - fruit fly (Drosophila melano...   255  7e-67
```

The first item on the line is the database and corresponding entry number (separated by ||) in this case, the PIR (Protein Identification Resource) entry I45557. It is the *Drosophila* homologue *eyeless*. The number 255 is a score for the match detected, and the significance of this match is measured by $E = 7 \times 10^{-67}$. E is related to the probability that the observed degree of similarity could have arisen by chance: E is the number of sequences that would be expected to match as well or better than the one being considered, if the same database were probed with random sequences. $E = 7 \times 10^{-67}$ means that it is *extremely* unlikely that even *one* random sequence would match as well as the *Drosophila* homologue. Values of E below about 0.05 would be considered significant; at least they might be worth considering. For borderline cases, you would ask: Are the mismatches conservative? Is there any pattern or are the matches and mismatches distributed randomly through the sequences? There is an elusive concept, the *texture* of an alignment, that you will become sensitive to. The court of last resort is whether the structures are similar, but often this information is not available.

→

→

Note that if there are many sequences in the databank that are very similar to the probe sequence, they will head the list. In this example, there are many very similar PAX genes in other vertebrates. You may have to scan far down the list to find a distant relative that you consider interesting.

Even in the case of *Drosophila* eyeless, a very close relative of the probe sequence, the program reports only a local match to a portion of the sequences. The full alignment is shown in the Box, Complete pairwise sequence alignment of human PAX-6 protein and *Drosophila melanogaster* eyeless, page 37.

Et in terra PAX hominibus, muscisque . . .

The eyes of the human, fly and octopus are very different in structure. Conventional wisdom, noting the immense selective advantage conferred by the ability to see, held that eyes arose independently in different phyla. It therefore came as a great surprise that a gene controlling human eye development has a homologue governing eye development in *Drosophila*.

The PAX-6 gene was first cloned in the mouse and human. It is a master regulatory gene, controlling a complex cascade of events in eye development. Mutations in the human gene cause the clinical condition *aniridia*, a developmental defect in which the iris of the eye is absent or deformed. The PAX-6 homologue in *Drosophila*—called the *eyeless* gene—has a similar function of control over eye development. Flies mutated in this gene develop without eyes; conversely, expression of this gene in a fly's wing, leg, or antenna produces ectopic (= out of place) eyes. (The *Drosophila eyeless* mutant was first described in 1915. Little did anyone then suspect a relation to a mammalian gene.)

Not only are the insect and mammalian genes similar in sequence, they are so closely related that their function crosses species boundaries. Expression of the mouse PAX-6 gene in the fly causes ectopic eye development just as expression of the fly's own *eyeless* gene does.

PAX-6 has homologues in other phyla, including flatworms, ascidians, sea urchins and nematodes. The observation that rhodopsins—a family of proteins containing retinal as a common chromophore—function as light-sensitive pigments in different phyla is supporting evidence for a common origin of different photoreceptor systems. The genuine structural differences in the macroscopic anatomy of different eyes reflect the divergence and independent development of higher-order structure.

Results of PSI-BLAST search for human PAX-6 protein

Five iterations of PSI-BLAST were run using Human PAX-6 as the query sequence, searching the nonredundant (nr) database. The NCBI nr database is a set of unique sequences selected from the full databases to eliminate multiple hits. The output contains a list of similar sequences identified in the database. It also contains pairwise alignments of well-matching regions from the query sequence and the retrieved sequences. Three selected alignments are shown here: PAX-6 from *Danio rerio*, $E = 10^{-134}$; *Drosophila eyeless*, $E = 7\times10^{-67}$ and another *Drosophila* protein, 'even skipped', with $E = 0.001$, for which the matching is both shorter and less perfect:

```
Score = 46.9 bits (110), Expect = 0.001
```

```
Reference: Altschul, Stephen F., Thomas L. Madden, Alejandro A. Schaffer,
Jinghui Zhang, Zheng Zhang, Webb Miller, and David J. Lipman (1997),
"Gapped BLAST and PSI-BLAST: a new generation of protein database search
programs", Nucleic Acids Res. 25:3389-3402.
RID: 1081182496-14481-96333547442.BLASTQ3
Query= sp|P26367|PAX6_HUMAN Paired box protein Pax-6(Oculorhombin)
(Aniridia, type II protein) - Homo sapiens (Human).
        (422 letters)

Database: All non-redundant GenBank CDS
translations+PDB+SwissProt+PIR+PRF
          2,738,511 sequences; 768,166,133 total letters

Results of PSI-Blast iteration 5
Sequences with E-value BETTER than threshold
Sequences producing significant alignments:                       (bits) Value

gi|2696972|dbj|BAA24025.1| PAX6 SL [Cynops pyrrhogaster]             499   e-140
gi|189353|gb|AAA59962.1| oculorhombin                                496   e-139
gi|4505615|ref|NP_000271.1| paired box gene 6 isoform a; Paired box h...  495   e-139
gi|27469846|gb|AAH41712.1| MGC52531 protein [Xenopus laevis]         495   e-139
gi|6981334|ref|NP_037133.1| paired box gene 6; paired box homeotic ge...  494   e-138
gi|2696971|dbj|BAA24024.1| PAX6 LL [Cynops pyrrhogaster]             491   e-137
gi|7305369|ref|NP_038655.1| paired box gene 6; small eye; Dickie's sm...  491   e-137
gi|42592306|emb|CAF29075.1| putative pax6 isoform 5a [Rattus norvegicus]  491   e-137
gi|1685047|gb|AAB36681.1| paired-type homeodomain Pax-6 protein [Xeno...  491   e-137
gi|26389393|dbj|BAC25729.1| unnamed protein product [Mus musculus]   490   e-137
gi|4580424|ref|NP_001595.2| paired box gene 6 isoform b; Paired box h...  490   e-137
gi|8132383|gb|AAF73271.1| paired domain transcription factor variant ...  490   e-137
gi|5758939|gb|AAD50903.1| paired-box transcription factor -+ isoform ...  489   e-137
gi|18138028|emb|CAC80516.1| paired box protein [Mus musculus]        489   e-137
gi|44922124|gb|AAS48919.1| paired box 6 isoform 5a [Rattus norvegicus]  489   e-137
gi|45384210|ref|NP_990397.1| PAX6 protein [Gallus gallus]            489   e-137
gi|383296|prf||1902328A PAX6 gene                                    488   e-137
gi|1488322|gb|AAB05932.1| Xpax6 [Xenopus laevis]                     488   e-136
gi|34364871|emb|CAE45868.1| hypothetical protein [Homo sapiens]      488   e-136
gi|2369655|emb|CAA68838.1| PAX-6 protein [Astyanax mexicanus]        487   e-136
```

. . . additional 'hits' deleted . . .
. . . three selected alignments follow. . .

→

Results of PSI-BLAST search for human PAX-6 protein *(continued)*

```
                              Alignments
>gi|18859209|ref|NP_571379.1| paired box gene 6a; paired box homeotic
gene 6 [Danio rerio] Length = 451

 Score =  478 bits (1231), Expect = e-134
 Identities = 404/436 (92%), Positives = 409/436 (93%), Gaps = 18/436 (4%)

Query: 1   MQNSHSGVNQLGGVFVNGRPLPDSTRQKIVELAHSGARPCDISRILQ------------ 47
            MQNSHSGVNQLGGVFVNGRPLPDSTRQKIVELAHSGARPCDISRILQ
Sbjct: 20  MQNSHSGVNQLGGVFVNGRPLPDSTRQKIVELAHSGARPCDISRILQTHADAKVQVLDNE 79

Query: 48  -VSNGCVSKILGRYYETGSIRPRAIGGSKPRVATPEVVSKIAQYKRECPSIFAWEIRDRL 106
             VSNGCVSKILGRYYETGSIRPRAIGGSKPRVATPEVV KIAQYKRECPSIFAWEIRDRL
Sbjct: 80  NVSNGCVSKILGRYYETGSIRPRAIGGSKPRVATPEVVGKIAQYKRECPSIFAWEIRDRL 139

Query: 107 LSEGVCTNDNIPSVSSINRVLRNLASEKQQMGADGMYDKLRMLNGQTGSWGTRPGWYPGT 166
            LSEGVCTNDNIPSVSSINRVLRNLASEKQQMGADGMY+KLRMLNGQTG+WGTRPGWYPGT
Sbjct: 140 LSEGVCTNDNIPSVSSINRVLRNLASEKQQMGADGMYEKLRMLNGQTGTWGTRPGWYPGT 199

Query: 167 SVPGQPTQDGCQQQEGGGENTNSISSNGEDSDEAQMRLQLKRKLQRNRTSFTQEQIEALE 226
            SVPGQP QDGCQQ +GGGENTNSISSNGEDSDE QMRLQLKRKLQRNRTSFTQEQIEALE
Sbjct: 200 SVPGQPNQDGCQQSDGGGENTNSISSNGEDSDETQMRLQLKRKLQRNRTSFTQEQIEALE 259

Query: 227 KEFERTHYPDVFARERLAAKIDLPEARIQVWFSNRRAKWRREEKLRNQRRQASNTPSHIP 286
            KEFERTHYPDVFARERLAAKIDLPEARIQVWFSNRRAKWRREEKLRNQRRQASN+ SHIP
Sbjct: 260 KEFERTHYPDVFARERLAAKIDLPEARIQVWFSNRRAKWRREEKLRNQRRQASNSSSHIP 319

Query: 287 ISSSFSTSVYQPIPQPTTPVSSFTSGSMLGRTDTALTNTYSALPPMPSFTMANNLPMQPP 346
            ISSSFSTSVYQPIPQPTTPV SFTSGSMLGR+DTALTNTYSALPPMPSFTMANNLPMQP
Sbjct: 320 ISSSFSTSVYQPIPQPTTPV-SFTSGSMLGRSDTALTNTYSALPPMPSFTMANNLPMQP- 377

Query: 347 VPSQTSSYSCMLPTSPSVNGRSYDTYTPPHMQTHMNSQPMGTSGTTSTGLISPGVSVPVQ 406
              SQTSSYSCMLPTSPSVNGRSYDTYTPPHMQ HMNSQ M  SGTTSTGLISPGVSVPVQ
Sbjct: 378 --SQTSSYSCMLPTSPSVNGRSYDTYTPPHMQAHMNSQSMAASGTTSTGLISPGVSVPVQ 435

Query: 407 VPGSEPDMSQYWPRLQ 422
            VPGSEPDMSQYWPRLQ
Sbjct: 436 VPGSEPDMSQYWPRLQ 451


>pir||I45557 eyeless, long form - fruit fly (Drosophila melanogaster)
 emb|CAA56038.1| (X79493) transcription factor [Drosophila melanogaster]
          Length = 838

 Score =  255 bits (644), Expect = 7e-67
 Identities = 124/132 (93%), Positives = 128/132 (96%)

Query: 5   HSGVNQLGGVFVNGRPLPDSTRQKIVELAHSGARPCDISRILQVSNGCVSKILGRYYETG 64
            HSGVNQLGGVFV GRPLPDSTRQKIVELAHSGARPCDISRILQVSNGCVSKILGRYYETG
Sbjct: 38  HSGVNQLGGVFVGGRPLPDSTRQKIVELAHSGARPCDISRILQVSNGCVSKILGRYYETG 97

Query: 65  SIRPRAIGGSKPRVATPEVVSKIAQYKRECPSIFAWEIRDRLLSEGVCTNDNIPSVSSIN 124
            SIRPRAIGGSKPRVAT EVVSKI+QYKRECPSIFAWEIRDRLL E VCTNDNIPSVSSIN
Sbjct: 98  SIRPRAIGGSKPRVATAEVVSKISQYKRECPSIFAWEIRDRLLQENVCTNDNIPSVSSIN 157

Query: 125 RVLRNLASEKQQ 136
            RVLRNLA++K+Q
Sbjct: 158 RVLRNLAAQKEQ 169
```

→

```
>gi|24652276|ref|NP_523670.2| CG2328-PA [Drosophila melanogaster]
 gi|12644102|sp|P06602|HMEV_DROME Segmentation protein even-skipped
 gi|7958|emb|CAA28784.1| even skipped [Drosophila melanogaster]
 gi|7303818|gb|AAF58865.1| CG2328-PA [Drosophila melanogaster]
          Length = 376

 Score = 46.9 bits (110), Expect = 0.001
 Identities = 33/95 (34%), Positives = 51/95 (53%)

Query: 174 QDGCQQQEGGGENTNSISSNGEDSDEAQMRLQLKRKLQRNRTSFTQEQIEALEKEFERTH 233
            Q G  Q      N    S +   +      +     ++R RT+FT++Q+  LEKEF + +
Sbjct: 34  QYGKPQTPPPSPNECLSSPDNSLNGSRGSEIPADPSVRRYRTAFTRDQLGRLEKEFYKEN 93

Query: 234 YPDVFARERLAAKIDLPEARIQVWFSNRRAKWRRE 268
            Y     R  LAA+++LPE+ I+VWF NRR K +R+
Sbjct: 94  YVSRPRRCELAAQLNLPESTIKVWFQNRRMKDKRQ 128
```

Complete pairwise sequence alignment of human PAX-6 protein and *Drosophila melanogaster* eyeless

```
PAX6_human      ---------------------------------MQNSHSGVNQLGGVFVNGRPLPDSTRQ 27
eyeless         MFTLQPTPTAIGTVVPPWSAGTLIERLPSLEDMAHKGHSGVNQLGGVFVGGRPLPDSTRQ 60
                                                  ::.************.**********

PAX6_human      KIVELAHSGARPCDISRILQVSNGCVSKILGRYYETGSIRPRAIGGSKPRVATPEVVSKI 87
eyeless         KIVELAHSGARPCDISRILQVSNGCVSKILGRYYETGSIRPRAIGGSKPRVATAEVVSKI 120
                *****************************************************.******

PAX6_human      AQYKRECPSIFAWEIRDRLLSEGVCTNDNIPSVSSINRVLRNLASEKQQ----------- 136
eyeless         SQYKRECPSIFAWEIRDRLLQENVCTNDNIPSVSSINRVLRNLAAQKEQQSTGSGSSSTS 180
                :*******************.*.*********************::*:*

PAX6_human      ------------MG-------------------------------------------ADG 141
eyeless         AGNSISAKVSVSIGGNVSNVASGSRGTLSSSTDLMQTATPLNSSESGGATNSGEGSEQEA 240
                            :*                                            :.

PAX6_human      MYDKLRMLNGQTGS--------------------WGTRP--------------------- 160
eyeless         IYEKLRLLNTQHAAGPGPLEPARAAPLVGQSPNHLGTRSSHPQLVHGNHQALQQHQQQSW 300
                :*:***:** * .:                     ***.

PAX6_human      -------GWYPG-------TSVP------------------------------GQP---- 172
eyeless         PPRHYSGSWYPTSLSEIPISSAPNIASVTAYASGPSLAHSLSPPNDIKSLASIGHQRNCP 360
                       .***        :*.*                              *:

PAX6_human      ----------TQDGCQQQEGG---GENTNSISSNGEDSDEAQMRLQLKRKLQRNRTSFTQ 219
eyeless         VATEDIHLKKELDGHQSDETGSGEGENSNGGASNIGNTEDDQARLILKRKLQRNRTSFTN 420
                            ** *.:* *   ***:*. :**  :::: * ** *************:

PAX6_human      EQIEALEKEFERTHYPDVFARERLAAKIDLPEARIQVWFSNRRAKWRREEKLRNQRRQAS 279
eyeless         DQIDSLEKEFERTHYPDVFARERLAGKIGLPEARIQVWFSNRRAKWRREEKLRNQRRTPN 480
                :**::********************.**.**************************** ..

PAX6_human      NTPSHIPISSSFSTSVYQPIPQPTTPVSSFTSGSMLG----------------------- 316
eyeless         STGASATSSSTSATASLTDSPNSLSACSSLLSGSAGGPSVSTINGLSSPSTLSTNVNAPT 540
                .* :  . **: :*:     *:. :. **: ***  *

PAX6_human      ------------------------------------------------------------
eyeless         LGAGIDSSESPTPIPHIRPSCTSDNDNGRQSEDCRRVCSPCPLGVGGHQNTHHIQSNGHA 600

PAX6_human      ----------------------------RTDTALTNTYSALPPMPSFTMANNLPMQPPVP 348
eyeless         QGHALVPAISPRLNFNSGSFGAMYSNMHHTALSMSDSYGAVTPIPSFNHSAVGPLAPPSP 660
                                            :*  :::::*.*:.*:***. :   *: ** *

PAX6_human      S-------QTSSYSCMLPTSP---------------------------------SVNGRS 368
eyeless         IPQQGDLTPSSLYPCHMTLRPPPMAPAHHHIVPGDGGRPAGVGLGSGQSANLGASCSGSG 720
                         :* *.* :.  *                                 * .* .

PAX6_human      YDTYTP-----------------------------PHMQTHMNSQP----------MGTS 389
eyeless         YEVLSAYALPPPPMASSSAADSSFSAASSASANVTPHHTIAQESCPSPCSSASHFGVAHS 780
                *:. :.                             **      :* *          :. *

PAX6_human      GTTSTGLISPGVS----------------VPVQVPGS----EPDMSQYWPRLQ----- 422
eyeless         SGFSSDPISPAVSSYAHMSYNYASSANTMTPSSASGTSAHVAPGKQQFFASCFYSPWV 838
                .  *:. ***.**                .* ...*:     *. .*::.
```

Species recognized by PSI-BLAST 'hits' to probe sequence human PAX-6

```
Achaearanea tepidariorum        Helobdella triserialis
Acipenser baerii                Hemicentrotus pulcherrimus
Acropora millepora              Herdmania curvata
Acropora palmata                Heterodontus francisci
Ambystoma mexicanum             Homo sapiens
Anopheles gambiae               Hydra littoralis
Archaster typicus               Hydra magnipapillata
Archegozetes longisetosus       Hydra vulgaris
Artemia franciscana             Hydractinia symbiolongicarpus
Astatoreochromis alluaudi       Ilyanassa obsoleta
Astatotilapia brownae           Labidochromis caeruleus
Asterina miniata                Lampetra fluviatilis
Astyanax mexicanus              Lampetra planeri
Bos taurus                      Latimeria menadoensis
Branchiostoma belcheri          Lethenteron japonicum
Branchiostoma floridae          Leucopsarion petersii
Branchiostoma lanceolatum       Lineus sanguineus
Caenorhabditis briggsae         Lingula anatina
Caenorhabditis elegans          Loligo opalescens
Canis familiaris                Lytechinus variegatus
Carassius auratus               Mesocricetus auratus
Cassiopea xamachana             Monodelphis domestica
Chrysaora quinquecirrha         Mus musculus
Ciona intestinalis              Nematostella vectensis
Ciona savignyi                  Nereis virens
Cladonema californicum          Oreochromis aureus
Coturnix coturnix               Oreochromis niloticus
Crocodylus niloticus            Oryctolagus cuniculus
Ctenophorus ornatus             Oryzias latipes
Cupiennius salei                Paracentrotus lividus
Cynops pyrrhogaster             Paralichthys olivaceus
Cyprinus carpio                 Patella vulgata
Danio rerio                     Petromyzon marinus
Discocelis tigrina              Phallusia mammilata
Drosophila ananassae            Phascolion strombi
Drosophila mauritiana           Platynereis dumerilii
Drosophila melanogaster         Pleurodeles waltl
Drosophila sechellia            Podocoryne carnea
Drosophila simulans             Pristina leidyi
Drosophila virilis              Ptychodera flava
Dugesia japonica                Raja eglanteria
Echinococcus granulosus         Rattus norvegicus
Eleutheria dichotoma            Related structures
Eleutherodactylus coqui         Saccoglossus kowalevskii
Emys orbicularis                Schistocerca americana
Ephydatia fluviatilis           Scyliorhinus canicula
Ephydatia muelleri              Spongilla lacustris
Equus caballus                  Stichopus japonicus
Eremothecium gossypii           Strongylocentrotus purpuratus
Euprymna scolopes               Sus scrofa
Gallotia stehlini               Takifugu rubripes
Gallus gallus                   Tetranychus urticae
Gasterosteus aculeatus          Theromyzon trizonare
Girardia tigrina                Tribolium castaneum
Gryllus bimaculatus             Trichoplax adhaerens
Haliotis asinina                Tripedalia cystophora
Haliotis rufescens              Triturus alpestris
Halocynthia roretzi             Tropheus duboisi
Heliocidaris erythrogramma      Xenopus laevis
Heliocidaris tuberculata        Xenopus tropicalis
Helobdella robusta              Xiphophorus hellerii
```

PERL Example 1.3 What species contain homologues of human PAX-6 detectable by PSI-BLAST?

PSI-BLAST reports the species in which the identified sequences occur (see Box, Results of PSI-BLAST search for human PAX-6 protein). These appear, embedded in the text of the output, in square brackets; for instance:

```
emb|CAA56038.1| (X79493) transcription factor [Drosophila melanogaster]
```

(In the section reporting *E*-values, the species names may be truncated.)

→

→

The following PERL program extracts species names from the PSI-BLAST output.

```
#!/usr/bin/perl
#extract species from psiblast output

# Method:
#   For each line of input, check for a pattern of form [Drosophila melanogaster]
#   Use each pattern found as the index in an associative array
#   The value corresponding to this index is irrelevant
#   By using an associative array, subsequent instances of the same
#      species will overwrite the first instance, keeping only a unique set
#   After processing of input complete, sort results and print.


while (<>) {                                # read line of input
    if (/\[([A-Z][a-z]+ [a-z]+)\]/) {       # select lines containing strings of form
                                            #     [Drosophila melanogaster]
     $species{$1} = 1;                      # make or overwrite entry in
   }                                        #         associative array
}

foreach (sort(keys(%species))){             # in alphabetical order,
    print "$_\n";                           #    print species names
}
```

There are 122 species found (see Box: Species recognized by PSI-BLAST 'hits' to probe sequence human PAX-6).

The program makes use of PERL's rich pattern recognition resources to search for character strings of the *form* `[Drosophila melanogaster]`. We want to specify the following pattern:

- a square bracket,
- followed by a word beginning with an upper case letter followed by a variable number of lower case letters,
- then a space between words,
- then a word all in lowercase letters,
- then a closing square bracket.

This kind of pattern is called a **regular expression** and appears in the PERL program in the following form: `[([A-Z][a-z]+ [a-z]+)]`.
Building blocks of the pattern specify ranges of characters:

`[A-Z]` = any letter in the range A, B, C, . . . Z

`[a-z]` = any letter in the range a, b, c, . . . z

We can specify repetitions:

`[A-Z]` = *one* upper case letter

`[a-z]+` = *one or more* lower case letters

and combine the results:

`[A-Z][a-z]+[a-z]+` = an upper case letter followed by one or more lower case letters (the genus name), followed by a blank, followed by one or more lower case letters (the species name).

→

PERL Example 1.3 *(continued)*

Enclosing these in parentheses: `([A-Z][a-z]+ [a-z]+)` tells PERL to save the material that matched the pattern for future reference. In PERL this matched material is designated by the variable `$1`. Thus if the input line contained `[Drosophila melanogaster]`, the statement

```
$species{$1} = 1;
```

would effectively be:

```
$species{"Drosophila melanogaster"} = 1;
```

Finally, we want to include the brackets surrounding the genus and species name, but brackets signify character ranges. Therefore we must precede the brackets by backslashes: `\[ . . . \]`, to give the final pattern: `\[([A-Z][a-z]+ [a-z]+)\]`.

The use of the associative array to retain only a unique set of species is another instructive aspect of the program. Recall that an associative array is a generalization of an ordinary array or vector, in which the elements are not indexed by integers but by arbitrary strings. A second reference to an associative array with a previously-encountered index string may possibly change the value in the array but not the list of index strings. In this case we do not care about the value but just use the index strings to compile a unique list of species detected. Multiple references to the same species will merely overwrite the first reference, *not* make a repetitive list.

Newer versions of PSI-BLAST report the taxonomic distribution of the hits. However, the program in this example would be useful if one wanted to retrieve the alignments, or do other kinds of analysis of the results.

Introduction to protein structure

With protein structures we leave behind the one-dimensional world of nucleotide and amino acid sequences and enter the three-dimensional world of molecular structures. Some of the facilities for archiving and retrieving molecular biological information survive this change pretty well intact, some must be substantially altered, and others do not make it at all.

Biochemically, proteins play a variety of roles in life processes: there are structural proteins (for example, viral coat proteins, the horny outer layer of human and animal skin, and proteins of the cytoskeleton); proteins that catalyse chemical reactions (the enzymes); transport and storage proteins (haemoglobin, ferritin); regulatory proteins, including hormones and receptor/signal transduction proteins; proteins that control gene transcription; and proteins involved in recognition, including cell adhesion molecules, and antibodies and other proteins of the immune system.

Proteins are large molecules. In many cases only a small part of the structure—an **active site**—is directly functional, the rest existing primarily to create and fix the spatial relationship among the active site residues. Proteins evolve by structural changes, produced by mutations in the amino acid sequence and genetic rearrangements, that bring together different combinations of structural subunits.

Approximately 30 000 protein structures are now known. Most were determined by X-ray crystallography or Nuclear Magnetic Resonance (NMR). From these we have derived our understanding both of the functions of individual proteins—for example, the chemical explanation of catalytic activity of enzymes—and of the general principles of protein structure and folding.

Chemically, protein molecules are long polymers typically containing several thousand atoms, composed of a uniform repetitive **backbone** (or **mainchain**) with a particular **sidechain** attached to each residue (see Fig. 1.6). The amino acid sequence of a protein records the succession of sidechains.

The polypeptide chain folds into a curve in space; the course of the chain defining a **folding pattern**. Proteins show a great variety of folding patterns. Underlying these are a number of common structural features. These include the recurrence of explicit structural paradigms—for example, α-helices and β-sheets (Fig. 1.7)—and common principles or features such as the dense packing of the atoms in protein interiors. Folding may be thought of as a kind of intramolecular condensation or crystallization. (See Chapter 5.)

The hierarchical nature of protein architecture

The Danish protein chemist K. U. Linderstrøm-Lang described the following levels of protein structure: The amino acid sequence—the set of primary chemical bonds—is called the **primary structure**. The assignment of helices and sheets—the hydrogen-bonding pattern of the mainchain—is called the **secondary structure**. The assembly and interactions of the helices and sheets is called the **tertiary structure**. For proteins composed of more than one subunit, J. D. Bernal called the assembly of the monomers the **quaternary structure**. In some cases, evolution can merge proteins—changing quaternary to tertiary structure. For example, five separate enzymes in the bacterium *E. coli*, that catalyse successive steps in the pathway of biosynthesis of aromatic amino acids, correspond to five regions of a single protein in the fungus *Aspergillus nidulans*. Sometimes homologous monomers form oligomers in different ways; for instance, globins form tetramers in mammalian haemoglobins, and dimers—using a different interface—in the ark clam *Scapharca inaequivalvis*.

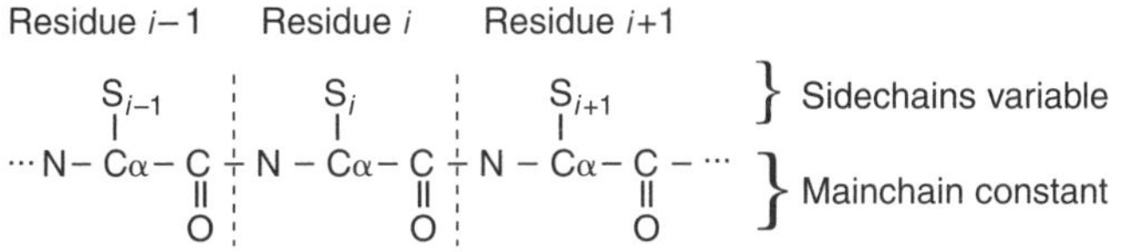

Residue $i-1$ Residue i Residue $i+1$

S_{i-1} S_i S_{i+1} } Sidechains variable

$\cdots$N– Cα– C(=O) – N – Cα– C(=O) – N – Cα– C(=O) – $\cdots$ } Mainchain constant

Fig. 1.6 The polypeptide chains of proteins have a mainchain of constant structure and sidechains that vary in sequence. Here S_{i-1}, S_i and S_{i+1} represent sidechains. The sidechains may be chosen, independently, from the set of 20 standard amino acids. It is the sequence of the sidechains that gives each protein its individual structural and functional characteristics.

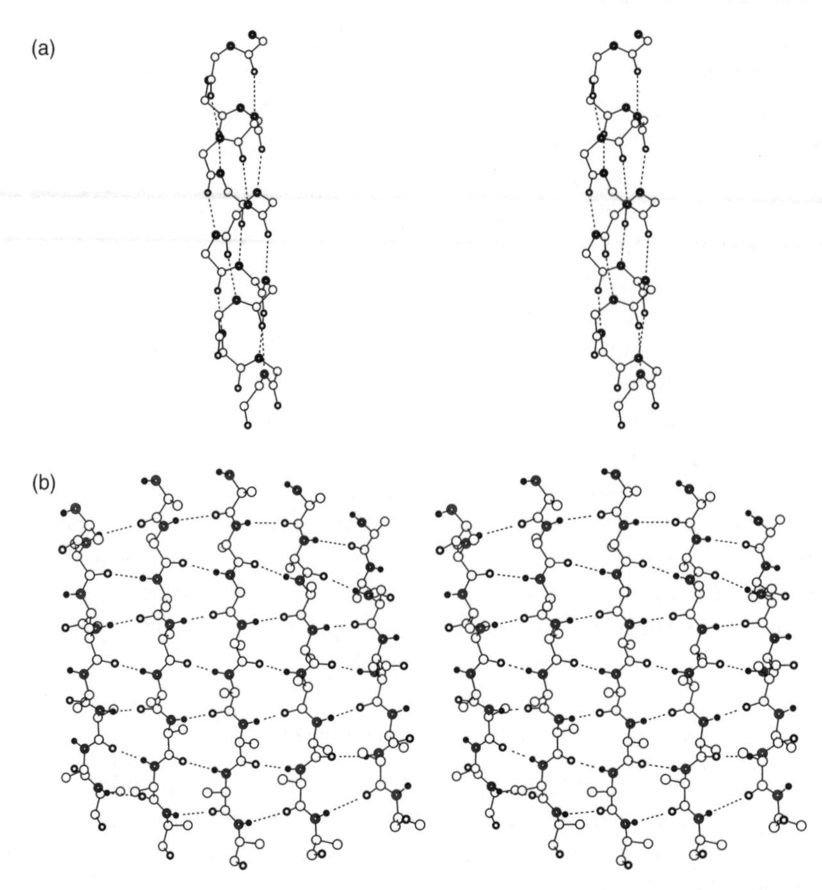

Fig. 1.7 Standard secondary structures of proteins. (a) α-helix. Hydrogen atoms not shown. (b) β-sheet. Hydrogen atoms shown. (b) illustrates a parallel β-sheet, in which all strands point in the same direction. Antiparallel β-sheets, in which all pairs of adjacent strands point in opposite directions, are also common. In fact, β-sheets can be formed by any combination of parallel and antiparallel strands.

It has proved useful to add additional levels to the hierarchy:

- **Supersecondary structures** Proteins show recurrent patterns of interaction between helices and sheets close together in the sequence. These *supersecondary structures* include the α-helix hairpin, the β-hairpin, and the β-α-β unit (Fig. 1.8).
- **Domains** Many proteins contain compact units within the folding pattern of a single chain, that look as if they should have independent stability. These are called domains. (Do not confuse domains as substructures of proteins with domains as general classes of living things: archaea, bacteria and eukarya.) The RNA-binding protein L1 has features typical of multidomain proteins: the binding site appears in a cleft between the two domains, and the relative geometry of the two domains is flexible, allowing for ligand-induced conformational changes (Fig. 1.9). In the hierarchy, domains fall between supersecondary structures and the tertiary structure of a complete monomer.

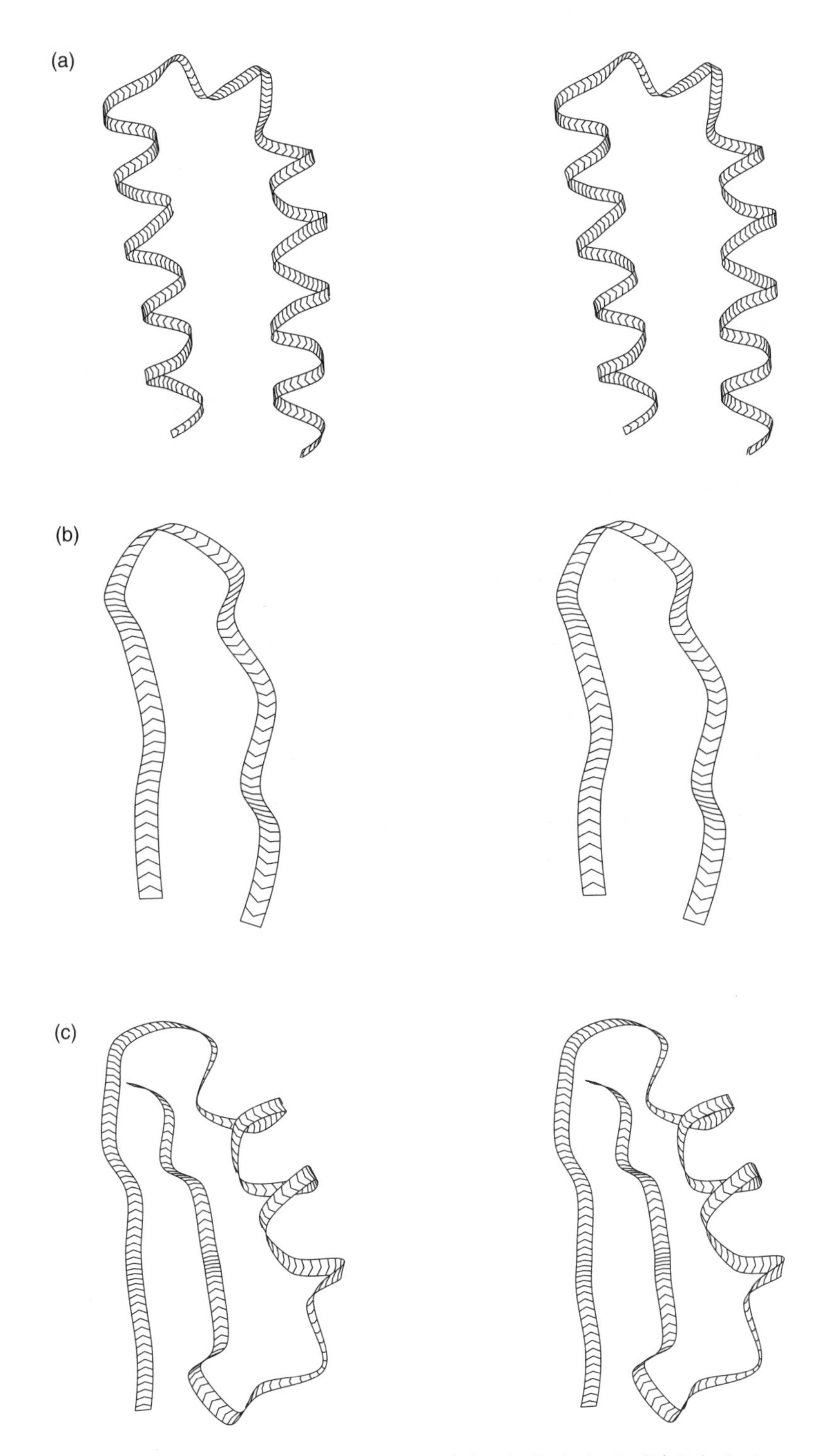

Fig. 1.8 Common supersecondary structures. (a) α-helix hairpin (b) β-hairpin, (c) β-α-β unit. The chevrons indicate the direction of the chain.

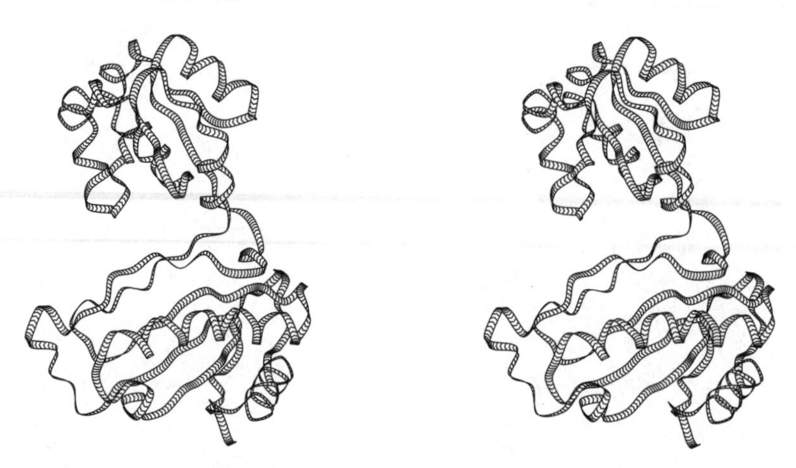

Fig. 1.9 Ribosomal protein L1 from *Methanococcus jannaschii* [1CJS]. ([1CJS] is the Protein Data Bank identification code of the entry.)

- **Modular Proteins** Modular proteins are multidomain proteins which often contain many copies of closely-related domains. Domains recur in many proteins in different structural contexts; that is, different modular proteins can 'mix and match' sets of domains. For example, fibronectin, a large extracellular protein involved in cell adhesion and migration, contains 29 domains including multiple tandem repeats of three types of domains called F1, F2, and F3. It is a linear array of the form: $(F1)_6(F2)_2(F1)_3(F3)_{15}(F1)_3$. Fibronectin domains also appear in other modular proteins. (See http://www.bork.embl-heidelberg.de/Modules/ for pictures and nomenclature.)

Classification of protein structures

The most general classification of families of protein structures is based on the secondary and tertiary structures of proteins.

Class	Characteristic
α-helical	Secondary structure exclusively or almost exclusively α-helical
β-sheet	secondary structure exclusively or almost exclusively β-sheet
α + β	α-helices and β-sheets separated in different parts of the molecule; absence of β-α-β supersecondary structure
α/β	helices and sheets assembled from β-α-β units
α/β-linear	line through centres of strands of sheet roughly linear
α/β-barrels	line through centres of strands of sheet roughly circular
little or no secondary structure	

Within these broad categories, protein structures show a variety of folding patterns. Among proteins with similar folding patterns, there are families that share enough features of structure, sequence and function to suggest evolutionary relationship. However, unrelated proteins often show similar structural themes.

Classification of protein structures occupies a key position in bioinformatics, not least as a bridge between sequence and function. We shall return to this theme, to describe results and relevant web sites. Meanwhile, the following album of small structures provides opportunities for practising visual analysis and recognition of the important spatial patterns (Fig. 1.10).

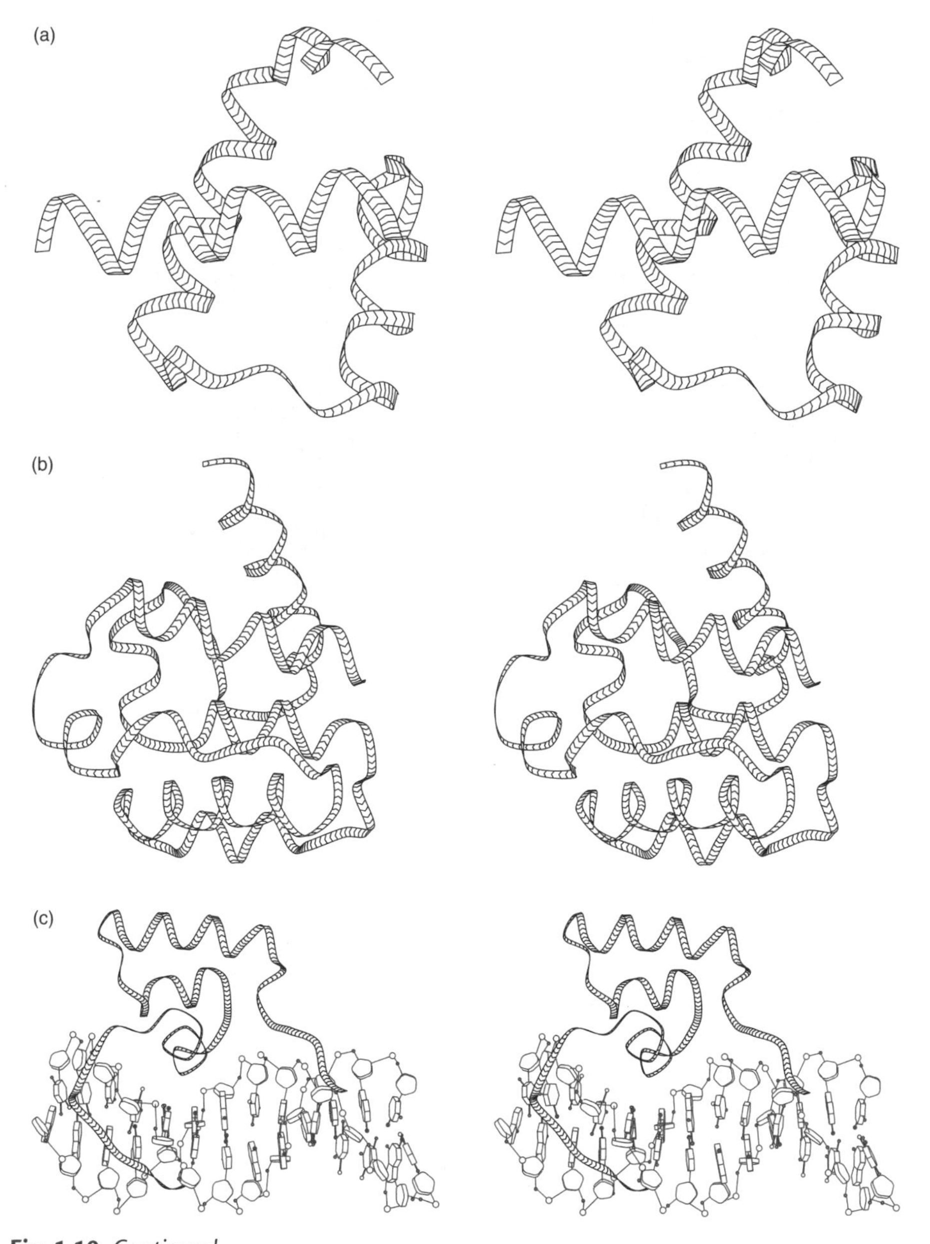

Fig. 1.10 *Continued*

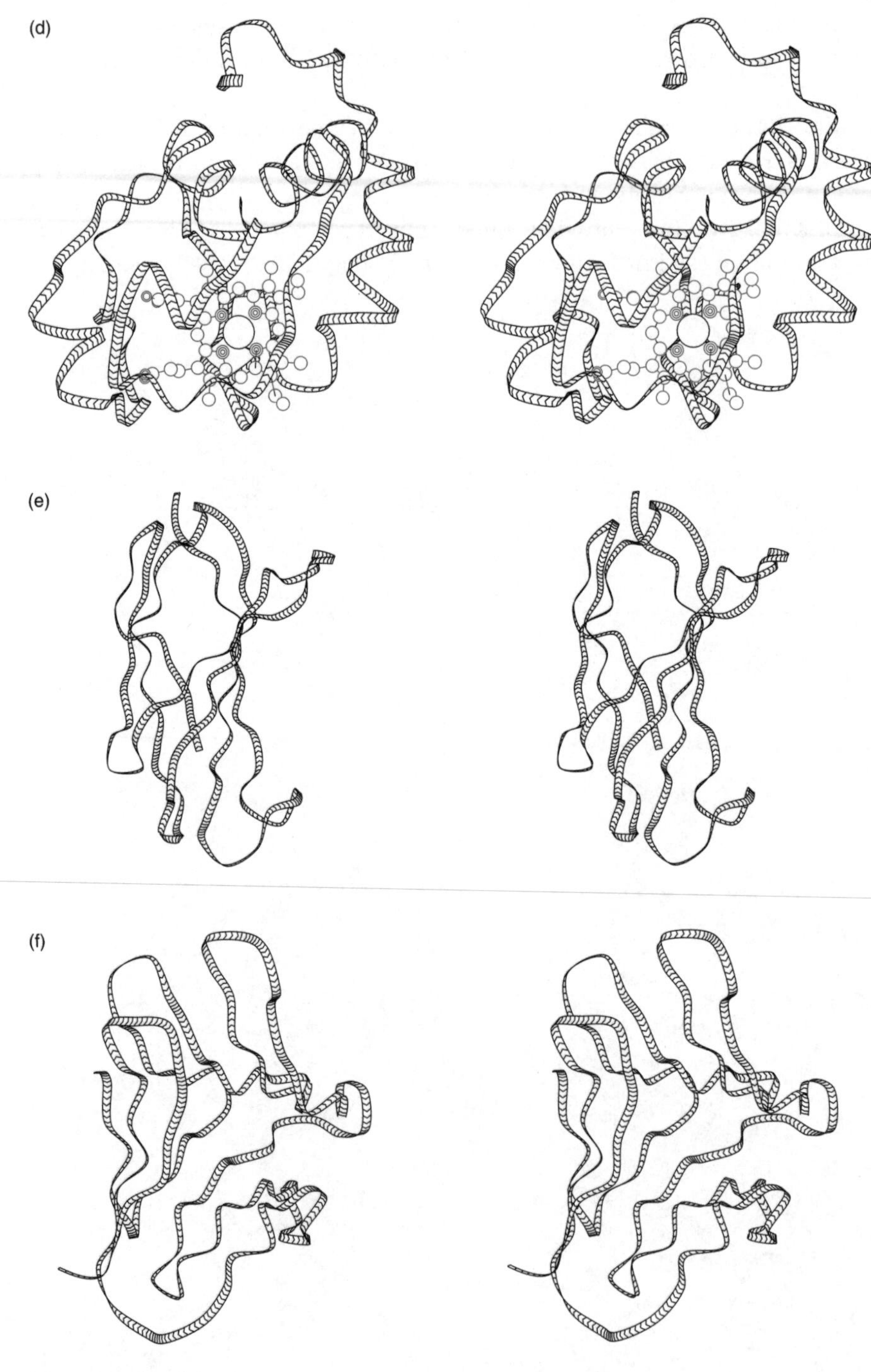

Fig. 1.10 *Continued*

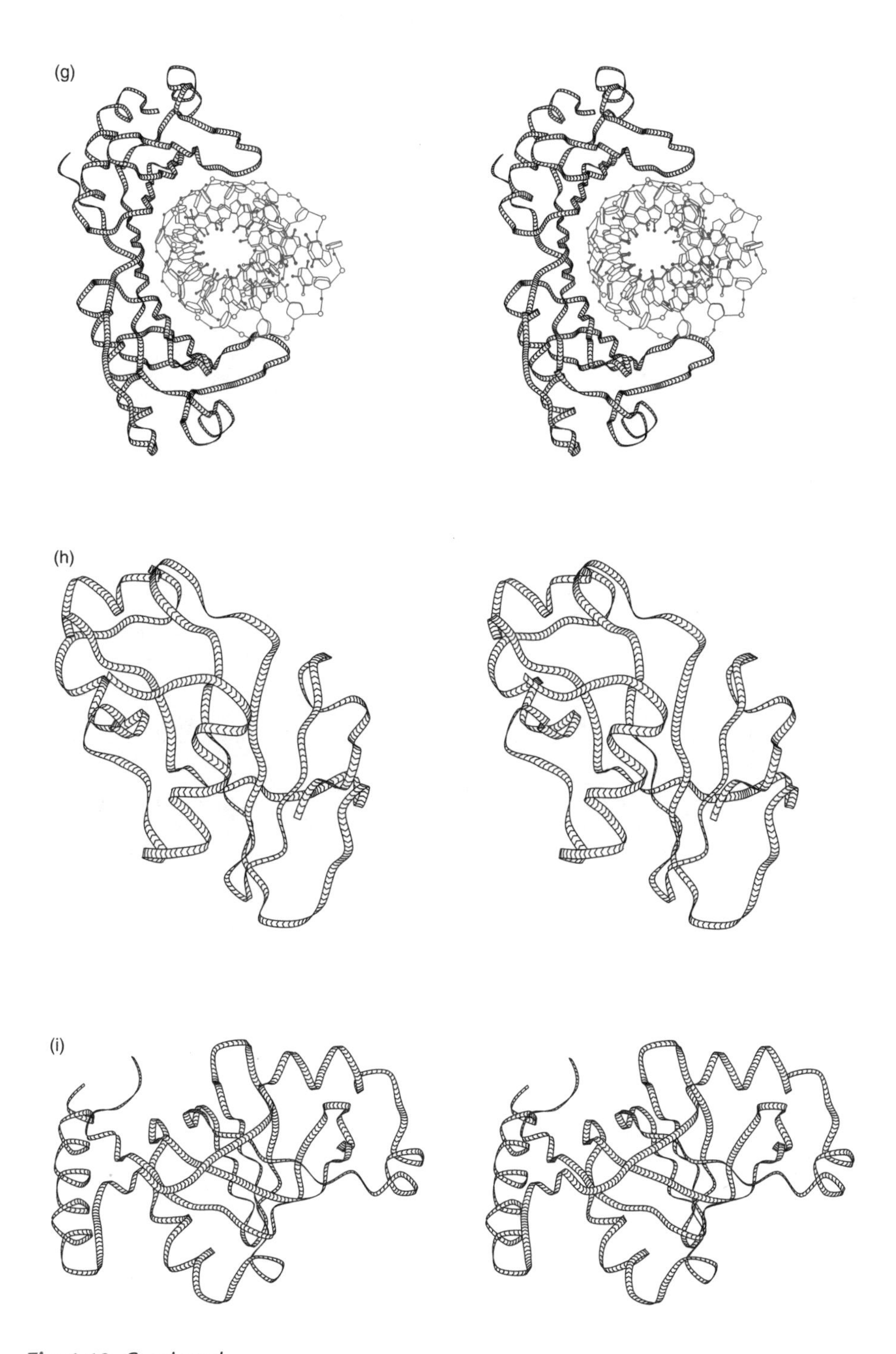

Fig. 1.10 *Continued*

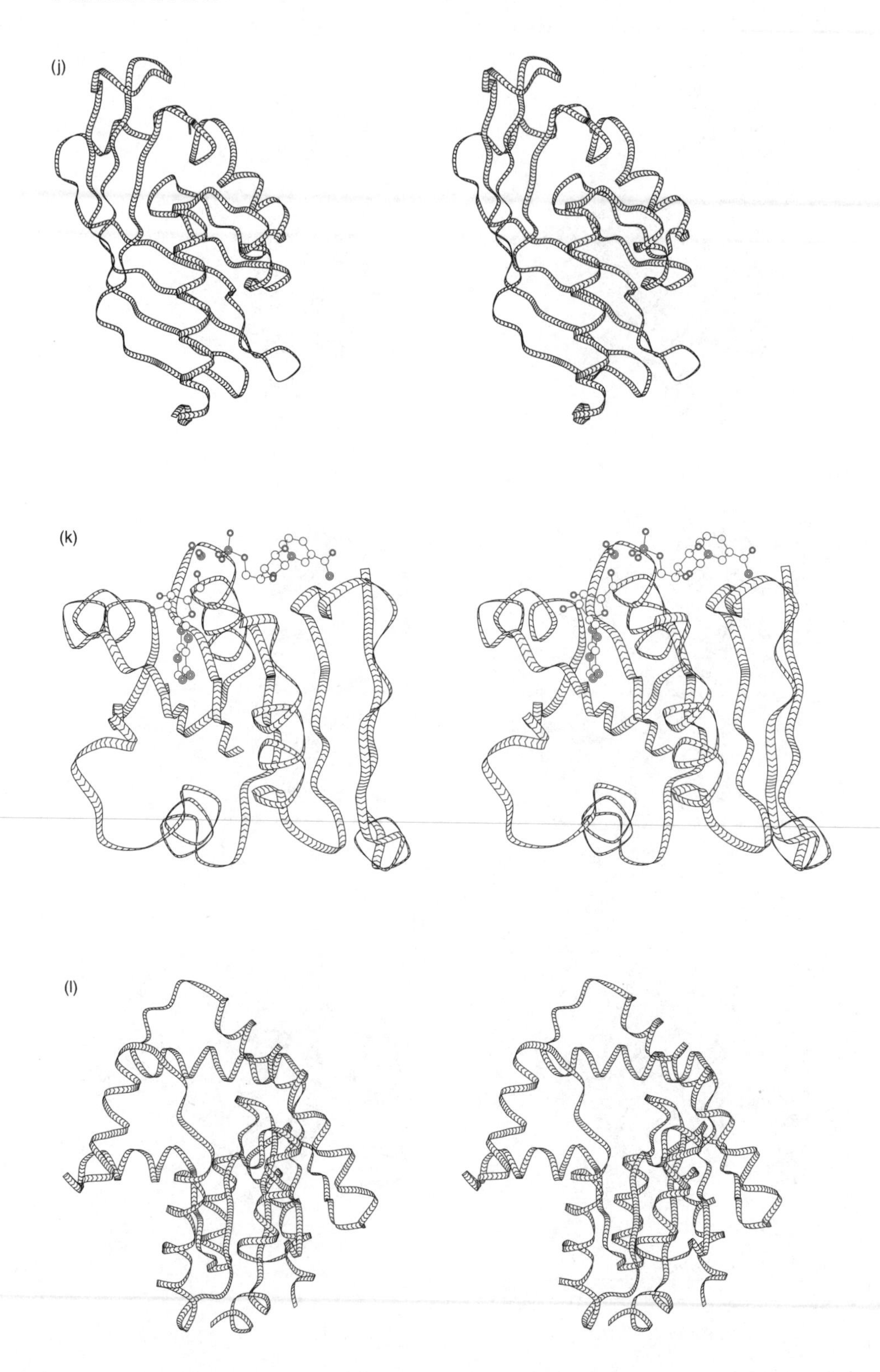

Fig. 1.10 *Continued*

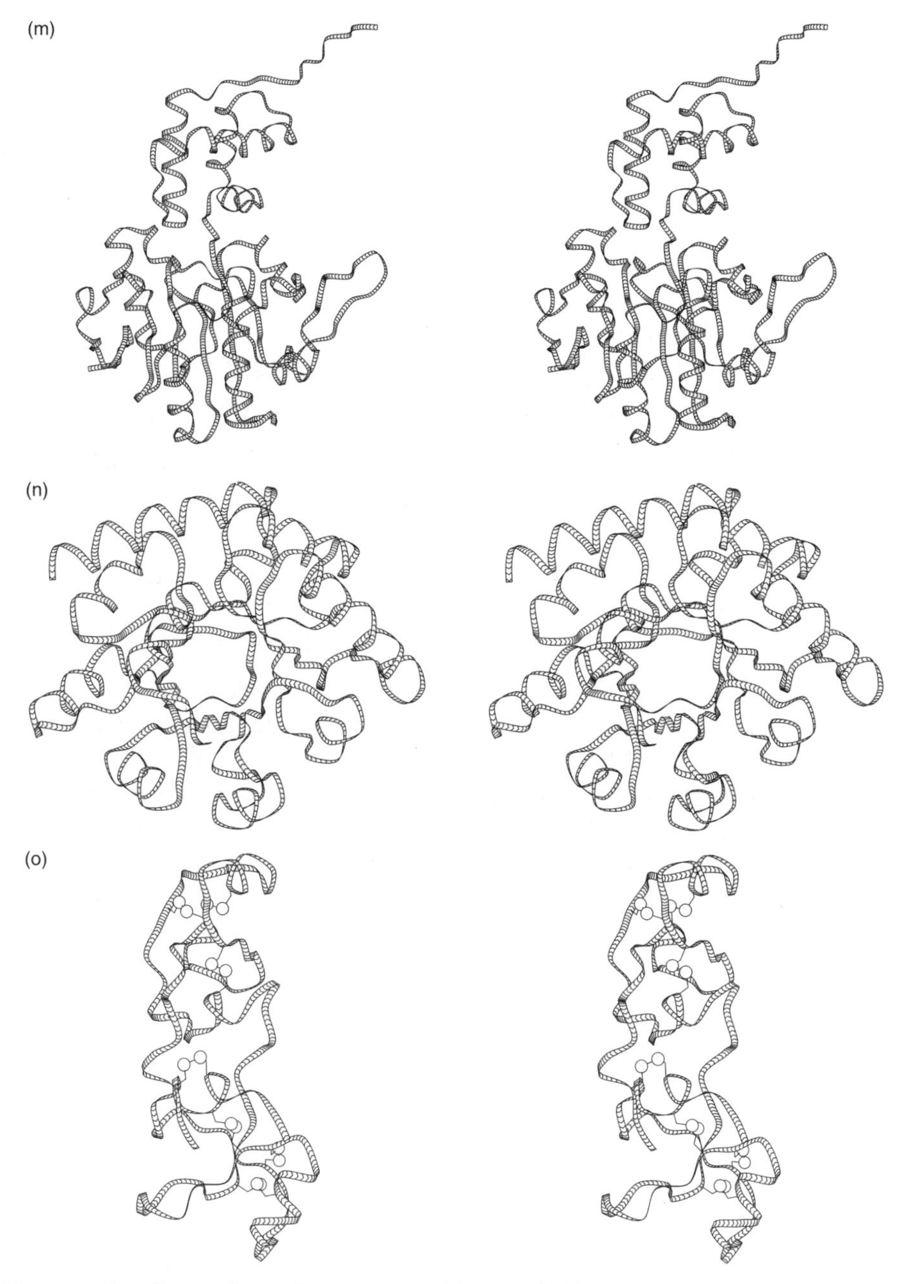

Fig. 1.10 An album of protein structures. (a) engrailed homeodomain [1ENH]. (b) second calponin homology domain from utrophin [1BHD]. (c) HIN recombinase, DNA-binding domain (protein black, DNA red) [1HCR]. (d) rice embryo cytochrome *c* [1CCR]. (e) fibronectin cell-adhesion module type III-10 [1FNA]. (f) mannose-specific agglutinin (lectin) [1NPL]. (g) TATA-box-binding protein core domain (black) binding DNA (red) [1CDW]. (h) barnase [1BRN]. (i) lysyl-tRNA synthetase [1BBW]. (j) scytalone dehydratase [3STD]. (k) alcohol dehydrogenase, NAD-binding domain. NAD in red. [1EE2]. (l) adenylate kinase [3ADK]. (m) chemotaxis receptor methyltransferase [1AF7]. (n) thiamin phosphate synthase [2TPS]. (o) porcine pancreatic spasmolytic polypeptide [2PSP].

Web resources: Macromolecular structures

The Worldwide PDB (wwPDB) is a collaboration between three primary archival projects to integrate the archiving and distribution of biological macromolecular structures:

- The Protein Data Bank (PDB) (USA)
- The Macromolecular Structure Database (MSD) (at the European Bioinformatics Institute (EBI), Hinxton, UK)
- The Protein Data Bank/Japan (Osaka, Japan)

The wwPDB sites accept depositions, process new entries, and maintain the archives. Other databanks reorganize, and provide access to the data, including:

- Structural Classification of Proteins (SCOP) is a carefully-curated database of all protein domains, classified according to structure, function, and evolution.
- The Molecular Modeling DataBase (MMDB) is the project within the US National Center for Biotechnology Information (NCBI) ENTREZ system, treating experimentally-determined macromolecular structures.

These and many other sites provide search facilities to identify structures of interest. For instance, to locate a protein of interest in SCOP the user can traverse the structural hierarchy, or search via keywords, such as protein name, PDB code, function (including Enzyme Commission number), name of fold (for instance, barrel). For each structure, SCOP provides textual information (including the full text of the entry), pictures, and links to other databases.

Naturally, there is considerable overlap between the sites. Each has its own strengths, based in many cases on the research interests of the contributing scientists. For instance, the Macromolecular Structure Database at the European Bioinformatics Institute maintains the Protein Quaternary Structure site, which gives the probable state of assembly of multi-chain proteins in their biologically active forms. Different sites differ also in their 'look and feel', and users will discover their own preferences.

Trace the chains visually, picking out helices and sheets. (The chevrons indicate the direction of the chain.) Can you see supersecondary structures? Into which general classes do these structures fall? (See Exercises 1.13 and 1.14, and Problem 1.2.) Many other examples appear in *Introduction to Protein Architecture: The Structural Biology of Proteins*, and in *Introduction to Protein Science: Architecture, Function and Genomics*.

Protein structure prediction and engineering

The amino acid sequence of a protein dictates its three-dimensional structure. In a medium of suitable solvent and temperature conditions, such as provided by a cell interior, proteins fold spontaneously to their active states. Chaperones help proteins to fold properly, but they catalyse the process rather than direct it.

If amino acid sequences contain sufficient information to specify three-dimensional structures of proteins, it should be possible to devise an algorithm to predict protein structure from amino acid sequence. This has proved elusive. In consequence, in addition to pursuing the fundamental problem of a priori prediction of protein structure from amino acid sequence, scientists have defined less-ambitious goals:

1. **Secondary structure prediction** Which segments of the sequence form helices and which form strands of sheet?
2. **Fold recognition** Given a library of known protein structures and their amino acid sequences, and the amino acid sequence of a protein of unknown structure, can we find the structure in the library that is most likely to have a folding pattern similar to that of the protein of unknown structure?
3. **Homology modelling** Suppose a target protein, of known amino acid sequence but unknown structure, is related to one or more proteins of known structure. Then we expect that much of the structure of the target protein will resemble that of the known protein, and it can serve as a basis for a model of the target structure. The completeness and quality of the result depend crucially on how similar the sequences are. As a rule of thumb, if the sequences of two homologous proteins have 50% or more identical residues in an optimal alignment, the structures are likely to have similar conformations over more than 90% of the model. (This is a conservative estimate, as the following illustration shows.)

Here are the aligned sequences, and superposed structures, of two related proteins, hen egg white lysozyme (black in structure diagram) and baboon α-lactalbumin (red in structure diagram). The sequences are closely related (37% identical residues in the aligned sequences), and the structures are very similar. Each protein could serve as a good model for the other, at least as far as the course of the mainchain is concerned.

```
Chicken lysozyme        KVFGRCELAAAMKRHGLDNYRGYSLGNWVCAAKFESNFNTQATNRNTDGS
Baboon α-lactalbumin    KQFTKCELSQNLY--DIDGYGRIALPELICTMFHTSGYDTQAIVEND-ES

Chicken lysozyme        TDYGILQINSRWWCNDGRTPGSRNLCNIPCSALLSSDITASVNCAKKIVS
Baboon α-lactalbumin    TEYGLFQISNALWCKSSQSPQSRNICDITCDKFLDDDITDDIMCAKKILD

Chicken lysozyme        DGN-GMNAWVAWRNRCKGTDVQA-WIRGCRL-
Baboon α-lactalbumin    I--KGIDYWIAHKALC-TEKL-EQWL--CE-K
```

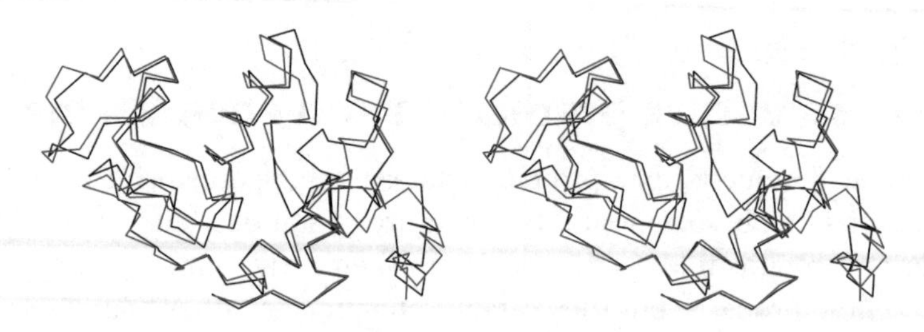

Critical Assessment of Structure Prediction (CASP)

Judging of techniques for predicting protein structures requires blind tests. To this end, J. Moult initiated biennial CASP (Critical Assessment of Structure Prediction) programmes. Crystallographers and NMR spectroscopists in the process of determining a protein structure are invited to (1) publish the amino acid sequence several months before the expected date of completion of their experiment, and (2) commit themselves to keeping the results secret until an agreed date. Predictors submit models, which are held until the deadline for release of the experimental structure. Then the predictions and experiments are compared—to the delight of a few and the chagrin of most.

The results of CASP evaluations record progress in the effectiveness of predictions, which has occurred partly because of the growth of the databanks but also because of improvements in the methods. We shall discuss protein structure prediction in Chapter 5.

Protein engineering

Molecular biologists used to be like astronomers—we could observe our subjects but not modify them. This is no longer true. In the laboratory we can manipulate nucleic acids and proteins at will. We can probe them by exhaustive mutation to see the effects on function. We can endow old proteins with new functions, as in the development of catalytic antibodies. We can even create new ones.

Many rules about protein structure were derived from observations of natural proteins. These rules do not *necessarily* apply to engineered proteins. Natural proteins have features required by general principles of physical chemistry, and by the mechanism of protein evolution. Engineered proteins must obey the laws of physical chemistry but not the constraints of evolution. With engineered proteins we can explore new territory.

Proteomics

The proteome, in analogy with the genome, is the set of proteins of an organism. Proteomics combines the census, distribution, interactions, dynamics, and expression patterns of the proteins within living systems. R. Simpson has drawn the analogy: if the genome is a list of the instruments in an orchestra, the proteome is the orchestra in the process of playing a symphony. It is a data-intensive subject, depending on high-throughput measurements. These include DNA microarrays, and mass spectrometry.

DNA microarrays

See Box: Applications of DNA microarrays.

DNA microarrays, or DNA chips, are devices for checking a sample *simultaneously* for the presence of many sequences. DNA microarrays can be used (1) to determine expression patterns of different proteins by detection of mRNAs; or (2) for genotyping, by detection of different variant gene sequences, including but not limited to single-nucleotide polymorphisms (SNPs). It is possible to measure simple presence or absence, or to quantify relative abundance.

From the point of view of bioinformatics, DNA arrays are yet another prolific stream of data creation. They demand effective design of archives and information retrieval systems. One advantage is that the data are all so new that the field is not encumbered with data structures and formats based on older generations of hardware and programs.

Applications of DNA microarrays

- **Identifying genetic individuality in tissues or organisms** In humans and animals, this permits correlation of genotype with susceptibility to disease. In bacteria, this permits identifying mechanisms of development of drug resistance by pathogens.
- **Investigating cellular states and processes** Patterns of expression that change with cellular state or growth conditions can give clues to the mechanism of sporulation, or to the change between aerobic to anaerobic metabolism.
- **Diagnosis of disease** Testing for the presence of mutations can confirm the diagnosis of a suspected genetic disease. Detection of carriers can help in counselling prospective parents.
- **Genetic warning signs** Some diseases are not determined entirely and irrevocably by genotype, but the probability of their development is correlated with genes or their expression patterns. A person aware of an enhanced risk of developing a condition can in some cases improve his or her prospects by adjustments in lifestyle, or in some cases even prophylactic surgery.
- **Drug selection** Detection of genetic factors that govern responses to drugs, that in some patients render treatment ineffective and in others cause unusual serious adverse reactions.
- **Specialized diagnosis of disease** Different types of leukemia can be identified by different patterns of gene expression. Knowing the exact type of the disease is important for prognosis and for selecting optimal treatment.
- **Target selection for drug design** Proteins showing enhanced transcription in particular disease states might be candidates for attempts at pharmacological intervention.

→

→

Applications of DNA Microarrays *(continued)*

- **Pathogen resistance** Comparisons of genotypes or expression patterns, between bacterial strains susceptible and resistant to an antibiotic, point to the proteins involved in the mechanism of resistance.
- **Following temporal variations in protein expression** permits timing the course of (1) responses to pathogen infection, (2) responses to environmental change, and (3) changes during the cell cycle.

Mass spectrometry

Mass spectrometry is a physical technique that characterizes molecules by measurements of the masses of their ions. Applications to proteomics include:

- Rapid identification of the components of a complex mixture of proteins.
- Sequencing of proteins and nucleic acids.
- Analysis of post-translational modifications, or substitutions relative to an expected sequence.
- Measuring extents of hydrogen-deuterium exchange, to reveal the solvent exposure of individual sites. This provides information about static conformation, dynamics—including folding, and interactions.

Systems biology

The watchword of systems biology is integration. Molecular biologists have spent a century taking cells apart—purifying individual proteins and measuring their properties in isolation. Our job now is to put it all back together.

Integration has two aspects. One is the study of patterns *within a cell or an organism:* patterns of protein-protein and protein-nucleic acid interaction, patterns of metabolic pathways and control cascades, and patterns of protein expression. Patterns have both static and dynamic aspects. Identification of pairs of proteins that bind to each other, and the assembly of pairwise interactions into a network, produces a static pattern. The flow of metabolites through a network of enzymes, or the flow of information down a control cascade, is a dynamic pattern.

The other aspect of integration is the comparison of occurrence, activities and interactions of genes and proteins *across different species*. The reason why the comparative approach is so powerful in biology is that the systems we are trying to understand arose through processes of evolution. Different species illuminate one another. To understand what it means to be human we must appreciate both what we have in common with other species, and how we differ.

High-throughput methods of genomics and proteomics provide data about sequences, expression patterns and interactions. From genome sequences we can infer the amino acid sequences of an organism's complement of proteins. Proteomics tells us how expression patterns of these proteins vary, within the organism, how they change during development or in response to changes in conditions, and how they cooperate with one another. Systems biology takes these data as pieces of a jigsaw puzzle that extends in both space and time. To understand the complex and delicate instrument that is the living cell, we must fit the pieces into their frame.

Clinical implications

There is consensus that the sequencing of the human and other genomes will lead to improvements in the health of mankind. Even discounting some of the more outrageous claims—hype springs eternal—categories of applications include:

1. **Diagnosis of disease and disease risks** DNA sequencing can detect the absence of a particular gene, or a mutation. Identification of specific gene sequences associated with diseases will permit fast and reliable diagnosis of conditions (a) when a patient presents with symptoms, (b) in advance of appearance of symptoms, as in tests for inherited late-onset conditions such as Huntington disease (see Box, page 56), (c) for *in utero* diagnosis of potential abnormalities such as cystic fibrosis, and (d) for genetic counselling of couples contemplating having children.

   In many cases our genes do not irrevocably condemn us to contract a disease, but raise the probability that we will. An example of a risk factor detectable at the genetic level involves α_1-antitrypsin, a protein that normally functions to inhibit elastase in the alveoli of the lung. People homozygous for the Z mutant of α_1-antitrypsin (342Glu→Lys) express only a dysfunctional protein. They are at risk of emphysema, because of damage to the lungs from endogenous elastase unchecked by normal inhibitory activity, and also of liver disease, because of accumulation of a polymeric form of α_1-antitrypsin in hepatocytes where it is synthesized. Smoking makes the development of emphysema all but certain. In these cases the disease is brought on by a *combination* of genetic and environmental factors.

   'Genetics loads the gun and environment pulls the trigger' – J. Stern

   Often the relationship between genotype and disease risk is much more difficult to pin down. Some diseases such as asthma depend on interactions of many genes, as well as environmental factors. In other cases a gene may be all present and correct, but a mutation elsewhere may alter its level of expression or distribution among tissues. Such abnormalities must be detected by measurements of protein activity. Analysis of protein expression patterns is also an important way to measure response to treatment.

2. **Genetics of reponses to therapy—customized treatment** Because people differ in their ability to metabolize drugs, different patients with the same condition may require different dosages. Sequence analysis permits selecting drugs and dosages optimal for individual patients, a fast-growing field called **pharmacogenomics**. Physicians can thereby avoid experimenting with different therapies, a procedure that is dangerous in terms of side effects—often even fatal—and in any case is expensive. Treatment of patients for adverse reactions to prescribed drugs demands billions of dollars in health-care costs.

For example, the very toxic drug 6-mercaptopurine is used in the treatment of childhood leukaemia. A small fraction of patients used to die from the treatment, because they lacked the enzyme thiopurine methyltransferase, needed to metabolize the drug. Testing of patients for this enzyme identifies those at risk.

Conversely, it may become possible to use drugs that are safe and effective in a minority of patients, but which have been rejected before or during clinical trials because of inefficacy or severe side effects in the majority.

Huntington disease

Huntington disease is an inherited neurodegenerative disorder affecting approximately 30 000 people in the USA. Its symptoms are quite severe, including uncontrollable dance-like (choreatic) movements, mental disturbance, personality changes, and intellectual impairment. Death usually follows within 10–15 years after the onset of symptoms. The gene arrived in New England during the colonial period, in the seventeenth century. It may have been responsible for some accusations of witchcraft. The gene has not been eliminated from the population, because the age of onset—30–50 years—is after the typical reproductive period.

Formerly, members of affected families had no alternative but to face the uncertainty and fear, during youth and early adulthood, of not knowing whether they had inherited the disease. The discovery of the gene for Huntington disease in 1993 made it possible to identify affected individuals. The gene contains expanded repeats of the trinucleotide CAG, corresponding to polyglutamine blocks in the corresponding protein, *huntingtin*. (Huntington disease is one of a family of neurodegenerative conditions resulting from trinucleotide repeats.) The larger the block of CAGs, the earlier the onset and more severe the symptoms. The normal gene contains 11–28 CAG repeats. People with 29–34 repeats are unlikely to develop the disease, and those with 35–41 repeats may develop only relatively mild symptoms. However, people with >41 repeats are almost certain to suffer full Huntington disease.

The inheritance is marked by a phenomenon called anticipation: the repeats grow longer in successive generations, progressively increasing the severity of the disease and reducing the age of onset. For some reason this effect is greater in paternal than in maternal genes. Therefore, even people in the borderline region, who might bear a gene containing 29–41 repeats, should be counselled about the risks to their offspring.

3. **Identification of drug targets** A **target** is a protein the function of which can be selectively modified by interaction by a drug, to affect the symptoms or underlying causes of a disease. Identification of a target provides the focus for subsequent steps in the drug design process. Among drugs now in use, the targets of about half are receptors, about a quarter enzymes, and about a quarter hormones. Approximately 7% act on unknown targets.

The growth in bacterial resistance to antibiotics is creating a crisis in disease control. There is a very real possibility that our descendants will look back at the second half of the twentieth century as a narrow window during which bacterial infections could be controlled, and before and after which they could not.

The urgency of finding new drugs is mitigated by the increasing availability of data on which to base their development. Genomics can suggest targets. Differential genomics, and comparison of protein expression patterns, between drug-sensitive and -resistant strains of pathogenic bacteria, can pinpoint the proteins responsible for drug resistance. The study of genetic variation between tumour and normal cells can identify differentially expressed proteins as potential targets for anticancer drugs.

4. **Gene therapy** If a gene is missing or defective, we'd like to replace it or at least supply its product. If a gene is overactive, we'd like to turn it off.

   Direct supply of proteins is possible for many diseases, of which insulin replacement for diabetes and Factor VIII for a common form of haemophilia are perhaps the best known.

   Gene transfer has succeeded in animals, for production of human proteins in the milk of sheep and cows. In human patients, gene replacement therapy for cystic fibrosis using adenovirus has shown encouraging results.

   One approach to blocking genes is called 'antisense therapy'. The idea is to introduce a short stretch of DNA or RNA that binds in a sequence-specific manner to a region of a gene. Binding to endogenous DNA can interfere with transcription; binding to mRNA can interfere with translation. Antisense therapy has shown some efficacy against cytomegalovirus and Crohn's disease.

   Antisense therapy is very attractive, because going directly from target sequence to blocker short-circuits many stages of the drug-design process.

The future

The new century will see a revolution in health-care development and delivery. Barriers between 'blue-sky' research and clinical practice are tumbling down. It is possible that a reader of this book will discover a cure for a disease that would otherwise kill him or her. Indeed it is extremely likely that Szent-Gyorgi's quip, 'Cancer supports more people than it kills' will come true. One hopes that this happens because the research establishment has succeeded in developing therapeutic or preventative measures against tumours rather than merely imitating their uncontrolled growth.

Recommended reading

A glimpse of the future?

Blumberg, B. S. (1996), Medical research for the next millenium, *The Cambridge Review*, **117**, 3–8. [A fascinating prediction of things to come, some of which are already here.]

The intellectual setting

Mayr, E., *What Makes Biology Unique? Considerations on the Autonomy of a Scientific Discipline*. (Cambridge: Cambridge University Press, 2004). [Perspectives, from a self-described 'dirty-fingernails biologist' with an unequalled clarity of mind.]

The overall biological context

Doolittle, W. F. (2000), Uprooting the tree of life, *Sci. Am.*, **282(2)**, 90–95. [Implications of analysis of sequences for our understanding of the relationships between living things.]

Genomic sequence determination

Green, E. D. (2001), Strategies for systematic sequencing of complex organisms, *Nature Reviews (Genetics)*, **2**, 573–583. [A clear discussion of possible approaches to large-scale sequencing projects. Includes list of, and links to, ongoing projects for sequencing of multicellular organisms.]

Sulston, J. & Ferry, G., *The Common Thread: a story of science, politics, ethics and the human genome* (New York: Bantam, 2002). [A first-hand account.]

Discussions of databases and information retrieval

Altschul, S. F., Madden, T. L., Schäffer, A. A., Zhang, J., Zhang, Z., Miller, W. & Lipman, D. J. (1997), Gapped BLAST and PSI-BLAST: a new generation of protein database search programs, *Nucleic Acids Res.*, **25**, 3389–3402.

Frishman, D., Heumann, K., Lesk, A. & Mewes, H. -W. (1998), Comprehensive, comprehensible, distributed and intelligent databases: Current status, *Bioinformatics*, **14**, 551–561.

Wheeler, D. L., Church, D. M., Federhen, S., Lash, A. E., Madden, T. L., Pontius, J. U., Schuler, G. D., Schriml, L. M., Sequeira, E., Tatusova, T. A. & Wagner, L. (2003), Database resources of the National Center for Biotechnology, *Nucleic Acids Res.*, **31**, 28–33.

Lesk, A. M. and 25 co-authors, (2000), Quality control in databanks for molecular biology, *BioEssays*, **22**, 1024–1034.

Stein, L. (2001), Genome annotation: from sequence to biology, *Nature Reviews (Genetics)*, **2**, 493–503.

Lesk, A. M., Editor, *Database Annotation in Molecular Biology: Principles and Practice* (Chichester: J. Wiley and Sons, 2004). [A set of articles describing what databases require in the way of annotation, and what annotators require in the way of skills.]

Proteins

Branden, C. I. & Tooze, J., *Introduction to Protein Structure* (2nd edn. New York: Garland, 1999). [A fine introductory text.]

Lesk, A. M., *Introduction to Protein Architecture: The Structural Biology of Proteins* (Oxford: Oxford University Press, 2000).

Lesk, A. M., *Introduction to Protein Science: Architecture, Function and Genomics* (Oxford: Oxford University Press, 2004).

[Companion volumes to *Introduction to Bioinformatics*, with a focus on protein structure, function and evolution.]

The transition to electronic publishing

Lesk. M., *Understanding Digital Libraries* (2nd edn. San Francisco: Morgan Kaufmann, 2004). [Introduction to the transition from traditional libraries to information provision by computer.]

Berners-Lee, T. (with Mark Fischetti), *Weaving the Web: The Original Design and Ultimate Destiny of the World Wide Web* (New York: HarperBusiness, 2000).

Berners-Lee, T. & Hendler, J. (2001), Publishing on the semantic web, *Nature*, **410**, 1023–1024. [From the inventor of the web.]

Butler, D. & Campbell, P. (2001), Future e-access to the primary literature, *Nature*, **410**, 613. [Describes developments in electronic publishing of scientific journals.]

Malakoff, D. (2003), Scientific publishing Opening the books on open access, *Science*, **302**, 550–554. [Description of the journals published by the Public Library of Science.]

Spedding V. (2003), Great data, but will it last? *Research Information* (Spring 2003) 16–20. [Problems of preservation of digital information. This journal has many articles of interest to scientists whose research depends on the quality and computer accessibility of data.]

Legal aspects

Human Genome Project Information Website: Genetics and Patenting http://www.ornl.gov/hgmis/elsi/patents.html

Bobrow, M. & Thomas, S. (2002), Patenting DNA, *Curr. Opin. Mol. Ther.*, **4**, 542–547.

F. S. Kieff, ed. (2003), Perspectives on Properties of the Human Genome Project, *Adv. Genetics*, vol. **50**. [A collection of articles, discussing legal aspects of genomics and bioinformatics.]

Exercises, Problems, and Weblems

Exercises

1.1 (a) The Sloan Digital Sky Survey is a mapping of the Northern sky over a 5-year period. The total raw data will amount to about 15 terabytes (1 byte = 1 character; 1TB = 10^{12} bytes). How many human genome equivalents does this amount to? (b) The Earth Observing System/Data Information System (EOS/DIS)—a series of long-term global observations of the Earth—is estimated to require 15 petabytes of storage (1 petabyte = 10^{15} bytes). How many human genome equivalents will this amount to? (c) Compare the data storage required for EOS/DIS with that required to store the complete DNA sequences of every inhabitant of the United States of America. (Ignore savings available using various kinds of storage compression techniques. Assume that each person's DNA sequence requires 1 byte/nucleotide.)

1.2 (a) How many floppy disks would be required to store the entire human genome? (b) How many CDs would be required to store the entire human genome? (c) How many DVDs would be required to store the entire human genome? (In all cases assume that the sequence is stored as 1 byte/per character, uncompressed.)

1.3 Suppose you were going to prepare the Box on Huntington disease (page 56) for a web site. For which words or phrases would you provide links?

1.4 The end of the human β-haemoglobin gene has the nucleotide sequence:

```
. . . ctg gcc cac aag tat cac taa
```

(a) What is the translation of this sequence into an amino acid sequence? (b) Write the nucleotide sequence of a single base change producing a silent mutation in this region. (A silent mutation is one that leaves the amino acid sequence unchanged.) (c) Write the nucleotide sequence, and the translation to an amino acid sequence, of a single base change producing a missense mutation in this region. (d) Write the nucleotide sequence, and the translation to an amino acid sequence, of a single base change producing a mutation in this region that would lead to premature truncation of the protein. (e) Write the nucleotide sequence of a single base change producing a mutation in this region that would lead to improper chain termination resulting in extension of the protein.

1.5 On a photocopy of the Box Complete pairwise sequence alignment of human PAX-6 protein and *Drosophila melanogaster* eyeless, indicate with a highlighter the regions aligned by PSI-BLAST.

1.6 On a copy of the alignment of Human PAX-6 protein and *Drosophila melanogaster* eyeless (page 37), highlight the regions aligned to even-skipped by PSI-BLAST (page 37).

1.7 (a) What cut-off value of E would you use in a PSI-BLAST search if all you want to know is whether your sequence is already in a databank? (b) What cut-off value of E would you use in a PSI-BLAST search if you want to locate distant homologues of your sequence?

1.8 In designing an antisense sequence, estimate the minimum length required to avoid exact complementarity to many random regions of the human genome.

1.9 It is suggested that all living humans are descended from a common ancestor called Eve, who lived approximately 140 000–200 000 years ago. (a) Assuming six generations per century, how many generations have there been between Eve and the present? (b) If a bacterial cell divides every 20 minutes, how long would be required for the bacterium to go through that number of generations?

1.10 Name an amino acid that has physicochemical properties similar to (a) leucine, (b) aspartic acid, (c) threonine? We expect that such substitutions would in most cases have relatively little effect on the structure and function of a protein. Name an amino acid that has physicochemical properties very different from (d) leucine, (e) aspartic acid, (f) threonine? Such substitutions might have severe effects on the structure and function of a protein, especially if they occur in the interior of the protein structure.

1.11 In Fig. 1.7 (a), does the direction of the chain from N-terminus to C-terminus point up the page or down the page? In Fig. 1.7 (b), does the direction of the chain from N-terminus to C-terminus point up the page or down the page?

1.12 From inspection of Fig. 1.9, how many times does the chain pass between the domains of *M. jannaschii* ribosomal protein L1?

1.13 On a photocopy of Fig. 1.10(k and l), indicate with highlighter the helices (in red) and strands of sheet (in blue). On a photocopy of Fig. 1.10(g and m), divide the protein into domains.

1.14 Which of the structures shown in Fig. 1.10 contains the following domain?

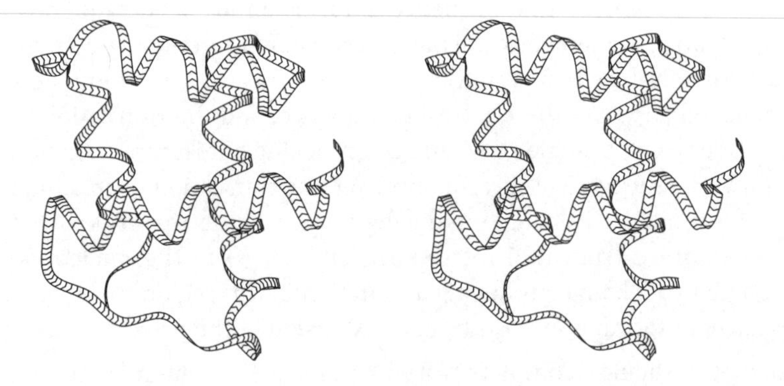

1.15 On a photocopy of the superposition of Chicken lysozyme and Baboon α-lactalbumin structures, indicate with a highlighter two regions in which the conformation of the mainchain is different.

1.16 In the PERL program on page 20, estimate the fraction of the text of the program that contains comment material. (Count full lines and half lines.)

1.17 Modify the PERL program that extracts species names from PSI-BLAST output so that it would also accept names given in the form `[D. melanogaster]`.

1.18 Modify the PERL program that extracts species names from PSI-BLAST output so that it would count the number of sequences from each species occurring in the list.

1.19 What is the nucleotide sequence of the molecule shown in Plate I?

Problems

1.1 The following table contains a multiple alignment of partial sequences from a family of proteins called ETS domains. Each line corresponds to the amino acid sequence from one protein, specified as a sequence of letters each specifying one amino acid. Looking down any column shows the amino acids that appear at that position in each of the proteins in the family. In this way patterns of preference are made visible.

```
TYLWEFLLKLLQDR.EYCPRFIKWTNREKGVFKLV..DSKAVSRLWGMHKN.KPD
VQLWQFLLEILTD..CEHTDVIEWVG.TEGEFKLT..DPDRVARLWGEKKN.KPA
IQLWQFLLELLTD..KDARDCISWVG.DEGEFKLN..QPELVAQKWGQRKN.KPT
IQLWQFLLELLSD..SSNSSCITWEG.TNGEFKMT..DPDEVARRWGERKS.KPN
IQLWQFLLELLTD..KSCQSFISWTG.DGWEFKLS..DPDEVARRWGKRKN.KPK
IQLWQFLLELLQD..GARSSCIRWTG.NSREFQLC..DPKEVARLWGERKR.KPG
IQLWHFILELLQK..EEFRHVIAWQQGEYGEFVIK..DPDEVARLWGRRKC.KPQ
VTLWQFLLQLLRE..QGNGHIISWTSRDGGEFKLV..DAEEVARLWGLRKN.KTN
ITLWQFLLHLLLD..QKHEHLICWTS.NDGEFKLL..KAEEVAKLWGLRKN.KTN
LQLWQFLVALLDD..PTNAHFIAWTG.RGMEFKLI..EPEEVARLWGIQKN.RPA
IHLWQFLKELLASP.QVNGTAIRWIDRSKGIFKIE..DSVRVAKLWGRRKN.RPA
RLLWDFLQQLLNDRNQKYSDLIAWKCRDTGVFKIV..DPAGLAKLWGIQKN.HLS
RLLWDYVYQLLSD..SRYENFIRWEDKESKIFRIV..DPNGLARLWGNHKN.RTN
IRLYQFLLDLLRS..GDMKDSIWWVDKDKGTFQFSSKHKEALAHRWGIQKGNRKK
LRLYQFLLGLLTR..GDMRECVWWVEPGAGVFQFSSKHKELLARRWGQQKGNRKR
```

On a photocopy of this page:

(a) Using coloured highlighter, mark, in each sequence, the residues in different classes in different colours:

small residues:	G A S T
medium-sized nonpolar residues:	C P V I L
large nonpolar residues:	F Y M W
polar residues:	H N Q
positively-charged residues:	K R
negatively-charged residues:	D E

(b) For each position containing the same amino acid in every sequence, write the letter symbolizing the common residue in upper case below the column. For each position containing the same amino acid in all but one of the sequences, write the letter symbolizing the preferred residue in lower case below the column.

(c) What patterns of periodicity of conserved residues suggest themselves?

(d) What secondary structure do these patterns suggest in certain regions?

(e) What distribution of conservation of charged residues do you observe? Propose a reasonable guess about what kind of molecule these domains interact with.

1.2 Classify the structures appearing in Fig. 1.10 in the following categories: α-helical, β-sheet, $\alpha + \beta$, α/β linear, α/β-barrels, little or no secondary structure.

1.3 Generalize the PERL program on page 18 to print the translations of a DNA sequence in all six possible reading frames.

1.4 Write a PERL program to read a CLUSTAL-W alignment, such as the alignment of mitochondrial cytochromes *b* from elephants and mammoths, and to count the number of sequence mismatches between all pairs of proteins.

1.5 For which of the following sets of fragment strings does the PERL program on page 20 work correctly?

(a) Would it correctly recover:

> Kate, when France is mine and I am
> yours, then yours is France and you are mine.

from:

> Kate, when France
> France is mine
> is mine and
> and I am\nyours
> yours then
> then yours is France
> France and you are mine\n

(b) Would it correctly recover:

One woman is fair, yet I am well; another is wise, yet I am well; another virtuous, yet I am well; but till all graces be in one woman, one woman shall not come in my grace.

from:

> One woman is
> woman is fair,
> is fair, yet I am
> yet I am well;
> I am well; another
> another is wise, yet I am well;
> yet I am well; another virtuous,
> another virtuous, yet I am well;
> well; but till all
> all graces be
> be in one woman,
> one woman, one
> one woman shall
> shall not come in my grace.

(c) Would it correctly recover:

That he is mad, 'tis true: 'tis true 'tis pity;
And pity 'tis 'tis true.

from:

That he is
is mad, 'tis
'tis true
true: 'tis true 'tis
true 'tis
'tis pity;\n
pity;\n And pity
pity 'tis
'tis 'tis
'tis true.\n

In (c), would it work if you deleted all punctuation marks from these strings?

1.6 Generalize the PERL program on page 20 so that it will correctly assemble all the fragments in the previous problem. (Warning–this is not an easy problem.)

1.7 Write a PERL program to find motif matches as illustrated in the Box on page 26. (a) Demand exact matches. (b) Allowing one mismatch, not necessarily at the first position as in the examples, but no insertions or deletions.

1.8 PERL is capable of great concision. Here is an alternative version of the program to assemble overlapping fragments (see page 20):

```
#!/usr/bin/perl

$/ = "";
@fragments = split("\n",<DATA>);

foreach (@fragments) { $firstfragment{$_} = $_; }

foreach $i (@fragments) {
    foreach $j (@fragments) { unless ($i eq $j) {
        ($combine = $i . "XXX" . $j) =~ /([\S ]{2,})XXX\1/;
        (length($1) <= length($successor{$i})) || { $successor{$i} = $j };
    }                             }
    undef $firstfragment{$successor{$i}};
}

$test = $outstring = join "", values(%firstfragment);
while ($test = $successor{$test}) { ($outstring .= "XXX" . $test) =~ s/([\S ]+)XXX\1/\1/; }

$outstring =~ s/\\n/\n/g; print "$outstring\n";

__END__
the men and women merely players;\n
one man in his time

All the world's
their entrances,\nand one man
stage,\nAnd all the men and women
They have their exits and their entrances,\n
world's a stage,\nAnd all
their entrances,\nand one man
in his time plays many parts.
merely players;\nThey have
```

(This is a good example of what to avoid. Anyone who produces code like this should be fired immediately. The absence of comments, and the tricky coding and useless brevity, make it difficult understand what the program is doing. A program written in this way is difficult to debug and virtually impossible to maintain. Someday you may succeed someone in a job and be presented with such a program to work on. You will have my sympathy.)

(a) Photocopy the concise program listed in this problem and the original version on page 20 so that they appear side-by-side on a page. Wherever possible, map each line of the concise program into the corresponding set of lines of the long one.

(b) Prepare a version of the concise program with enough comments to clarify what it is doing (for this you could consider adapting the comments from the original program) and how it is doing it. Do not change any of the executable statements (back to the original version or to anything else); just add comments.

Weblems

1.1 Identify the source of all quotations from Shakespeare's plays in the Box on alignment on page 26.

1.2 Identify web sites that give *elementary* tutorial explanations and/or on-line demonstrations of (a) the Polymerase Chain Reaction (PCR), (b) Southern blotting, (c) restriction maps, (d) suffix tree, (e) heapsort. Report their URLs. Write one-paragraph explanations of these terms based on these sites.

1.3 To which phyla do the following species belong? (a) Starfish (b) Lamprey (c) Tapeworm (d) Ginko tree (e) Scorpion (f) Jellyfish (g) Sea anemone.

1.4 What are the common names of the following species? (a) *Acer rubrum* (b) *Orycteropus afer* (c) *Beta vulgaris* (d) *Pyractomena borealis* (e) *Macrocystis pyrifera*

1.5 A typical British breakfast consists of: eggs (from chickens) fried in lard, bacon, kippered herrings, grilled cup mushrooms, fried potatoes, grilled tomatoes, baked beans, toast, and tea with milk. Write the complete taxonomic classfication of the organisms from which these are derived.

1.6 Recover and align the mitochondrial cytochrome *b* sequences from horse, whale and kangaroo. (a) Compare the degree of similarity of each pair of sequences with the results from comparison of the pancreatic ribonuclease sequences from these species in Case Study 1.2. Are the conclusions from the analysis of mitochondrial cytochromes *b* sequences consistent with those from analysis of the pancreatic ribonucleases? (b) Compare the *relative* similarity of these sequences with the results from comparison of the pancreatic ribonuclease sequences from these species in Case Study 1.2. Are the conclusions from the analysis of mitochondrial cytochromes *b* sequences consistent with those from analysis of the pancreatic ribonucleases?

1.7 Recover and align the pancreatic ribonuclease sequences from sperm whale, horse, and hippopotamus. Are the results consistent with the relationships shown by the SINES?

1.8 We observed that the amino acid sequences of cytochrome *b* from elephants and the mammoth are very similar. One hypothesis to explain this observation is that a functional cytochrome *b* might *require* so many conserved residues that cytochromes *b* from all animals are as similar to one another as the elephant and mammoth proteins are. Test this hypothesis by retrieving cytochrome *b* sequences from other mammalian species, and check whether the cytochrome *b* amino acid sequences from more distantly-related species are as similar to the elephant sequences as the elephant and mammoth sequences are to each other.

1.9 Recover and align the cytochrome *c* sequences of human, rattlesnake, and monitor lizard. Which pair appears to be the most closely related? Is this surprising to you? Why or why not?

1.10 Send the sequences of pancreatic ribonucleases from horse, minke whale, and red kangaroo (Case Study 1.2) to the T-coffee multiple-alignment server: `http://www.ch.embnet.org/software/TCoffee.html`. Is the resulting alignment the same as that shown in Case Study 1.2 produced by CLUSTAL-W? If not, where do they differ?

1.11 Create a multiple sequence alignment of the mitochondrial cytochrome *b* *genes* from African and Indian elephants, and Siberian mammoths. How many mismatches are there between each pair of sequences? Is the result consistent with the conclusion from comparison of protein sequences that the mammoth is more closely related to the African than to the Indian elephant?

1.12 Linnaeus divided the animal kingdom into six classes: mammals, birds, amphibia (including reptiles), fishes, insects and worms. This implies, for instance, that he considered crocodiles and salamanders more closely related than crocodiles and birds. Thomas Huxley, on the other hand, in the nineteenth century, grouped reptiles and birds together. For three suitable proteins with homologues in crocodiles, salamanders and birds, determine the similarity between the homologous sequences. Which pair of animal groups appears most closely related? Who was right, Linnaeus or Huxley?

1.13 When was the last new species of primate discovered?

1.14 In how many more species have PAX-6 homologues been discovered since the table on page 38 was compiled?

1.15 What are the classifications in SCOP of the proteins in Fig. 1.10 (a), (e), (g), (i), (l) and (o)?

1.16 Identify three modular proteins in addition to fibronectin itself that contain fibronectin III domains.

1.17 Find six examples of diseases other than diabetes and haemophilia that are treatable by administering missing protein directly. In each case, what protein is administered?

1.18 To what late-onset disease are carriers of a variant apolipoprotein E gene at unusually high risk? What variant carries the highest risk? What is known about the mechanism by which this variant influences the development of the disease?

1.19 For approximately 10% of Europeans, the painkiller codeine is ineffective because the patients lack the enzyme that converts codeine into the active molecule, morphine. What is the most common mutation that causes this condition?

1.20 Find the page in SCOP headed by Protein: Thermopin from *Thermobifida fusca*. (a) What is the PDB code of this protein? (b) To what superfamily does this protein belong? (c) For what homologue of Thermopin in the chicken is the structure known? (d) By clicking on embedded links, get to a page that displays the abstract of the article describing the structure determination. How many clicks are there in the shortest path that you found? What were the URLs of the intermediate sites on this path?

1.21 Monotremes, of which the best known is probably the platypus (*Ornithorhynchus anatinus*), form an order in the class *mammalia*. Only a few species of monotremes are known. (a) Which if any of them are critically endangered? (b) Where are they found in the wild? (c) What is the nature of the current threat to their continued survival? (d) What gene sequences, if any, are known from endangered monotreme species? (e) Suppose you wanted to sequence a gene from an endangered monotreme species. Identify a zoo that harbours a specimen.

CHAPTER 2

Genome organization and evolution

Chapter contents

Genomes and proteomes 68
Genes 69
Proteomes 71
Eavesdropping on the transmission of genetic information 72
Mappings between the maps 77
High-resolution maps 78
Picking out genes in genomes 80
Genomes of prokaryotes 81
The genome of the bacterium *Escherichia coli* 82
The genome of the archaeon *Methanococcus jannaschii* 85
The genome of one of the simplest organisms: *Mycoplasma genitalium* 86
Genomes of eukaryotes 87
The genome of *Saccharomyces cerevisiae* (baker's yeast) 89
The genome of *Caenorhabditis elegans* 93
The genome of *Drosophila melanogaster* 94
The genome of *Arabidopsis thaliana* 95
The genome of *Homo sapiens* (the human genome) 96
Protein coding genes 97
Repeat sequences 99
RNA 100
Single-nucleotide polymorphisms (SNPs) 101
Genetic diversity in anthropology 102
Genetic diversity and personal identification 103
Genetic analysis of cattle domestication 104
Evolution of genomes 104
Please pass the genes: horizontal gene transfer 108
Comparative genomics of eukaryotes 109
Recommended reading 111
Exercises, Problems, and Weblems 112

Learning goals

1. To know the basic sizes and organizing principles of simple and complex genomes.
2. To understand ideas of proteomics—the patterns of expression of proteins—and techniques such as microarrays used to study them.
3. To know how genomes are analysed, and the relation of gene sequences to phenotypic features, including inherited diseases.
4. To absorb a general idea of the contents of particular genomes, and how the genomes of prokaryotes and eukaryotes differ systematically.
5. To recognize both the importance and the difficulty of deriving from a complete genome sequence the amino acid sequences of the proteins encoded, and assigning functions to these proteins.
6. To realize that published genomes record the characteristics of only a single individual, and that there is considerable variation within populations, and great variation between separated populations of organisms belonging to the same species.
7. To appreciate the power of comparative genomics to identify features responsible for differences between species—for instance, what is it that makes us human?

Genomes and proteomes

The genome of a typical bacterium comes as a single DNA molecule, which, if extended, would be about 2 mm long. (The cell itself has a diameter of about 0.001 mm). The DNA of higher organisms is organized into chromosomes—normal human cells contain 23 chromosome pairs. The total amount of genetic information per cell—the sequence of nucleotides of DNA—is very nearly constant for all members of a species, but varies widely between species (see Box for longer list):

Organism	**Genome size (base pairs)**
Epstein-Barr virus	0.172×10^6
Bacterium (*E. coli*)	4.6×10^6
Yeast (*S. cerevisiae*)	12.5×10^6
Nematode worm (*C. elegans*)	100.3×10^6
Thale cress (*A. thaliana*)	115.4×10^6
Fruit fly (*D. melanogaster*)	128.3×10^6
Human (*H. sapiens*)	3223×10^6

Not all DNA codes for proteins. Conversely, some genes exist in multiple copies. Therefore the amount of protein sequence information in a cell cannot easily be estimated from the genome size.

Genes

A single gene coding for a particular protein corresponds to a sequence of nucleotides along one or more regions of a molecule of DNA. The DNA sequence is collinear with the protein sequence. In species for which the genetic material is double-stranded DNA, genes may appear on either strand. Bacterial genes are continuous regions of DNA. Therefore the functional unit of genetic sequence information from a bacterium is a string of 3*N* nucleotides encoding a string of *N* amino acids, or a string of *N* nucleotides encoding a structural RNA molecule. Such a string, equipped with annotations, would form a typical entry in one of the genetic sequence archives.

Genome Sizes

Organism	Number of base pairs	Number of genes	Comment
φX-174	5 386	10	virus infecting *E. coli*
Human mitochondrion	16 569	37	subcellular organelle
Epstein-Barr virus (EBV)	172 282	80	cause of mononucleosis
Mycoplasma pneumoniae	816 394	680	cause of cyclic pneumonia epidemics
Rickettsia prowazekii	1 111 523	834	bacterium, cause of epidemic typhus
Treponema pallidum	1 138 011	1 039	bacterium, cause of syphilis
Borrelia burgdorferi	1 471 725	1 738	bacterium, cause of Lyme disease
Aquifex aeolicus	1 551 335	1 749	bacterium from hot spring
Thermoplasma acidophilum	1 564 905	1 509	archaeal prokaryote, lacks cell wall
Campylobacter jejuni	1 641 481	1 708	frequent cause of food poisoning
Helicobacter pylori	1 667 867	1 589	chief cause of stomach ulcers
Methanococcus jannaschii	1 664 970	1 783	archaeal prokaryote, thermophile
Haemophilus influenzae	1 830 138	1 738	bacterium, cause of middle ear infections
Thermotoga maritima	1 860 725	1 879	marine bacterium
Archaeoglobus fulgidus	2 178 400	2 437	another archaeon
Deinococcus radiodurans	3 284 156	3 187	radiation-resistant bacterium
Synechocystis	3 573 470	4 003	cyanobacterium, 'blue-green alga'

→

Genome Sizes (*continued*)

Vibrio cholerae	4 033 460	3 890	cause of cholera
Mycobacterium tuberculosis	4 411 532	3 959	cause of tuberculosis
Bacillus subtilis	4 214 814	4 779	popular in molecular biology
Escherichia coli	4 639 221	4 377	molecular biologists' all-time favourite
Pseudomonas aeruginosa	6 264 403	5 570	one of largest prokaryote genomes sequenced
Saccharomyces cerevisiae	12 495 682	5 770	yeast, first eukaryotic genome sequenced.
Caenorhabditis elegans	103 006 709	20 598	the worm
Arabidopsis thaliana	115 409 949	25 498	flowering plant (angiosperm)
Drosophila melanogaster	128 343 463	13 525	the fruit fly
Takiugu rubripes	329×10^6	34 080?	puffer fish (fugu fish)
Human	3223×10^6	23 000?	
Wheat	16×10^9	30 000	
Salamander	10^{11}	?	
Psilotum nudum	2.5×10^{11}	?	whisk fern—a simple plant

In eukaryotes the nucleotide sequences that encode the amino acid sequences of individual proteins are organized in a more complex manner. Frequently one gene appears split into separated segments in the genomic DNA. An **exon** (expressed region) is a stretch of DNA retained in the mature messenger RNA that a ribosome translates into protein. An **intron** is an **intervening region** between two exons. Cellular machinery splices together the proper segments, in RNA transcripts, based on signal sequences flanking the exons in the sequences themselves. Many introns are very long—in some cases substantially longer than the exons.

Control information organizes the expression of genes. Control mechanisms may turn genes on and off (or more finely regulate gene expression) in response to concentrations of nutrients, or to stress, or to unfold complex programs of development during the lifetime of the organism. Many control regions of DNA lie near the segments coding for proteins. They contain **signal sequences** that serve as binding sites for the molecules that transcribe the DNA sequence, or sequences that bind regulatory molecules that can **block** transcription. Bacterial genomes contain examples of contiguous genes coding for several proteins that catalyse successive steps in an integrated sequence of reactions, all under the control of the same regulatory sequence. F. Jacob, J. Monod and E. Wollman named these **operons**. One can readily understand the utility of a parallel control mechanism. In animals, methylation of DNA provides the signals for tissue-specific expression of developmentally regulated genes.

Products of certain genes cause cells to commit suicide—a process called **apoptosis**. Defects in the apoptotic mechanism leading to uncontrolled growth are observed in some cancers, and stimulation of these mechanisms is a general approach to cancer therapy.

The conclusion is that to reduce genetic data to individual coding sequences is to disguise the very complex nature of the interrelationships among them, and to ignore the historical and integrative aspects of the genome. Robbins has expressed the situation unimprovably:

'. . . Consider the 3.2 gigabytes of a human genome as equivalent to 3.2 gigabytes of files on the mass-storage device of some computer system of unknown design. Obtaining the sequence is equivalent to obtaining an image of the contents of that mass-storage device. Understanding the sequence is equivalent to reverse engineering that unknown computer system (both the hardware and the 3.2 gigabytes of software) all the way back to a full set of design and maintenance specifications.

. . .

'Reverse engineering the sequence is complicated by the fact that the resulting image of the mass-storage device will not be a file-by-file copy, but rather a streaming dump of the bytes in the order they were entered into the device. Furthermore, the files are known to be fragmented. In addition, some of the device contains erased files or other garbage. Once the garbage has been recognized and discarded and the fragmented files reassembled, the reverse engineering of the codes can be undertaken with only a partial, and sometimes incorrect, understanding of the CPU on which the codes run. In fact, deducing the structure and function of the CPU is part of the project, since some of the 3.2 gigabytes are the binary specifications for the computer-assisted-manufacturing process that fabricates the CPU. In addition, one must also consider that the huge database also contains code generated from the result of literally millions of maintenance revisions performed by the worst possible set of kludge-using, spaghetti-coding, opportunistic hackers who delight in clever tricks like writing self-modifying code and relying upon undocumented system quirks.'

Robbins, R. J. Challenges in the Human genome project, *IEEE Engineering in Medicine and Biology*, **11**, 25–34 (1992) (©1992 IEEE).

Proteomes

The **proteome project** is a large-scale programme dealing in an integral way with patterns of expression of proteins in biological systems, in ways that complement and extend genome projects.

What kinds of data would we like to measure, and what mature experimental techniques exist to determine them? The basic goal is a spatio-temporal description of the deployment of proteins in the organism. The rates of synthesis of different proteins vary among different tissues and different cell types and states of activity. Methods are available for analysis of transcription patterns of genes. However, because proteins 'turn over' at different rates, it is also necessary to measure proteins directly. High-resolution two-dimensional polyacylamide gel electrophoresis (2D PAGE) shows the pattern of protein content in a sample. Mass-spectroscopic techniques identify the proteins into which the sample has been separated. We shall return to these topics in Chapter 6.

In principle, a database of amino acid sequences of proteins is inherent in the database of nucleotide sequences of DNA, by virtue of the genetic code. Indeed, new protein sequence data are now being determined almost exclusively by translation of DNA sequences, rather than by direct sequencing of proteins. (Historically the chemical problem of determining amino acid sequences of proteins directly was solved before the genetic code was established and before methods for determination of nucleotide sequences of DNA were developed. F. Sanger's sequencing of insulin in 1955 first proved that proteins had definite amino acid sequences, a proposition that until then was hypothetical.)

Should any distinction be made between amino acid sequences determined directly from proteins and those determined by translation from DNA? First, we must assume that it is possible correctly to identify within the DNA data stream the regions that encode proteins. The pattern-recognition programs that address this question are subject to three types of errors: a genuine protein sequence may be missed entirely, or an incomplete protein may be reported, or a gene may be incorrectly spliced. Several variations on the theme add to the complexity: Genes for different proteins may overlap, or genes may be assembled from exons in different ways in different tissues. Conversely, some genetic sequences that appear to code for proteins may in fact be defective or not expressed. *A protein inferred from a genome sequence is a hypothetical object until an experiment verifies its existence.*

Second, in many cases the expression of a gene produces a molecule that must be modified within a cell, to make a *mature* protein that differs significantly from the one suggested by translation of the gene sequence. In many cases the missing details of **post-translational modifications**—the molecular analogues of body piercing—are quite important. Post-translational modifications include addition of ligands (for instance the covalently-bound haem group of cytochrome *c*), glycosylation, methylation, excision of peptides, and many others. Patterns of disulphide bridges—primary chemical bonds between cysteine residues—cannot be deduced from the amino acid sequence. In some cases, mRNA is edited before translation, creating changes in amino acid sequences that are not inferrable from the genes.

Eavesdropping on the transmission of genetic information

How hereditary information is stored, passed on, and implemented is perhaps *the* fundamental problem of biology. Three types of maps have been essential (see Box):

1. Linkage maps of genes
2. Banding patterns of chromosomes
3. DNA sequences

These represent three very different types of data. Genes, as discovered by Mendel, were entirely abstract entities. Chromosomes are physical objects, the

Gene maps, chromosome maps, and sequence maps

1. A **gene map** is classically determined by observed patterns of heredity. Linkage groups and recombination frequencies can detect whether genes are on the same or different chromosomes, and, for genes on the same chromosome, how far apart they are. The principle is that the farther apart two linked genes are, the more likely they are to recombine, by crossing over during meiosis. Indeed, two genes on the same chromosome but very far apart will appear to be unlinked. The unit of length in a gene map is the Morgan, defined by the relation that 1 cM corresponds to a 1% recombination frequency. (We now know that $1 \text{ cM} \sim 1 \times 10^6$ bp in humans, but it varies with the location in the genome and with the distance between genes.)
2. **Chromosome banding pattern maps** Chromosomes are physical objects. Banding patterns are visible features on them. The nomenclature is as follows: In many organisms, chromosomes are numbered in order of size, 1 being the largest. The two arms of human chromosomes, separated by the centromere, are called the p (petite = short) arm and q (= queue) arm. Regions within the chromosome are numbered p1, p2, . . . and q1, q2 . . . outward from the centromere. Subsequent digits indicate subdivisions of bands. For example, certain bands on the q arm of human chromosome 15 are labelled 15q11.1, 15q11.2, 15q12. Originally bands 15q11 and 15q12 were defined; subsequently 15q11 was divided into 15q11.1 and 15q11.2. Deletions including this region are associated with Prader-Willi and Angleman syndromes. These syndromes have the interesting feature that the alternative clinical consequences depend on whether the affected chromosome is paternal or maternal. This observation of **genomic imprinting** shows that the genetic information in a fertilized egg is not simply the bare DNA sequences contributed by the parents. Chromosomes of paternal and maternal origin have different states of methylation, signals for differential expression of their genes. The process of modifying the DNA which takes place during differentiation in development is already present in the zygote.
3. **The DNA sequence itself** Physically a sequence of nucleotides in the molecule, computationally a string of characters A, T, G and C. Genes are regions of the sequence, in many cases interrupted by non-coding regions.

banding patterns their visible landmarks. Only with DNA sequences are we dealing directly with stored hereditary information in its physical form.

It was the very great achievement of the last century of biology to forge connections between these three types of data. The first steps—and giant strides they were indeed—proved that, for any chromosome, the maps are one-dimensional arrays, and indeed that they are collinear. Any schoolchild now knows that genes are strung out along chromosomes, and that each gene corresponds to a DNA sequence. But the proofs of these statements earned a large number of Nobel prizes.

Splitting a long molecule of DNA—for example, the DNA in an entire chromosome—into fragments of convenient size for cloning and sequencing requires additional maps to report the order of the fragments, so that the entire sequence can be reconstructed from the sequences of the fragments. A restriction endonuclease is an enzyme that cuts DNA at a specific sequence, usually about 6 bp long. Cutting DNA with several restriction enzymes with different specificities produces sets of overlapping fragments. From the sizes of the fragments it is possible to construct a **restriction map**, stating the order and distance between the restriction enzyme cleavage sites. A mutation in one of these cleavage sites will change the sizes of the fragments produced by the corresponding enzyme, allowing the mutation to be located in the map.

Restriction enzymes can produce fairly large pieces of DNA. Cutting the DNA into smaller pieces, which are cloned and ordered by sequence overlaps (like the example using text on pages 19–20) produces a finer dissection of the DNA called a **contig map**.

In the past, the connections between chromosomes, genes and DNA sequences have been essential for identifying the molecular deficits underlying inherited diseases, such as Huntington disease or cystic fibrosis. Sequencing of the human genome has changed the situation radically.

Given a disease attributable to a defective protein:

- If we know the protein involved, we can pursue rational approaches to therapy.
- If we know the gene involved, we can devise tests to identify sufferers or carriers.
- In many cases, knowledge of the chromosomal location of the gene is unnecessary for either therapy or detection; it is required only for identifying the gene, providing a bridge between the patterns of inheritance and the DNA sequence. (This is not true of diseases arising from chromosome abnormalities.)

For instance, in the case of sickle-cell anaemia, we know the protein involved. The disease arises from a single point mutation in haemoglobin. We can proceed directly to drug design. We need the DNA sequence only for genetic testing and counselling. In contrast, if we know neither the protein nor the gene, we must somehow go from the phenotype back to the gene, a process called **positional cloning** or **reverse genetics**. Positional cloning used to involve a kind of 'Tinker to Evers to Chance' cascade from the gene map to the chromosome map to the DNA sequence. Later we shall see how recent developments have short-circuited this process.

Patterns of inheritance identify the type of genetic defect responsible for a condition. They show, for example, that Huntington disease and cystic fibrosis are caused by single genes. To find the gene associated with cystic fibrosis it was necessary to begin with the gene map, using linkage patterns of heredity in affected families to localize the affected gene to a particular region of a particular chromosome. Knowing the general region of the chromosome, it was then possible to search the DNA of that region to identify candidate genes, and finally to pinpoint the particular gene responsible and sequence it (see Boxes, Identification of the cystic fibrosis gene, and Positional cloning: finding the cystic fibrosis gene.) In contrast, many diseases do not show simple inheritance, or, even if only a single gene is involved, heredity creates only a predisposition, the clinical consequences of which depend on environmental factors. The full human genome sequence, and measurements of expression patterns, will be essential to identify the genetic components of these more complex cases.

Identification of the cystic fibrosis gene

Cystic fibrosis, a disease known to folklore since at least the Middle Ages and to science for about 500 years, is an inherited recessive autosomal condition. Its symptoms include intestinal obstruction, reduced fertility including anatomical abnormalities (especially in males); and recurrent clogging and infection of lungs—the primary cause of death now that there are effective treatments for the gastrointestinal symptoms. Approximately half the sufferers die before age 32 years, and few survive beyond 50. Cystic fibrosis affects 1/2500 individuals in the American and European populations. Approximately 1/25 Caucasians carry a mutant gene, and 1/65 African-Americans. The protein that is defective in cystic fibrosis also acts as a receptor for uptake of *Salmonella typhi*, the pathogen that causes typhoid fever. Increased resistance to typhoid in heterozygotes—who do not develop cystic fibrosis itself but are carriers of the mutant gene—probably explains why the gene has not been eliminated from the population.

The pattern of inheritance showed that cystic fibrosis was the effect of a single gene. However, the actual protein involved was unknown. It had to be found *via* the gene.

Clinical observations provided the gene hunters with useful clues. It was known that the problem had to do with Cl^- transport in epithelial tissues. Folklore had long recognized that children with excessive salt in their sweat—tastable when kissing an infant on the forehead—were short-lived. Modern physiological studies showed that epithelial tissues of cystic fibrosis patients cannot reabsorb chloride. When closing in on the gene, the expected distribution among tissues of its expression, and the type of protein implicated, were useful guides.

In 1989 the gene for cystic fibrosis was isolated and sequenced. This gene—called CFTR (cystic fibrosis transmembrane conductance regulator)—codes for a 1480 amino acid protein that normally forms a cyclicAMP-regulated epithelial Cl^- channel. The gene, comprising 24 exons, spans a 250 kbp region. For 70% of mutant alleles, the mutation is a three-base-pair deletion,

→

→

Identification of the cystic fibrosis gene *(continued)*

deleting the residue 508Phe from the protein. The mutation is denoted del508. The effect of this deletion is defective translocation of the protein, which is degraded in the endoplasmic reticulum rather than transported to the cell membrane.

An *in utero* test for cystic fibrosis is based on recovery of foetal DNA. A PCR primer is designed to give a 154 bp product from the normal allele and a 151 bp product from the del508 allele.

Clinicians have taken advantage of the fact that the affected tissues of the airways are easily accessible, to develop gene therapy. Genetically engineered adenovirus sprayed into the respiratory passages can deliver the correct gene to epithelial tissues.

Positional cloning: finding the cystic fibrosis gene

The process by which the gene was found has been called *positional cloning* or *reverse genetics.*

- A search in family pedigrees for a linked marker showed that the cystic fibrosis gene was close to a known variable number tandem repeat (VNTR), DOCR-917. Somatic cell hybrids placed this on chromosome 7, band q3.
- Other markers found were linked more tightly to the target gene. It was thereby bracketed by a VNTR in the MET oncogene and a second VNTR, D7S8. The target gene lies 1.3 cM from MET and 0.9 cM from D7S8—localizing it to a region of approximately 1–2 million bp. A region this long could well contain 100–200 genes.
- The inheritance patterns of additional markers from within this region localized the target more sharply to within 500 kbp. A technique called chromosome jumping made the exploration of the region more efficient.
- A 300 kbp region at the right distance from the markers was cloned. Probes were isolated from the region, to look for active genes, characterized by an upstream CCGG sequence. (The restriction endonuclease HpaII is useful for this step; it cuts DNA at this sequence, but only when the second C is not methylated, that is, when the gene is active.)
- Identification of genes in this region by sequencing.
- Checking in animals for genes similar to the candidate genes turned up four likely possibilities. Checking these possibilities against a cDNA library from sweat glands of cystic fibrosis patients and healthy controls identified one probe with the right tissue distribution for the expected expression pattern of the gene responsible for cystic fibrosis. One long coding segment had the right properties, and indeed corresponded to an exon of the cystic fibrosis gene. Most cystic fibrosis patients have a common alteration

→

→

in the sequence of this gene—a three base-pair deletion, deleting the residue 508Phe from the protein.

Proof that the gene was correctly identified included:

- 70% of cystic fibrosis alleles have the deletion. It is not found in people who are neither sufferers nor likely to be carriers.
- expression of the wild-type gene in cells isolated from patients restores normal Cl^- transport.
- knockout of the homologous gene in mice produces the cystic fibrosis phenotype.
- The pattern of gene expression matches the organs in which it is expected.
- The protein encoded by the gene would contain a transmembrane domain, consistent with involvement in transport.

Mappings between the maps

A gene linkage map can be calibrated to chromosome banding patterns through observation of individuals with deletions or translocations of parts of chromosomes. The genes responsible for phenotypic changes associated with a deletion must lie within the deletion. Translocations are correlated with altered patterns of linkage and recombination.

There have been several approaches to coordinating chromosome banding patterns with individual DNA sequences of genes:

- In Fluorescent *In Situ* Hybridization (FISH) a probe sequence is labelled with fluorescent dye. The probe is hybridized with the chromosomes, and the chromosomal location where the probe is bound shows up directly in a photograph. (See Plate II.) Typical resolution is ~10^5 base pairs, but specialized new techniques can achieve high resolution, down to 1 kbp. Simultaneous FISH with two probes can detect linkage and even estimate genetic distances. This is important in species for which the generation time is long enough to make standard genetic approaches inconvenient. FISH can also detect chromosomal abnormalities.
- **Somatic cell hybrids** are rodent cells containing few, one, or even partial human chromosomes. (Chromosome fragments are produced by irradiating the human cells prior to fusion. Such lines are called **radiation hybrids**.) Hybridization of a probe sequence with a panel of somatic cell hybrids, detected by fluorescence, can identify which chromosome contains the probe. This approach has been superseded by use of clones of yeast or bacteria or phage containing fragments of human DNA in artificial chromosomes (YACs and BACs and PACs).

Of course, for fully-sequenced genomes these methods are obsolete. Given a DNA sequence one would just look it up (but mind the gaps).

High-resolution maps

Formerly, genes were the only visible portions of genomes. Now, markers are no longer limited to genes with phenotypically observable effects, which are anyway too sparse for an adequately high-resolution map of the human genome. Now that we can interrogate DNA sequences directly, any features of DNA that vary among individuals can serve as markers, including:

- **variable number tandem repeats** (VNTRs), also called minisatellites. VNTRs contain regions 10–100 base pairs long, repeated a variable number of times—same sequence, different number of repeats. In any individual, VNTRs based on the same repeat motif may appear only once in the genome; or several times, with different lengths on different chromosomes. The distribution of the sizes of the repeats is the marker. Inheritance of VNTRs can be followed in a family and mapped to a disease phenotype like any other trait. VNTRs were the first genetic sequence data used for personal identification—genetic fingerprints—in paternity and in criminal cases.

  Formerly, VNTRs were observed by producing **restriction fragment length polymorphisms** (RFLPs) from them. VNTRs are generally flanked by recognition sites for the same restriction enzyme, which will neatly excise them. The results can be spread out on a gel, and detected by Southern blotting. Note the distinction: VNTRs are characteristics of genome sequences; RFLPs are artificial mixtures of short stretches of DNA created in the laboratory in order to identify VNTRs.

  It is much easier and more efficient to measure the sizes of VNTRs by amplifying them with PCR, and this method has replaced the use of restriction enzymes.

- **short tandem repeat polymorphisms** (STRPs), also called microsatellites. STRPs are regions of only 2–5 bp but repeated many times; typically 10–30 consecutive copies. They have several advantages as markers over VNTRs, one of which is a more even distribution over the human genome.

There is no reason why these markers need lie within expressed genes, and usually they do not. (The CAG repeats in the gene for huntingtin and certain other disease genes are exceptions.)

Panels of microsatellite markers greatly simplify the identification of genes. It is interesting to compare a recent project to identify a disease gene now that the human genome sequence is available, with such classic studies as the identification of the gene for cystic fibrosis (see Box: Identification of a gene for Berardinelli-Seip syndrome).

Additional mapping techniques deal more directly with the DNA sequences, and can short-circuit the process of gene identification:

- A **contig**, or **contiguous clone map**, is a series of overlapping DNA clones of known order along a chromosome from an organism of interest—for instance, human—stored in yeast or bacterial cells as YACs (Yeast Artificial Chromosomes) or BACs (Bacterial Artificial Chromosomes). A contig map can produce a very fine mapping of a genome. In a YAC, human DNA is stably integrated into a

Identification of a gene for Berardinelli-Seip syndrome

Berardinelli-Seip syndrome (congenital generalized lipodystrophy) is an autosomal recessive disease. Its symptoms include absence of body fat, insulin-resistant diabetes, and enhanced rate of skeletal growth.

To determine the gene involved, a group led by J. Magré subjected DNA from members of affected families to linkage analysis and homozygosity mapping with a genome-wide panel containing ~400 microsatellite markers of known genetic location, with an average spacing of ~10 cM. In this procedure, a fixed panel of primers specific for the amplification and analysis of each marker was used to compare whole DNA of affected individuals with that of unaffected relatives. The measurements reveal the lengths of the repeats associated with each microsatellite. For every microsatellite, each observed length is an allele. Identifying microsatellite markers that are closely linked to the phenotype localizes the desired gene. The measurements are done efficiently and in parallel using commercial primer sets and instrumentation.

Two markers in chromosome band 11q13—D11S4191 and D11S987—segregated with the disease, and some affected individuals born from consanguineous families were homozygous for them. Finer probing, mapping with additional markers, localized the gene on chromosome 11 to a region of about 2.5 Mbp.

There are 27 genes in the implicated 2.5 Mbp region and its vicinity. Sequencing these genes in a set of patients identified a deletion of three exons in one of them. It was proved to be the disease gene by comparing its sequences in members of the families studied, and demonstrating a correlation between presence of the syndrome and abnormalities in the gene. None of the other 26 genes in the suspect interval showed such correlated alterations.

Previous studies had identified a different gene, *BSCL1*, at 9q34, in other families with the same syndrome. The gene *BSCL1* has not yet been identified. It is possible that abnormalities in these two genes produce the same effect because their products participate in a common pathway which can be blocked by dysfunction of either.

The gene on chromosome 11, *BSCL2*, contains 11 exons spanning > 14 kbp. It encodes a 398-residue protein, named seipin. Observed alterations in the gene include large and small deletions, and single amino acid substitutions. The effects are consistent with loss of functional protein, either by causing frameshifts or truncation, or a missense mutation Ala212 → Pro that credibly interferes with the stability of a helix or sheet in the structure.

Seipin has homologues in mouse and *Drosophila*. There are no clues to the function of any of the homologues, although they are predicted to contain transmembrane helices. What does provide suggestions about the aetiology of some aspects of the syndrome is the expression pattern, highest in brain and testis. This might be consistent with earlier endocrinological studies

→

Identification of a gene for Berardinelli-Seip syndrome *(continued)*

of Berardinelli-Seip syndrome that identified a problem in the regulation of release of pituitary hormones by the hypothalamus. Discovery of a protein of unknown function involved in the syndrome opens the way to investigation of what may well be a new biological pathway.

small extra chromosome in a yeast cell. A YAC can contain up to 10^6 base pairs. In principle, the entire human genome could be represented in 10 000 YAC clones. In a BAC, human DNA is inserted into a plasmid in an *E. coli* cell. (A plasmid is a small piece of double-stranded DNA found in addition to the main genome, usually but not always circular.) A BAC can carry about 250 000 bp. Despite their smaller capacities, BACs are preferred to YACs because of their greater stability and ease of handling.

- A **sequence tagged site** (STS) is a short, sequenced region of DNA, typically 200–600 bp long, that appears in a unique location in the genome. It need not be polymorphic. An STS can be mapped into the genome by using PCR to test for the presence of the sequence in the cells containing a contig map.

  One type of STS arises from an **expressed sequence tag** (EST), a piece of cDNA (complementary DNA; that is, a DNA sequence derived from the messenger RNA of an expressed gene). The sequence contains only the exons of the gene, spliced together to form the sequence that encodes the protein. cDNA sequences can be mapped to chromosomes using FISH, or located within contig maps.

How do contig maps and sequence tagged sites facilitate identifying genes? If you are working with an organism for which the full-genome sequence is not known, but for which full contig maps are available for all chromosomes, you would identify STS markers tightly linked to your gene, then locate these markers in the contig maps.

Picking out genes in genomes

Computer programs for genome analysis identify **open reading frames** or ORFs. An ORF is a region of DNA sequence that begins with an initiation codon (ATG) and ends with a stop codon. An ORF is a potential protein-coding region.

Approaches to identifying protein-coding regions choose from or combine two possible approaches:

(1) **Detection of regions similar to known coding regions from other organisms.** These regions may encode amino acid sequences similar to known proteins, or may be similar to Expressed Sequence Tags (ESTs). Because ESTs are derived from messenger RNA, they correspond to genes known to be transcribed. It is necessary to sequence only a few hundred initial bases of cDNA to give enough information to identify a gene: characterization of genes by ESTs is like indexing poems or songs by their first lines.

(2) **Ab initio methods, that seek to identify genes from the properties of the DNA sequences themselves.** Computer-assisted annotation of genomes is more complete and accurate for bacteria than for eukaryotes. Bacterial genes are relatively easy to identify because they are contiguous—they lack the introns characteristic of eukaryotic genomes—and the intergene spaces are small. In higher organisms, identifying genes is harder. Identification of exons is one problem, assembling them is another. Alternative splicing patterns present a particular difficulty.

A framework for ab initio gene identification in eukaryotic genomes includes the following features:

- The initial (5′) exon starts with a transcription start point, preceded by a core promotor site such as the TATA box typically ~30 bp upstream. It is free of in-frame stop codons, and ends immediately before a GT splice signal. (Occasionally a noncoding exon precedes the exon that contains the initiator codon.)
- Internal exons, like initial exons, are free of in-frame stop codons. They begin immediately after an AG splice signal and end immediately before a GT splice signal.
- The final (3′) exon starts immediately after an AG splice signal and ends with a stop codon, followed by a polyadenylation signal sequence. (Occasionally a non-coding exon follows the exon that contains the stop codon.)

All coding regions have nonrandom sequence characteristics, based partly on codon usage preferences. Empirically, it is found that statistics of hexanucleotides perform best in distinguishing coding from noncoding regions. Starting from a set of known genes from an organism as a training set, pattern recognition programs can be tuned to particular genomes.

Accurate gene detection is a crucial component of genome sequence analysis. This problem is an important focus of current research.

Genomes of prokaryotes

Most prokaryotic cells contain their genetic material in the form of a large single circular piece of double-stranded DNA, usually less than 5 Mbp long. In addition they may contain plasmids.

The protein-coding regions of bacterial genomes do not contain introns. In many prokaryotic genomes the protein-coding regions are partially organized into *operons*—tandem genes transcribed into a single messenger RNA molecule, under common transcriptional control. In bacteria, the genes of many operons code for proteins with related functions. For instance, successive genes in the trp operon of *E. coli* code for proteins that catalyse successive steps in the biosynthesis of tryptophan (see Fig. 2.1). In archaea, a metabolic relationship between genes in operons is less frequently observed.

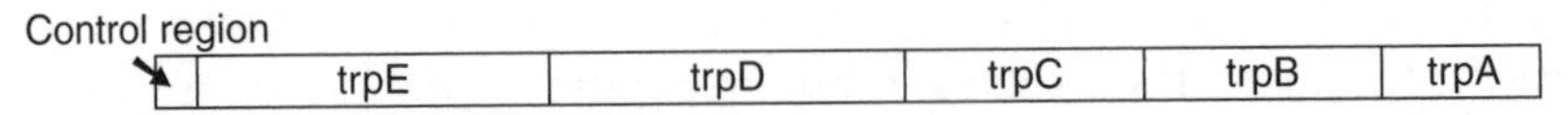

Fig. 2.1 The trp operon in *E. coli* begins with a control region containing promoter, operator and leader sequences. Five structural genes encode proteins that catalyse successive steps in the synthesis of the amino acid tryptophan from its precursor chorismate:

chorismate $\xrightarrow{(1)}$ anthranilate $\xrightarrow{(2)}$ phosphoribosyl-anthranilate $\xrightarrow{(3)}$ indoleglycerol-phosphate $\xrightarrow{(4)}$ indole $\xrightarrow{(5)}$ tryptophan

Reaction step (1): trpE and trpD encode two components of anthranilate synthase. This tetrameric enzyme, comprising two copies of each subunit, catalyses the conversion of chorismate to anthranilate. Reaction step (2): The protein encoded by trpD also catalyses the subsequent phosphoribosylation of anthranilate. Reaction step (3): trpC encodes another bifunctional enzyme, phosphoribosylanthranilate isomerase – indoleglycerolphosphate synthase. It converts phosphoribosyl anthranilate to indoleglycerolphosphate, through the intermediate, carboxyphenylaminodeoxyribulose phosphate. Reaction steps (4) and (5): trpB and trpA encode the β and α subunits, respectively, of a third bifunctional enzyme, tryptophan synthase (an $\alpha_2\beta_2$ tetramer). A tunnel within the structure of this enzyme delivers, without release to the solvent, the intermediate produced by the α subunit – indoleglycerolphosphate → indole – to the active site of the β subunit – which converts indole → tryptophan.

A separate gene, trpR, not closely linked to this operon, codes for the trp repressor. The repressor can bind to the operator sequence in the DNA (within the control region) only when binding tryptophan. Binding of the repressor blocks access of RNA polymerase to the promoter, turning the pathway off when tryptophan is abundant. Further control of transcription in response to tryptophan levels is exerted by the attenuator element in the mRNA, within the leader sequence. The attenuator region (1) contains two tandem trp codons and (2) can adopt alternative secondary structures, one of which terminates transcription. Levels of tryptophan govern levels of trp-tRNAs, which govern the rate of progress of the tandem trp codons through the ribosome. Stalling on the ribosome at the tandem trp codons in response to low tryptophan levels reduces the formation of the mRNA secondary structure that terminates transcription.

The typical prokaryotic genome contains only a relatively small amount of noncoding DNA (in comparison with eukaryotes), distributed throughout the sequence. In *E. coli* only ~11% of the DNA is noncoding.

The genome of the bacterium *Escherichia coli*

E. coli, strain K-12, has long been the workhorse of molecular biology. The genome of strain MG1655, published in 1997 by the group of F. Blattner at the University of Wisconsin, contains 4 639 221 bp in a single circular DNA molecule, with no plastids. Approximately 89% of the sequence codes for proteins or structural RNAs. An inventory reveals:

- 4284 protein-coding genes
- 122 structural RNA genes
- noncoding repeat sequences
- regulatory elements
- transcription/translation guides
- transposases
- prophage remnants
- insertion sequence elements
- patches of unusual composition, likely to be foreign elements introduced by horizontal transfer.

Analysis of the genome sequence required identification and annotation of protein-coding genes and other functional regions. Many *E. coli* proteins were known before the sequencing was complete, from the many years of intensive investigation: 1853 proteins had been described before publication of the genome sequence. Other genes could be assigned functions from identification of homologues by searching in sequence databanks. The narrower the range of specificity of the function of the homologues, the more precise could be the assignment. Currently, over 80% of proteins can be assigned at least a general function (compare Box: Distribution of *E. coli* proteins among 22 functional groups). Other regions of the genome are recognized as regulatory sites, or mobile genetic elements, also on the basis of similarity to homologous sequences known in other organisms.

Distribution of *E. coli* proteins among 22 functional groups

Functional class	Number	%
Regulatory function	45	1.05
Putative regulatory proteins	133	3.10
Cell structure	182	4.24
Putative membrane proteins	13	0.30
Putative structural proteins	42	0.98
Phage, transposons, plasmids	87	2.03
Transport and binding proteins	281	6.55
Putative transport proteins	146	3.40
Energy metabolism	243	5.67
DNA replication, recombination, modification, and repair	115	2.68
Transcription, RNA synthesis, metabolism, and modification	55	1.28
Translation, post-translational protein modification	182	4.24

→

→

Distribution of *E. coli* proteins among 22 functional groups *(continued)*

Cell processes (including adaptation, protection)	188	4.38
Biosynthesis of cofactors, prosthetic groups, and carriers	103	2.40
Putative chaperones	9	0.21
Nucleotide biosynthesis and metabolism	58	1.35
Amino acid biosynthesis and metabolism	131	3.06
Fatty acid and phospholipid metabolism	48	1.12
Carbon compound catabolism	130	3.03
Central intermediary metabolism	188	4.38
Putative enzymes	251	5.85
Other known genes (gene product or phenotype known)	26	0.61
Hypothetical, unclassified, unknown	1632	38.06

From Blattner *et al.* (1997), The complete genome sequence of *Escherichia coli* K12, *Science*, **277**, 1453–1462.

The distribution of protein-coding genes over the genome of *E. coli* does not seem to follow any simple rules, either along the DNA or on different strands. Indeed, comparison of strains suggests that genes are mobile.

The *E. coli* genome is relatively gene dense. Genes coding for proteins or structural RNAs occupy ~89% of the sequence. The average size of an open reading frame is 317 amino acids. If the genes were evenly distributed, the average intergenic region would be 130 bp; the observed average distance between genes is 118 bp. However, the sizes of intergenic regions vary considerably. Some intergenic regions are large. These contain sites of regulatory function, and repeated sequences. The longest intergenic region, 1730 bp, contains noncoding repeat sequences.

Approximately three-quarters of the transcribed units contain only one gene; the rest contain several consecutive genes, or operons. It is estimated that the *E. coli* genome contains 630–700 operons. Operons vary in size, although few contain more than five genes. Genes within operons tend to have related functions.

In some cases, the same DNA sequence encodes parts of more than one polypeptide chain. One gene codes for both the τ and γ subunits of DNA polymerase III. Translation of the entire gene forms the τ subunit. The γ subunit corresponds approximately to the N-terminal two-thirds of the τ subunit. A frameshift on the ribosome at this point leads to chain termination 50% of the time, causing a 1:1 ratio of expressed τ and γ subunits. There do not appear to be any overlapping genes in which different reading frames both code for expressed proteins.

In other cases, the same polypeptide chain appears in more than one enzyme. A protein that functions on its own as lipoate dehydrogenase is also an essential subunit of pyruvate dehydrogenase, 2-oxoglutarate dehydrogenase and the glycine cleavage complex.

Having the complete genome, we can examine the protein repertoire of *E. coli*. The largest class of proteins are the enzymes—appoximately 30% of the total genes. Many enzymatic functions are shared by more than one protein. Some of these sets of functionally similar enzymes are very closely related, and appear to have arisen by duplication, either in *E. coli* itself or in an ancestor. Other sets of functionally similar enzymes have very dissimilar sequences, and differ in specificity, regulation, or intracellular location.

Several features of *E. coli's* generous endowment of enzymes give it an extensive and versatile metabolic competence, which allow it to grow and compete under varying conditions:

- It can synthesize all components of proteins and nucleic acids (amino acids and nucleotides), and cofactors.
- It has metabolic flexibility: Both aerobic and anaerobic growth are possible, utilizing different pathways of energy capture. It can grow on many different carbon sources. Not all metabolic pathways are active at any time, the versatility allowing response to changes in conditions.
- Even for specific metabolic reactions there are many cases of multiple enzymes. These provide redundancy, and contribute to an ability to tune metabolism to varying conditions.
- However, *E. coli* does not possess a complete range of enzymatic capacity. It cannot fix CO_2 or N_2.

We have described here some of the *static* features of the *E. coli* genome and its protein repertoire. Current research is moving into dynamic aspects, to investigations of protein expression patterns in time and space.

The genome of the archaeon *Methanococcus jannaschii*

S. Luria once suggested that to determine common features of all life one should not try to survey everything, but rather identify the organism most different from us and see what we have in common with it. The assumption was that the way to do this would be to find an organism adapted to the most different environment.

Deep-sea exploration has revealed environments as far from the familiar as those portrayed in science fiction. Hydrothermal vents are underwater volcanoes emitting hot lava and gases through cracks in the ocean floor. They create niches for communities of living things disconnected from the surface, depending on the minerals exuded from the vent for their sources of inorganic nutrients. They support living communities of micro-organisms that are the only known forms of life not dependent on sunlight, directly or indirectly, as their energy source.

The micro-organism *Methanococcus jannaschii* was collected from a hydrothermal vent 2600 m deep off the coast of Baja California, Mexico, in 1983. It is a thermophilic organism, surviving at temperatures from 48–94 °C, with an optimum at 85 °C. It is a strict anaerobe, capable of self-reproduction from inorganic components. Its overall metabolic equation is to synthesize methane from H_2 and CO_2.

M. jannaschii belongs to the archaea, one of the three major divisions of life along with the bacteria and eukarya (see Fig. 1.2). The archaea comprise groups of prokaryotes, including organisms adapted to extreme environmental conditions such as high temperature and pressure, or high salt concentration.

The genome of *M. jannaschii* was sequenced in 1996 by The Institute for Genomic Research (TIGR). It was the first archaeal genome sequenced. It contains a large chromosome containing a circular double-stranded DNA molecule 1 664 976 bp long, and two extrachromosomal elements of 58 407 and 16 550 bp respectively. There are 1743 predicted coding regions, of which 1682 are on the chromosome, and 44 and 12 are on the large and small extrachromosomal elements respectively. Some RNA genes contain introns. As in other prokaryotic genomes there is little noncoding DNA.

M. jannaschii would appear to satisfy Luria's goal of finding our most distant extant relative. Comparison of its genome sequence with others shows that it is distantly related to other forms of life. Only 38% of the open reading frames could be assigned a function, on the basis of homology to proteins known from other organisms. However, to everyone's great surprise, archaea are in some ways more closely related to eukaryotes than to bacteria! They are a complex mixture. In archaea, proteins involved in transcription, translation and regulation are more similar to those of eukaryotes. Archaeal proteins involved in metabolism are more similar to those of bacteria.

The genome of one of the simplest organisms: *Mycoplasma genitalium*

Mycoplasma genitalium is an infectious bacterium, the cause of nongonococcal uretheitis. Its genome was sequenced in 1995 by a collaboration of groups at TIGR, The Johns Hopkins University and The University of North Carolina. The genome is a single DNA molecule containing 580 070 base pairs. This is the smallest cellular genome yet sequenced. So far, *M. genitalium* is the closest we have to a **minimal organism**, the smallest capable of independent life. (Viruses, in contrast, require the cellular machinery of their hosts.)

The genome is dense in coding regions. 468 genes have been identified as expressed proteins. Some regions of the sequence are gene rich, others gene poor, but overall 85% of the sequence is coding. The average length of a coding region is 1040 bp. As in other bacterial genomes, the coding regions do not contain introns. Further compression of the genome is achieved by overlapping genes. It appears that many of these have arisen through loss of stop codons.

The gene repertoire of *M. genitalium* includes some that encode proteins essential for independent reproduction, such as those involved in DNA replication, transcription, and translation, plus ribosomal and transfer RNAs. Other genes are specific for the infectious activity, including adhesins that mediate binding to infected cells, other molecules for defence against the host's immune system, and many transport proteins. As adaptation to the parasitic lifestyle of the organism, there has been widespread loss of metabolic enzymes, including those responsible for amino acid biosynthesis—indeed, one amino acid is absent from all *M. genitalium* proteins (see Weblem 2.7).

Genomes of eukaryotes

It is rare in science to encounter a completely new world containing phenomena entirely unsuspected. The complexity of the eukaryotic genome is such a world (see Box: Inventory of a eukaryotic genome).

Inventory of a eukaryotic genome

Moderately repetitive DNA

- Functional
  - dispersed gene families
    - e.g. actin, globin
  - tandem gene family arrays
    - rRNA genes (250 copies)
    - tRNA genes (50 sites with 10–100 copies each in human)
    - histone genes in many species
- Without known function
  - short interspersed elements (SINEs)
    - Alu is an example
    - 200–300 bp long
    - 100 000s of copies (300 000 Alu)
    - scattered locations (not in tandem repeats)
  - long interspersed elements (LINEs)
    - 1–5 kbp long
    - 10–10 000 copies per genome
  - pseudogenes

Highly repetitive DNA

- minisatellites
  - composed of repeats of 14–500 bp segments
  - 1–5 kbp long
  - many different ones
  - scattered throughout the genome
- microsatellites
  - composed of repeats of up to 13 bp

Inventory of a eukaryotic genome *(continued)*

  - ~100s of kbp long
  - ~10^6 copies/genome
  - most of the heterochromatin around the centromere
- telomeres
  - contain a short repeat unit (typically 6 bp: TTAGGG in human genome, TTGGGG in *Paramecium*, TAGGG in trypanosomes, TTTAGGG in *Arabidopsis*)
  - 250–1000 repeats at the end of each chromosome

In eukaryotic cells, the majority of DNA is in the nucleus, separated into bundles of nucleoprotein, the chromosomes. Each chromosome contains a single double-stranded DNA molecule. Smaller amounts of DNA appear in organelles—mitochondria and chloroplasts. The organelles originated as intracellular parasites. Organelle genomes usually have the form of circular double-stranded DNA, but are sometimes linear and sometimes appear as multiple circles. The genetic code by which organelle genes are translated differs from that of nuclear genes.

Nuclear genomes of different species vary widely in size (see pages 68–69). The correlation between genome size and complexity of the organism is very rough. It certainly does not support any preconception that humans stand on a pinnacle. In many cases differences in genome size reflect different amounts of simple repetitive sequences.

In addition to variation in DNA content, eukaryotic species vary in the number of chromosomes and distribution of genes among them. Some differences in the distribution of genes among chromosomes involve translocations, or chromosome fragmentations or joinings. For instance, humans have 23 pairs of chromosomes; chimpanzees have 24. Human chromosome 2 is equivalent to a fusion of chimpanzee chromosomes 12 and 13 (see Fig. 2.2). The difficulty of chromosome pairing during mitosis in a zygote after such an event can contribute to the reproductive isolation required for species separation.

Other differences in chromosome complement reflect duplication or hybridization events. The wheat first used in agriculture, in the Middle East at least 10 000–15 000 years ago, is a diploid called *einkorn* (*Triticum monococcum*), containing 14 pairs of chromosomes. *Emmer* wheat (*T. dicoccum*), also cultivated since palaeolithic times, and *durum* wheat (*T. turgidum*), are merged hybrids of relatives of einkorn with other wild grasses, to form tetraploid species. Additional hybridizations, to different wild wheats, gave hexaploid forms, including *spelt* (*T. spelta*), and modern common wheat *T. aestivum*. *Triticale*, a robust crop developed in modern agriculture and currently used primarily for animal feed, is an artificial genus arising from crossing durum wheat (*Triticum turgidum*) and rye (*Secale cereale*). Most triticale varieties are hexaploids.

Variety of wheat	Classification	Chromosome complement
einkorn	*Triticum monococcum*	AA
emmer wheat	*Triticum dicoccum*	AABB
durum wheat	*Triticum turgidum*	AABB
spelt	*Triticum spelta*	AABBDD
common wheat	*Triticum aestivum*	AABBDD
triticale	*Triticosecale*	AABBRR

A = genome of original diploid wheat or a relative, B = genome of a wild grass *Aegilops speltoides* or *Triticum speltoides* or a relative, D = genome of another wild grass, *Triticum tauschii* or a relative, R = genome of rye *Secale cereale*.

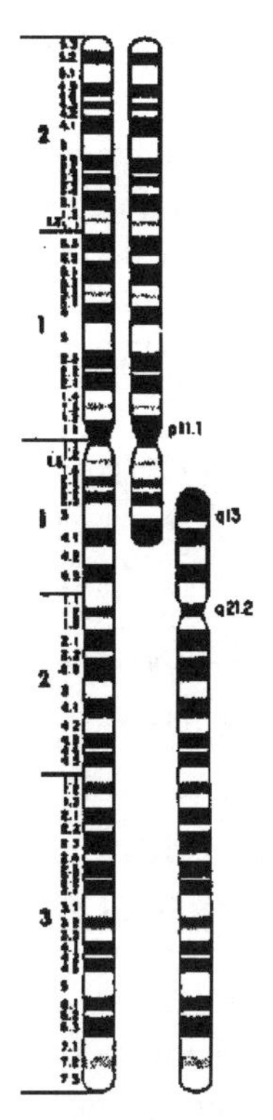

Fig. 2.2 Left: human chromosome 2. Right: matching chromosomes from a chimpanzee. (From: Yunis, J. J., Sawyer, J. R. & Dunham, K. (1980), The striking resemblance of high-resolution G-banded chromosomes of man and chimpanzee, *Science*, **208**, 1145–1148.)

All these species are still cultivated—some to only minor extents—and have their individual uses in cooking. Spelt, or *farro* in Italian, is the basis of a well-known soup; pasta is made from durum wheat; and bread from *T. aestivum*.

Even within single chromosomes, gene families are common in eukaryotes. Some family members are **paralogues**—related genes that have diverged to provide separate functions in the same species. (**Orthologues**, in contrast, are homologues that perform the same function in different species. For instance, human α and β globin are paralogues, and human and horse myoglobin are orthologues.) Other related sequences may be pseudogenes, which may have arisen by duplication, or by retrotransposition from messenger RNA, followed by the accumulation of mutations to the point of loss of function. The human globin gene cluster is a good example (see Box, page 90: The human haemoglobin gene cluster).

The genome of *Saccharomyces cerevisiae* (baker's yeast)

Yeast is one of the simplest known eukaryotic organisms. Its cells, like our own, contain a nucleus and other specialized intracellular compartments. The

The human haemoglobin gene cluster

Human haemoglobin genes and pseudogenes appear in clusters on chromosomes 11 and 16. The normal adult human synthesizes primarily three types of globin chains: α- and β-chains, which assemble into haemoglobin $\alpha_2\beta_2$ tetramers; and myoglobin, a monomeric protein found in muscle. Other forms of haemoglobin, encoded by different genes, are synthesized in the embryonic and foetal stages of life.

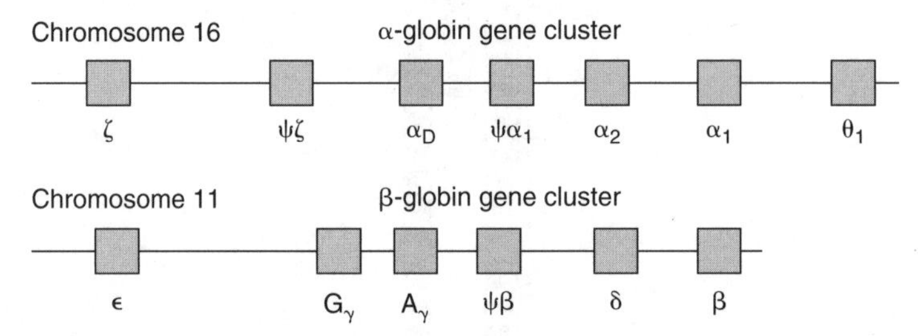

The α gene cluster on chromosome 16 extends over 28 kbp. It contains three functional genes: ζ, and 2 α genes identical in their coding sequences and a third expressed gene; two pseudogenes, ψζ and $\psi\alpha_1$ and another homologous gene the function of which is obscure, θ_1. The β gene cluster on chromosome 11 extends over 50 kbp. It includes five functional genes: ε, two γ genes (G_γ and A_γ), which differ in one amino acid, δ, and β; and one pseudogene, ψβ. The gene for myoglobin is unlinked from both of these clusters.

All human haemoglobin and myoglobin genes have the same intron/exon structure. They contain three exons separated by two introns:

Here E = exon and I = intron. The lengths of the regions in this diagram reflect the human β-globin gene. This exon/intron pattern is conserved in most expressed vertebrate globin genes, including haemoglobin α and β chains and myoglobin. In contrast, the genes for plant globins have an additional intron, genes for *Paramecium* globins one fewer intron, and genes for insect globins contain none. The gene for human neuroglobin, a recently-discovered homologue expressed at low levels in the brain, contains three introns, like plant globin genes.

Haemoglobin genes and pseudogenes are distributed on their chromosomes in a way that appears to reflect their evolution via duplication and divergence. (See diagram, next page.) The expression of these genes follows a strict developmental pattern. In the embryo (up to 6 weeks after conception) two haemoglobin chains are primarily synthesized—ζ and ε—which form a $\zeta_2\epsilon_2$ tetramer. Between 6 weeks after conception until about 8 weeks after birth, foetal haemoglobin—$\alpha_2\gamma_2$—is the predominant species. This is succeeded by adult haemoglobin—$\alpha_2\beta_2$.

→

→

The human haemoglobin gene cluster *(continued)*

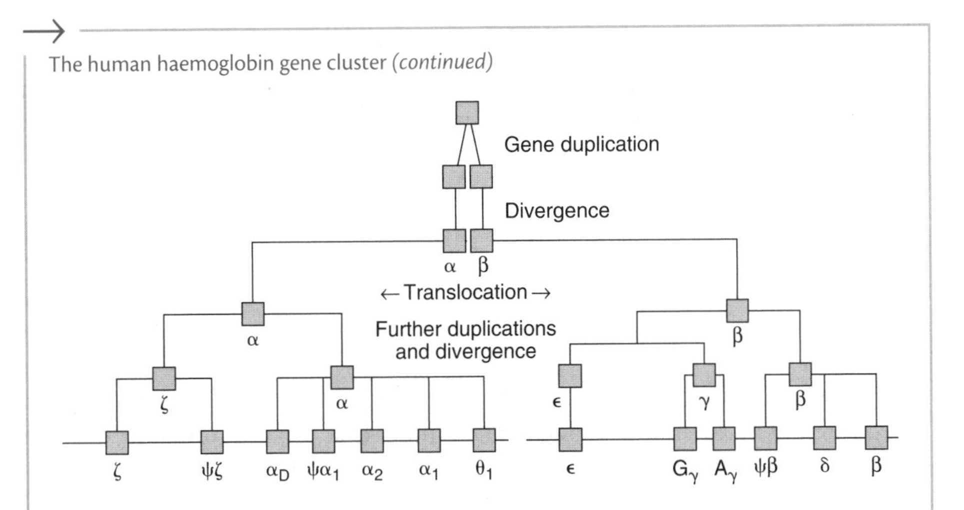

Thalassaemias are genetic diseases associated with defective or deleted haemoglobin genes. Most Caucasians have four genes for the α chain of normal adult haemoglobin, two alleles of each of the two tandem genes α_1 and α_2. Therefore α-thalassaemias can present clinically in different degrees of severity, depending on how many genes express normal α chains. Only deletions leaving fewer than two active genes present as symptomatic under normal conditions. Observed genetic defects include deletions of both genes (a process made more likely by the tandem gene arrangement and repetitive sequences which make crossing-over more likely); and loss of chain termination leading to transcriptional 'read through', creating extended polypeptide chains which are unstable.

β-thalassaemias are usually point mutations, including missense mutations (amino acid substitutions), or nonsense mutations (changes from a triplet coding for an amino acid to a stop codon) leading to premature termination and a truncated protein, mutations in splice sites, or mutations in regulatory regions. Certain deletions including the normal termination codon and the intergenic region between δ and β genes create δ–β fusion proteins.

sequencing of its genome, by an unusually effective international consortium involving ~100 laboratories, was completed in 1992. The yeast genome contains 12 057 500 bp of nuclear DNA, distributed over 16 chromosomes. The chromosomes range in size over an order of magnitude, from the 1352 kbp chromosome IV to the 230 kbp chromosome I.

The yeast genome contains 5885 predicted protein-coding genes, ~140 genes for ribosomal RNAs, 40 genes for small nuclear RNAs, and 275 transfer RNA genes. In two respects, the yeast genome is denser in coding regions than the known genomes of the more complex eukaryotes *Caenorhabditis elegans*, *Drosophila melanogaster*, and human: (1) Introns are relatively rare, and relatively small. Only

231 genes in yeast contain introns. (2) There are fewer repeat sequences compared with more complex eukaryotes.

A duplication of the entire yeast genome appears to have occurred ~150 million years ago. This was followed by translocations of pieces of the duplicated DNA and loss of one of the copies of most (~92%) of the genes.

Of the 5885 potential protein-coding genes, 3408 correspond to known proteins. About 1000 more contain some similarity to known proteins in other species. Another ~800 are similar to ORFs in other genomes that correspond to unknown proteins. Many of these homologues appear in prokaryotes. Only approximately one-third of yeast proteins have identifiable homologues in the human genome.

In taking censuses of genes, it has been useful to classify their functions into broad categories. The following classification of yeast protein functions is taken from http://www.mips.biochem.mpg.de/proj/yeast/catalogues/funcat/:

- Metabolism
- Energy
- Cell growth, cell division and DNA synthesis
- Transcription
- Protein synthesis
- Protein destination
- Transport facilitation
- Cellular transport and transport mechanisms
- Cellular biogenesis
- Cellular communication/signal transduction
- Cell rescue, defence, cell death and ageing
- Ionic homeostasis
- Cellular organization
- Transposable elements, viral and plasmid proteins
- Unclassified

Yeast is a testbed for development of methods to assign functions to gene products. The search for homologues has been exhaustive and continues. Collections of mutants exist containing a knockout of every gene. (A unique sequence 'bar code' introduced into each mutant facilitates identification of the ones that grow under selected conditions.) A surprisingly-large fraction of these single-gene knockout strains are viable—80%. Cellular localization and expression patterns are being investigated. Several types of measurements, including those based on activation of transcription by pairs of proteins that can form dimers—the **two-hybrid method**—are producing catalogues of interprotein interactions.

The genome of *Caenorhabditis elegans*

The nematode worm *Caenorhabditis elegans* entered biological research in the 1960s, at the express invitation of S. Brenner. He recognized its potential as a sufficiently complex organism to be interesting but simple enough to permit complete analysis, at the cellular level, of its development and neural circuitry.

The *C. elegans* genome, completed in 1998, provided the first full DNA sequence of a multicellular organism. The *C. elegans* genome contains ~97 Mbp of DNA distributed on paired chromosomes I, II, III, IV, V and X (see Box). There is no Y chromosome. Different genders in *C. elegans* appear in the XX genotype, a self-fertilizing hermaphrodite; and the XO genotype, a male.

Distribution of *C. elegans* genes

Chromosome	Size (Mbp)	Number of protein genes	Density of protein genes (kbp/gene)	Number of tRNA genes
I	13.9	2803	5.06	13
II	14.7	3259	3.65	6
II	12.8	2508	5.40	9
IV	16.1	3094	5.17	7
V	20.8	4082	4.15	5
X	17.2	2631	6.54	3

The *C. elegans* genome is about 8 times larger than that of yeast, and its 19 099 predicted genes are approximately three times the number in yeast. The gene density is relatively low for a eukaryote, with ~1 gene/5 kbp of DNA. Exons cover ~27% of the genome; the genes contain an average of 5 introns. Approximately 25% of the genes are in clusters of related genes.

Many *C. elegans* proteins are common to other life forms. Others are apparently specific to nematodes. 42% of proteins have homologues outside the phylum, 34% are homologous to proteins of other nematodes and 24% have no known homologues outside *C. elegans* itself. Many of the proteins have been classified according to structure and function (see Box).

***C. elegans*: 20 commonest protein domains**

Type of domain	Number
Seven-transmembrane spanning chemoreceptor	650
Eukaryotic protein kinase domain	410
Two domain, C4 type zinc finger	240
Collagen	170

→

→

C. elegans: 20 commonest protein domains

Type of domain	Number
Seven-transmembrane spanning receptor (rhodopsin-family)	140
C2H2 type zinc finger	130
C-type lectin	120
RNA recognition motif	100
C3HC4 type (RING finger) zinc fingers	90
Protein tyrosine phosphatase	90
Ankyrin repeat	90
WD domain G-beta repeat	90
Homeobox domain	80
Neurotransmitter gated ion channel	80
Cytochrome P450	80
Conserved C-terminal helicase	80
Short chain alcohol dehydrogenases	80
UDP-glucoronosyl and UDP-glucosyl transferases	70
EGF-like domain	70
Immunoglobulin superfamily	70

Source: *C. elegans* genome consortium paper in the 11 December 1998, *Science.*

Several kinds of RNA genes have been identified. The *C. elegans* genome contains 659 genes for tRNA, almost half of them (44%) on the X chromosome. Spliceosomal RNAs appear in dispersed copies, often identical. (Spliceosomes are the organelles that convert pre-mRNA transcripts to mature mRNA by excising introns.) Ribosomal RNAs appear in a long tandem array at the end of chromosome I. 5S RNAs appear in a tandem array on chromosome V. Some RNA genes appear in introns of protein-coding genes.

The *C. elegans* genome contains many repeat sequences. Approximately 2.6% of the genome consists of tandem repeats. Approximately 3.6% of the genome contains inverted repeats; these appear preferentially within introns, rather than between genes. Repeats of the hexamer sequence TTAGGC appear in many places. There are also simple duplications, involving hundreds to tens of thousands of kbp.

The genome of *Drosophila melanogaster*

Drosophila melanogaster, the fruit fly, has been the subject of detailed studies of genetics and development for a century. Its genome sequence, the product of a collaboration between Celera Genomics and the Berkeley Drosophila Genome Project, was announced in 1999.

The chromosomes of *D. melanogaster* are nucleoprotein complexes. Approximately one-third of the genome is contained in heterochromatin, highly coiled and compact (and therefore densely staining) regions flanking the centromeres. The other two-thirds is euchromatin, a relatively uncoiled, less-compact form. Most of the active genes are in the euchromatin. The heterochromatin in *D. melanogaster* contains many tandem repeats of the sequence AATAACATAG, and relatively few genes.

The total chromosomal DNA of *Drosophila melanogaster* contains about ~180 Mbp. The released sequence consists of the euchromatic portion, about 120 Mbp.

The genome is distributed over five chromosomes: three large autosomes, a Y chromosome, and a fifth tiny chromosome containing only ~1 Mbp of euchromatin. The fly's 13 601 genes are approximately double the number in yeast, but are fewer than in *C. elegans*, perhaps a surprise. The average density of genes in the euchromatin sequence is 1 gene/9 kbp; much lower than the typical 1 gene/kbp densities of prokaryotic genomes.

Despite the fact that insects are not very closely related to mammals, the fly genome is useful in the study of human disease. It contains homologues of 289 human genes implicated in various diseases, including cancer, and cardiovascular, neurological, endocrinological, renal, metabolic, and haematological diseases. Some of these homologues have different functions in humans and flies. Other human disease-associated genes can be introduced into, and studied in, the fly. For instance, the gene for human spinocerebellar ataxia type 3, when expressed in the fly, produces similar neuronal cell degeneration. There are now fly models for Parkinson disease and malaria.

The noncoding regions of the *D. melanogaster* genome must contain regions controlling spatio-temporal patterns of development. The developmental biology of the fly has been studied very intensively. It is therefore an organism in which the study of the genomics of development will prove extremely informative.

The genome of *Arabidopsis thaliana*

As a flowering plant *Arabidopsis thaliana* is a very distant relative of the other higher eukaryotic organisms for which genome sequences are available. It invites comparative analysis to identify common and specialised features.

The *Arabidopsis thaliana* genome contains ~125 Mbp of DNA, of which 115.4 Mbp was reported in 2000 by The *Arabidopsis* Genome Initiative. There are five pairs of chromosomes, containing 25 498 predicted genes. The genome is relatively compact, with 1 gene/4.6 kbp on average. This figure is intermediate between prokaryotes and *Drosophila*, and roughly similar to *C. elegans*. The genes of *Arabidopsis* are relatively small. Exons are typically 250 bp long, and introns relatively small, with mean length 170 bp. Typical of plant genes is an enrichment of coding regions in GC content.

The *Arabidopsis thaliana* genome

	Chromosome					
	1	**2**	**3**	**4**	**5**	**Total**
Length(bp)	29 105 111	19 646 945	23 172 617	17 549 867	25 353 409	115 409 949
Number of genes	6 543	4 036	5 220	3 825	5 874	25 498
Density (kbp/gene)	4.0	4.9	4.5	4.6	4.4	
Mean gene length	2 078	1 949	1 925	2 138	1 974	

Most *Arabidopsis* proteins have homologues in animals, but some systems are unique, among higher organisms, to plants, including cell wall production and photosynthesis. Many proteins shared with animals have diverged widely since the last common ancestor. Typical of another difference between plants and animals, 25% of the nuclear genes have signal sequences governing their transport into organelles—mitochondria and chloroplasts—compared to 5% of mitochondrial-targeted nuclear genes in animals.

Traces of large- and small-scale duplication events appear in the *Arabidopsis* genome. 58% of the genome contains 24 duplicated segments, $\geq$ 100 kbp in length. The genes in the duplicated segments are usually *not* identifiable homologues. However, 4140 genes (17% of all genes) appear in 1528 tandem arrays of gene families, containing up to 23 members. The pattern of duplications suggests that an ancestor of *Arabidopsis* underwent a whole-genome duplication, to form a tetraploid species. That this happened a long time—estimated at 112 million years—ago is shown by the very high amount of loss of one copy of duplicated genes, and wide divergence of those that remain.

Higher plants must integrate the effects of three genomes—nuclear, chloroplast and mitochondrial. The organelle genomes are much smaller:

Gene distribution in *A. thaliana* between nucleus and organelles

	Nuclear	Chloroplast	Mitochondrion
Size (kbp)	125 100	154	367
Protein genes	25 498	79	58
Density (kbp/protein gene)	4.5	1.2	6.25

Many genes for proteins synthesized by nuclear genes and transported to organelles appear to have originated in the organelles and been transferred to the nucleus.

The genome of *Homo sapiens* (the human genome)

Notice

PERSONS attempting to find a motive in this narrative will be prosecuted; persons attempting to find a moral in it will be banished; persons attempting to find a plot in it will be shot.

Mark Twain, Preface to *The Adventures of Huckleberry Finn*

In February 2001, the International Human Genome Sequencing Consortium and Celera Genomics published, separately, drafts of the human genome. The sequence amounted to $\sim 3.2 \times 10^9$ bp, thirty times larger than the genomes of *C. elegans* or *D. melanogaster*. One reason for this disparity in size is that coding sequences form less than 5% of the human genome; repeat sequences over 50%. Perhaps the most surprising feature was the small number of genes identified. The finding of only about 20 000–25 000 genes suggests that alternative splicing patterns make a very significant contribution to our protein repertoire. It is estimated that ~50% of genes have alternative splicing patterns.

The published human genome is distributed over 22 chromosome pairs plus the X and Y chromosomes. The DNA contents of the autosomes range from 279 Mbp down to 48 Mbp. The X chromosome contains 163 Mbp and the Y chromosome only 51 Mbp.

The exons of human protein-coding genes are relatively small compared to those in other known eukaryotic genomes. The introns are relatively long. As a result many protein-coding genes span long stretches of DNA. For instance, the dystrophin gene, coding for a 3685 amino acid protein, is >2.4 Mbp long.

Protein coding genes

Analysis of the human protein repertoire implied by the genome sequence has proved difficult because of the problems in reliably detecting genes, and alternative splicing patterns.

The top categories in a functional classification are:

Function	Number	% of genome
Nucleic acid binding	2207	14.0
DNA binding	1656	10.5
DNA repair protein	45	0.2
DNA replication factor	7	0.0
Transcription factor	986	6.2
RNA binding	380	2.4
Structural protein of ribosome	137	0.8
Translation factor	44	0.2
Transcription factor binding	6	0.0
Cell cycle regulator	75	0.4
Chaperone	154	0.9
Motor	85	0.5
Actin binding	129	0.8
Defence/immunity protein	603	3.8
Enzyme	3242	20.6
Peptidase	457	2.9
Endopeptidase	403	2.5
Protein kinase	839	5.3
Protein phosphatase	295	1.8
Enzyme activator	3	0.0
Enzyme inhibitor	132	0.8
Apoptosis inhibitor	28	0.1

Function	Number	% of genome
Signal transduction	1790	11.4
Receptor	1318	8.4
Transmembrane receptor	1202	7.6
G-protein linked receptor	489	3.1
Olfactory receptor	71	0.4
Storage protein	7	0.0
Cell adhesion	189	1.2
Structural protein	714	4.5
Cytoskeletal structural protein	145	0.9
Transporter	682	4.3
Ion channel	269	1.7
Neurotransmitter transporter	19	0.1
Ligand binding or carrier	1536	9.7
Electron transfer	33	0.2
Cytochrome P450	50	0.3
Tumour suppressor	5	0.0
Unclassified	4813	30.6
Total	15683	100.0

From: http://www.ebi.ac.uk/proteome/
[under Functional classification of *H. sapiens* using Gene Ontology (GO): General statistics (InterPro proteins with GO hits)]

A classification based on structure revealed the most common types:

Most common types of proteins

Protein	Number
Immunoglobulin and major histocompatibility complex domain	591
Zinc finger, C2H2 type	499
Eukaryotic protein kinase	459
Rhodopsin-like GPCR superfamily	346
Serine/Threonine protein kinase family active site	285
EGF-like domain	259
RNA-binding region RNP-1 (RNA recognition motif)	214
G-protein beta WD-40 repeats	196
Src homology 3 (SH3) domain	194
Pleckstrin homology (PH) domain	188
EF-hand family	185
Homeobox domain	179

Protein	Number
Tyrosine kinase catalytic domain	173
Immunoglobulin V-type	163
RING finger	159
Proline rich extensin	156
Fibronectin type III domain	151
Ankyrin-repeat	135
KRAB box	133
Immunoglobulin subtype	128
Cadherin domain	118
PDZ domain (also known as DHR or GLGF)	117
Leucine-rich repeat	113
Serine proteases, trypsin family	108
Ras GTPase superfamily	103
Src homology 2 (SH2) domain	100
BTB/POZ domain	99
TPR repeat	92
AAA ATPase superfamily	92
Aspartic acid and asparagine hydroxylation site	91

From: http://www.ebi.ac.uk/proteome/

Repeat sequences

Repeat sequences comprise over 50% of the genome:

- Transposable elements, or interspersed repeats—almost half the entire genome! These include the LINEs and SINEs (see Box).

Type of transposable elements in the human genome

Element	Size (bp)	Copy number	Fraction of genome (%)
Short Interspersed Nuclear Elements (SINEs)	100–300	1 500 000	13
Long Interspersed Nuclear Elements (LINEs)	6000–8000	850 000	21
Long Terminal Repeats	15 000–110 000	450 000	8
DNA Transposon fossils	80–3000	300 000	3

- Retroposed pseudogenes
- Simple 'stutters'—repeats of short oligomers. These include the minisatellites and microsatellites. Trinucleotide repeats such as CAG, corresponding to glutamine repeats in the corresponding protein, are implicated in numerous diseases.

- Segmental duplications, of blocks of ~10–300 kbp. Interchromosomal duplications appear on nonhomologous chromosomes, sometimes at multiple sites. Some intrachromosomal duplications include closely spaced duplicated regions many kbp long of very similar sequence implicated in genetic diseases; for example Charcot-Marie-Tooth syndrome type 1A, a progressive peripheral neuropathy resulting from duplication of a region containing the gene for peripheral myelin protein 22.
- Blocks of tandem repeats, including gene families

RNA

RNA genes in the human genome include:

1. 497 transfer RNA genes. One large cluster contains 140 tRNA genes within a 4Mbp region on chromosome 6.
2. Genes for 28S and 5.8S ribosomal RNAs appear in a 44 kbp tandem repeat unit of 150–200 copies. 5S RNA genes also appear in tandem arrays containing 200–300 genes, the largest of which is on chromosome 1.
3. Small nucleolar RNAs include two families of molecules that cleave and process ribosomal RNAs.
4. Splicesomal snRNAs, including the U1, U2, U4, U5, and U6 snRNAs, many of which appear in clusters of tandem repeats of nearly-identical sequences, or inverted repeats.

Web resources: Human genome information

Interactive access to DNA and protein sequences:
http://www.ensembl.org/
http://www.genome.ucsc.edu

Images of chromosomes, maps, loci:
http://www.ncbi.nlm.nih.gov/genome/guide/

Gene map 99:
http://www.ncbi.nlm.nih.gov/genemap99/

Overview of human genome structure:
http://hgrep.ims.u-tokyo.ac.jp

Single-nucleotide polymorphisms:
http://snp.cshl.org/

Human genetic disease:
http://www.ncbi.nlm.nih.gov/Omim/
http://www.geneclinics.org/profiles/all.html

Social, ethical, legal issues:
http://www.nhgri.nih.gov/ELSI/

Single-nucleotide polymorphisms (SNPs)

A single-nucleotide polymorphism or SNP (pronounced 'snip') is a genetic variation between individuals, limited to a single base pair which can be substituted, inserted or deleted. Sickle-cell anaemia is an example of a disease caused by a specific SNP: an A → T mutation in the β-globin gene changes a Glu→ Val, creating a sticky surface on the haemoglobin molecule that leads to polymerization of the deoxy form.

SNPs are distributed throughout the genome, occurring on average every 5000 base pairs. Although they arose by mutation, many positions containing SNPs have low mutation rates, and provide stable markers for mapping genes. A consortium of four major academic genome centres and eleven private companies is creating a high-quality, high-density human SNP map, with a commitment to open public access. A recently published version contains 1.8 million SNPs.

Not all SNPs are linked to diseases. Many are not within functional regions (although the density of SNPs is higher than the average in regions containing genes). Some SNPs that occur within exons are mutations to synonymous codons, or cause substitutions that do not significantly affect protein function. Other types of SNPs can cause more than local perturbation to a protein: (1) A mutation from a sense codon to a stop codon, or vice versa, will cause either premature truncation of protein synthesis or 'readthrough'. (2) A deletion or insertion will cause a phase shift in translation.

The A, B and O allelles of the genes for blood groups illustrate these possibilities. A and B alleles differ by four SNP substitutions. They code for related enzymes that add different saccharide units to an antigen on the surface of red blood cells.

Allele	Sequence	Saccharide
A	...gctggtgaccccctt...	N-acetylgalactosamine
B	...gctcgtcaccgcta...	galactose
O	...cgtggt-accccctt...	—

The O allele has undergone a mutation causing a phase shift, and produces no active enzyme. The red blood cells of type O individuals contain neither the A nor the B antigen. This is why people with type O blood are universal donors in blood transfusions. The loss of activity of the protein does not seem to carry any adverse consequences. Indeed, individuals of blood types B and O have greater resistance to smallpox.

Stong correlation of a disease with a specific SNP is advantageous in clinical work, because it is relatively easy to test for affected people or carriers. But if a disease arises from dysfunction of a specific protein, there ought to be many sites of mutations that could cause inactivation. A particular site may predominate if (1) all bearers of the gene are descendants of a single individual in whom the mutation occurred, and/or (2) the disease results from a *gain* rather than loss of a specific property, such as in the ability of sickle cell haemoglobin to polymerize, and/or (3) the mutation rate at a particular site is unusually high, as in the Gly380→Arg mutation in the fibroblast growth receptor gene FGFR3, associated with achondroplasia (a syndrome including short stature).

In contrast, many independent mutations have been detected in the BRCA1 and BRCA2 genes, loci associated with increased disposition to early-onset breast and ovarian cancer (see Chapter 6). The normal gene products function as tumour suppressors. Insertion or deletion mutants causing phase shifts generally produce a missing or inactive protein. But it cannot be deduced a priori whether a novel **substitution** mutant in BRCA1 or BRCA2 confers increased risk or not.

Treatments of diseases caused by defective or absent proteins include:

1. **Providing normal protein** We have mentioned insulin for diabetes, and Factor VIII in the most common type of haemophilia. Another example is the administration of human growth hormone in patients with an absence or severe reduction in normal levels. Use of recombinant proteins eliminates the risk of transmission of AIDS through blood transfusions or of Creutzfeld-Jakob disease from growth hormone isolated from crude pituitary extracts.
2. **Lifestyle adjustments that make the function unnecessary** Phenylketoneuria (PKU) is a genetic disease caused by deficiency in phenylalanine hydroxylase, the enzyme that converts phenylalanine to tyrosine. Accumulation of high levels of phenylalanine causes developmental defects, including mental retardation. The symptoms can be avoided by a phenylalanine-free diet. Screening of newborns for high blood phenylalanine levels is legally required in the United States and many other countries.
3. Gene therapy to replace absent proteins is an active field of research.

Other clinical applications of SNPs reflect correlations between genotype and reaction to therapy (pharmacogenomics). For example, a SNP in the gene for N-acetyl transferase (NAT-2) is correlated with peripheral neuropathy—weakness, numbness and pain in the arms, legs, hands or feet—as a side effect of treatment with isoniazid (isonicotinic acid hydrazide), a common treatment for tuberculosis (see Chapter 6). Patients who test positive for this SNP are given alternative treatment.

Genetic diversity in anthropology

SNP data are of great utility in anthropology, giving clues to historical variations in population size, and migration patterns.

Degrees of genetic diversity are interpretable in terms of the size of the founding population. Founders are the original set of individuals from whom an entire population is descended. Founders can be either original colonists, such as the Polynesians who first settled New Zealand, or merely the survivors of a near-extinction. Cheetahs show the effects of a population bottleneck, estimated to have occurred 10 000 years ago. All living cheetahs are as closely related to one another as typical siblings in other species. Extrapolations of mitochondrial DNA variation in contemporary humans suggest a single maternal ancestor who lived 14 000–200 000 years ago. Calling her Eve suggests that she was the first woman. But Mitochondrial Eve was in fact the founder of a surviving population following extinction of descendants of all her contemporaries.

Population-specific SNPs are informative about migrations. Mitochondrial sequences provide information about female ancestors and Y chromosome sequences provide information about male ancestors. For example, it has been suggested that the population of Iceland—first inhabited about 1100 years ago—is descended from Scandinavian males, and from females from both Scandinavia and the British Isles. Mediaeval Icelandic writings refer to raids on settlements in the British Isles.

A fascinating relationship between human DNA sequences and language families has been investigated by L. L. Cavalli-Sforza and colleagues. These studies have proved useful in working out interrelationships among American Indian languages. They confirm that the Basques, known to be a linguistically isolated population, have been genetically isolated also.

In the discovery of isolated populations, anthropological genetics provides data useful in medicine, for mapping disease genes is easier if the background variation is low. Genetically isolated populations in Europe include, in addition to the Basques, the Finns, Icelanders, Welsh, and Lapps.

Genetic diversity and personal identification

Variations in our DNA sequences give us individual fingerprints, useful for identification and for establishment of relationships, including but not limited to questions of paternity. The use of DNA analysis as evidence in criminal trials is now well-established.

Genetic fingerprinting techniques were originally based on patterns of variable number tandem repeats (VNTRs), but have been extended to include analysis of other features including mitochondrial DNA sequences.

For most of us, all our mitochondria are genetically identical, a condition called homoplasmy. However, some individuals contain mitochondria with different DNA sequences; this is heteroplasmy. Such sequence variation in a disease gene can complicate the observed inheritance pattern of the disease.

The most famous case of heteroplasmy involved Tsar Nicholas II of Russia. After the revolution in 1917 the Tsar and his family were taken to exile in Yekaterinburg in Central Russia. During the night of 16–17 July 1918, the Tsar, Tsarina Alexandra, at least three of their five children, plus their physician and three servants who had accompanied the family, were killed, and their bodies buried in a secret grave. When the remains were rediscovered, assembly of the bones and examination of the dental work suggested—and sequence analysis confirmed—that the remains included an expected family group. The identity of the remains of the Tsarina were proved by matching the mitochondrial DNA sequence with that of a maternal relative, Prince Philip, Chancellor of the University of Cambridge, Duke of Edinburgh, and grandnephew of the Tsarina.

However, comparisons of mitochondrial DNA sequences of the putative remains of Nicholas II with those of two maternal relatives revealed a difference at base 16169: the Tsar had a C and the relatives a T. The extreme political and even religious sensitivities mandated that no doubts were tolerable. Further tests showed that the Tsar was heteroplasmic; T was a minor component of his mitochondrial DNA at position 16169. To confirm the identity beyond any reasonable question, the body of Grand Duke Georgij, brother of the Tsar, was exhumed, and was shown to have the same rare heteroplasmy.

Genetic analysis of cattle domestication

Animal resources are an integral and essential aspect of human culture. Analysis of DNA sequences sheds light on their historical development and on the genetic variety characterizing modern breeding populations.

Contemporary domestic cattle include those familiar in Western Europe and North America, *Bos taurus*; and the zebu of Africa and India, *Bos indicus*. The most obvious difference in external appearance is the humpback of the zebu. It has been widely believed that the domestication of cattle occurred once, about 8000–10 000 years ago, and that the two species subsequently diverged.

Analysis of mitochondrial DNA sequences from European, African, and Asian cattle suggest, however, that (1) all European and African breeds are more closely related to each other than either is to Indian breeds, and (2) the two groups diverged about 200 000 years ago, implying recent independent domestications of different species. The similarity in physical appearance of African and Indian zebu (and other similarities at the molecular level; for instance, VNTR markers in nuclear DNA) must then be attributable to importation of male cattle from India to East Africa.

Evolution of genomes

The availability of complete information about genomic sequences has redirected research. A general challenge in analysis of genomes is to identify 'interesting events'. A background mutation rate in coding sequences is reflected in **synonymous** nucleotide substitutions: changes in codons that do not alter the amino acid. With this as a baseline one can search for instances in which there are significantly higher rates of **nonsynonymous** nucleotide substitutions: changes in codons that cause mutations in the corresponding protein. (Note, however, that synonymous changes are not necessarily selectively neutral.)

Given two aligned gene sequences, we can calculate K_s = the number of synonymous substitutions, and K_a = the number of nonsynonymous substitutions. (The calculation involves more than simple counting because of the need to estimate and correct for possible multiple changes.) A high ratio of K_s/K_a identifies pairs of sequences apparently showing positive selection, possibly even functional changes.

The new field of comparative genomics treats questions that can only now be addressed, such as:

- What *genes* do different phyla share? What genes are unique to different phyla? Do the arrangements of these genes in the genome vary from phylum to phylum?
- What homologous *proteins* do different phyla share? What proteins are unique to different phyla? Does the integration of the activities of these proteins vary from phylum to phylum? Do the mechanisms of control of expression patterns of these proteins vary from phylum to phylum?
- What *biochemical functions* do different phyla share? What biochemical functions are unique to different phyla? Does the integration of these biochemical functions vary from phylum to phylum? If two phyla share a function, and the protein that carries out this function in one phylum has a homologue in the other, does the homologous protein carry out the same function?

The same questions could be asked about different species within each phylum.

M. A. Andrade, C. Ouzounis, C. Sander, J. Tamames and A. Valencia compared the protein repertoire of species from the three major domains of life: *Haemophilus influenzae* represented the bacteria, *Methanococcus jannaschii* the archaea, and *Saccharomyces cerevisiae* (yeast) the eukarya. Their classification of protein functions contained as major categories processes involving energy, information, and communication and regulation:
General functional classes

- Energy
  - Biosynthesis of cofactors, amino acids
  - Central and intermediary metabolism
  - Energy metabolism
  - Fatty acids and phospholipids
  - Nucleotide biosynthesis
  - Transport
- Information
  - Replication
  - Transcription
  - Translation
- Communication and regulation
  - Regulatory functions
  - Cell envelope/cell wall
  - Cellular processes

The number of genes in the three species are:

Species	Number of genes
Haemophilus influenzae	1680
Methanococcus jannaschii	1735
Saccharomyces cerevisiae	6278

Are there, among these, shared proteins for shared functions? In the category of Energy, proteins are shared across the three domains. In the category of Communication and regulation, proteins are unique to each domain. In the Information category, archaea share some proteins with bacteria and others with eukarya.

Analysis of shared functions among all domains of life has led people to ask whether it might be possible to define a **minimal organism**—that is, an organism with the smallest gene complement consistent with independent life based on the central DNA→RNA→protein dogma (that is, excluding protein-free life forms based solely on RNA). The minimal organism must have the ability to reproduce, but not be required to compete in growth and reproductive rate with other organisms. One may reasonably assume a generous nutrient medium, relieving the organism of biosynthetic responsibility, and dispensing with stress-reponse functions including DNA repair.

The smallest known independent organism is *Mycoplasma genitalium*, with 468 predicted protein sequences. In 1996, A. R. Mushegian and E. V. Koonin compared the genomes of *M. genitalium* and *Haemophilus influenzae*. (At the time, these were the only completely sequenced bacterial genomes.) The last common ancestor of these widely diverged bacteria lived about two billion years ago. Of 1703 protein-coding genes of *H. influenzae*, 240 are homologues of proteins in *M. genitalium*. Mushegian and Koonin reasoned that all of these must be essential, but might not be sufficient for autonomous life—for some essential functions might be carried out by unrelated proteins in the two organisms. For instance, the common set of 240 proteins left gaps in essential pathways, which could be filled by adding 22 enzymes from *M. genitalium*. Finally, removing functional redundancy and parasite-specific genes gave a list of 256 genes as the proposed necessary *and* sufficient minimal set.

What is in the proposed minimal genome? Functional classes include:

- Translation, including protein synthesis
- DNA replication
- Recombination and repair—a second function of essential proteins involved in DNA replication.
- Transcription apparatus
- Chaperone-like proteins
- Intermediary metabolism—the glycolytic pathway
- No nucleotide, amino acid, or fatty acid biosynthesis
- Protein-export machinery
- Limited repertoire of metabolite transport proteins

It should be emphasized that the viability of an organism with these proteins has not been proven. Moreover, even if experiments proved that some minimal gene content—the proposed set or some other set—is necessary and sufficient, this does not answer the related question of identifying the gene complement of the common ancestor of *M. genitalium* and *H. influenzae*, or of the earliest cellular forms of life. For only 71% of the proposed set of 256 proteins have recognizable homologues among eukaryotic *or* archaeal proteins.

Nevertheless, identification of functions necessarily common to all forms of life allows us to investigate the extent to which different forms of life accomplish these functions in the same ways. Are similar reactions catalysed in different species by homologous proteins? Genome analysis has revealed families of proteins with homologues in archaea, bacteria, and eukarya. The assumption is that these have evolved from an individual ancestral gene through a series of speciation and duplication events, although some may be the effects of horizontal transfer. The challenge is to map common functions and common proteins.

Several thousand protein families have been identified with homologues in archaea, bacteria and eukarya. Different species contain different amounts of these common families: In bacteria, the range is from *Aquifex aeolicus*, 83% of the proteins of which have archaeal and eukaryotic homologues, to *Borrelia burgdorferi*, in which only 52% of the proteins have archaeal and eukaryotic homologues. Archaeal genomes have somewhat higher percentages (62%–71%) of proteins with bacterial and eukaryotic homologues. But only 35% of the proteins of yeast have bacterial and archaeal homologues.

Does the common set of proteins carry out the common set of functions? Among the proteins of the minimal set identified from *M. genitalium*, only ~30% have homologues in all known genomes. Other essential functions must be carried out by unrelated proteins, or in some cases possibly by unrecognized homologues. The protein families for which homologues carry out common functions in archaea, bacteria and eukarya are enriched in those involved in translation, and biosynthesis:

Protein functional class	Number of families appearing in all known genomes
Translation, including ribosome structure	53
Transcription	4
Replication, recombination, repair	5
Metabolism	9
Cellular processes: (chaperones, secretion, cell division, cell wall biosynthesis)	9

The picture is emerging that evolution has explored the vast potential of proteins to different extents for different types of functions. It has been most conservative in the area of protein synthesis.

Web resources: Databases of aligned gene families

Pfam: Protein families database
http://www.sanger.ac.uk/Software/Pfam/

COG: Clusters of orthologous groups
http://www.ncbi.nlm.nih.gov/COG/

HOBACGEN: Homologous Bacterial Genes Database
http://pbil.univ-lyon1.fr/databases/hobacgen.html

HOVERGEN: Homologous Vertebrate Genes Database
http://pbil.univ-lyon1.fr/databases/hovergen.html

TAED: The Adaptive Evolution Database
http://www.sbc.su.se/liberles/TAED.html

Many other sites contain data on individual families.

Please pass the genes: horizontal gene transfer

Learning that *S. griseus* trypsin is more closely related to bovine trypsin than to other microbial proteinases, Brian Hartley commented in 1970 that, '. . . the bacterium must have been infected by a cow.' It was a clear example of lateral or horizontal gene transfer—a bacterium picking up a gene from the soil in which it was growing, which an organism of another species had deposited there. The classic experiments on pneumococcal transformation by O. Avery, C. MacLeod and M. McCarthy that identified DNA as the genetic material are another example. In general, horizontal gene transfer is the acquisition of genetic material by one organism from another, by natural rather than laboratory procedures, through some means other than descent from a parent during replication or mating. Several mechanisms of horizonal gene transfer are known, including direct uptake, as in the pneumococcal transformation experiments, or via a viral carrier.

Analysis of genome sequences has shown that horizontal gene transfer is not a rare event, but has affected most genes in micro-organisms. It requires a change in our thinking from ordinary clonal or parental models of heredity. Evidence for horizontal transfer includes (1) discrepancies among evolutionary trees constructed from different genes, and (2) direct sequence comparisons between genes from different species:

- In *E. coli*, about 25% of the genes appear to have been acquired by transfer from other species.
- In microbial evolution, horizontal gene transfer is more prevalent among operational genes—those reponsible for 'housekeeping' activities such as biosynthesis—than among informational genes—those responsible for organizational activities such as transcription and translation. For example: *Bradyrhizobium japonicum*, a nitrogen-fixing bacterium symbiotic with higher plants, has two glutamine synthetase genes. One is similar to those of its

bacterial relatives; the other 50% identical to those of higher plants. Rubisco (ribulose-1,5-bisphosphate carboxylase/oxygenase), the enzyme that first fixes carbon dioxide at the entry to the Calvin cycle of photosynthesis, has been passed around between bacteria, mitochondria and algal plastids, as well as undergoing gene duplication. Many phage genes appearing in the *E. coli* genome provide further examples and point to a mechanism of transfer.

Nor is the phenomenon of horizontal gene transfer limited to prokaryotes. Both eukaryotes and prokaryotes are chimeras. Eukaryotes derive their informational genes primarily from an organism related to *Methanococcus*, and their operational genes primarily from proteobacteria with some contributions from cyanobacteria and methanogens. Almost all informational genes from *Methanococcus* itself are similar to those in yeast. At least 8 human genes appeared in the *M. tuberculosis* genome.

The observations hint at the model of a 'global organism', a genetic common market, or even a World-Wide DNA Web from which organisms download genes at will! How can this be reconciled with the fact that the discreteness of species has been maintained? The conventional explanation is that the living world contains ecological 'niches' to which individual species are adapted. It is the discreteness of niches that explains the discreteness of species. But this explanation depends on the stability of normal heredity to maintain the fitness of the species. Why would not the global organism break down the lines of demarcation between species, just as global access to pop culture threatens to break down lines of demarcation between national and ethnic cultural heritages? Perhaps the answer is that it is the informational genes, which appear to be less subject to horizontal transfer, that determine the identity of the species.

It is interesting that although evidence for the importance of horizontal gene transfer is overwhelming, it was dismissed for a long time as rare and unimportant. The source of the intellectual discomfort is clear: Parent-to-child transmission of genes is at the heart of the Darwinian model of biological evolution whereby selection (differential reproduction) of parental phenotypes alters gene frequencies in the next generation. For offspring to gain genes from elsewhere than their parents smacks of Lamarck and other discredited alternatives to the paradigm. The evolutionary tree as an organizing principle of biological relationship is a deeply ingrained concept: scientists display an environmentalist-like fervour in their commitment to trees, even when trees are not an appropriate model of a network of relationships. Perhaps it is well to recall that Darwin knew nothing of genes, and the mechanism that generated the variation on which selection could operate was a mystery to him. Maybe he would have accepted horizontal gene transfer more easily than his followers!

Comparative genomics of eukaryotes

A comparison of the genomes of yeast, fly, worm and human revealed 1308 groups of proteins that appear in all four. These form a conserved core of proteins for basic functions, including metabolism, DNA replication and repair, and translation.

These proteins are made up of individual protein domains, including single-domain proteins, oligomeric proteins, and modular proteins containing many domains (the biggest, the muscle protein titin, contains 250–300 domains.) The proteins of the worm and fly are built from a structural repertoire containing about three times as many domains as the proteins of yeast. Human proteins are built from about twice as many as those of the worm and fly. Most of these domains appear also in bacteria and archaea, but some are specific to (probably, invented by) vertebrates (see Table). These include proteins that mediate activities unique to vertebrates, such as defence and immunity proteins, and proteins in the nervous system; only one of them is an enzyme, a ribonuclease.

Distribution of probable homologues of predicted human proteins	
Vertebrates only	22%
Vertebrates and other animals	24%
Animals and other eukaryotes	32%
Eukaryotes and prokaryotes	21%
No homologues in animals	1%
Prokaryotes only	1%

To create new proteins, inventing new domains is a unusual event. It is far more common to create different combinations of existing domains in increasingly complex ways. A common mechanism is by accretion of domains at the ends of modular proteins (see Fig. 2.3). This process can occur independently, and take different courses, in different phyla.

Gene duplication followed by divergence is a mechanism for creating protein families. For instance, there are 906 genes + pseudogenes for olfactory receptors in the human genome. These are estimated to bind ~10 000 odour molecules. Homologues have been demonstrated in yeast and other fungi (some comparisons *are* odorous), but it is the need of vertebrates for a highly-developed sense of smell that multiplied and specialized the family to such a great extent. Eighty percent of the human olfactory receptor genes are in clusters. Compare the small size of the globin gene cluster (page 90), which did not require such great variety.

Web resources: Genome databases

Lists of completed genomes:

http://www.ebi.ac.uk/genomes
http://pir.georgetown.edu/pirwww/search/genome.html
http://www.nslij-genetics.org/seq/

Organism-specific databases:

http://www.hgmp.mrc.ac.uk/GenomeWeb/genome-db.html

http://www.bioinformatik.de/cgi-bin/browse/Catalog/Databases/Genome_Projects/

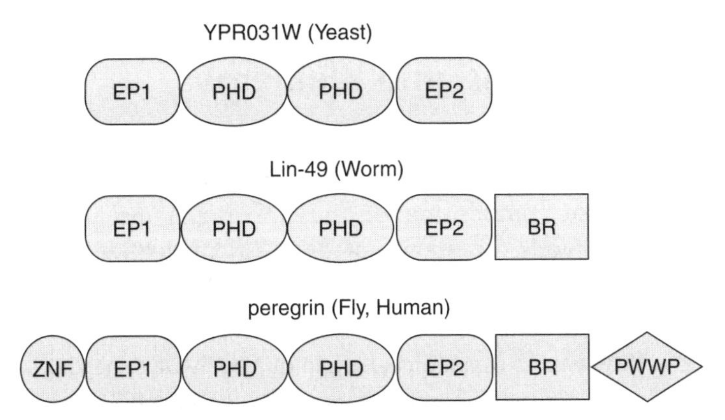

Fig. 2.3 Evolution by accretion of domains, of molecules related to peregrin, a human protein that probably functions in transcription regulation. The *C. elegans* homologue, lin-49, is essential for normal development of the worm. The yeast homologue is a component of the histone acetyltransferase complex. The proteins contain these domains: ZNF = C_2H_2 type zinc finger (not to be confused with acetylene; C and H stand for cysteine and histidine); EP1 and EP2 = Enhancer of polycomb 1 and 2, PHD = plant homeodomain, a repressor domain containing the $C_4H_3C_3$ type of zinc finger, BR = Bromo domain; PWWP = domain containing sequence motif Pro-Trp-Trp-Pro.

Recommended reading

C. elegans sequencing consortium (1999), How the worm was won. The *C. elegans* genome sequencing project, *Trends Genet.*, **15**, 51–58. [Description of the project in which high-throughput DNA sequencing was originally developed, and the results; the first metazoan genome to be sequenced.]

Ureta-Vidal, A., Ettwiller, L. & Birney, E. (2003), Comparative genomics: genome-wide analysis in metazoan eukaryotes, *Nature Reviews Genetics*, **4**, 251–262. [Introduction to the repertoire of solved genomes and how genomes from different species illuminate one another.]

Doolittle, W. F. (1999), Lateral genomics, *Trends Cell Biol.*, **9**, M5–M8. [How horizontal gene transfer upsets traditional views of evolution.]

Koonin, E. V. (2000), How many genes can make a cell: The minimal-gene-set concept, *Annu. Rev. Genomics Hum. Genet.*, **1**, 99–116. [A summary of work on comparative genomics.]

Kwok, P. -Y. & Gu, Z. (1999), SNP libraries: why and how are we building them? *Molecular Medicine Today*, **5**, 538–543. [Progress and rationale for databases of single-nucleotide polymorphisms.]

Publications of the drafts of the human genome sequences appeared in special issues of *Nature*, 15 Feb. 2001, containing the results of the publicly supported Human Genome Project, and *Science*, 16 Feb. 2001, containing the results produced by Celera Genomics. These are landmark issues.

The May 2001 issue of *Genome Research*, Volume **11**, Number 5, is devoted to the Human Genome.

The completion of the human genome in 2003 was announced in various press releases, and described in issues of *Nature* and *Science* magazines:

Collins, F. S., Green, E. D., Guttmacher, A. E. & Guyer, M. S. (2003), A vision for the future of genomics research, *Nature*, **422**, 835–847.

The ENCODE Project Consortium (2004), The ENCODE (ENCyclopedia of DNA Elements) Project, Science, **306**, 636–640. [Systematic comparative sequencing.]

Building on the DNA Revolution, Special section, *Science*, **300**, 11 Apr. 2003, p. 277 ff.

International Human Genome Sequencing Consortium (2004), Finishing the euchromatic sequence of the human genome, *Nature*, **431**, 931–945. [The end of the beginning, at the very least.]

Southan, C. (2004), Has the yo-yo stopped? An assessment of human protein-coding gene number, *Proteomics*, **4**, 1712–1726. [How many genes are there in the human genome? Estimates have been changing.]

Bentley, D. R. (2004), Genomes for medicine, *Nature*, **429**, 440–445. [Summary and discussion of the results of genomic sequencing and their applications.]

Exercises, Problems, and Weblems

Exercises

2.1 The overall base composition of the *E. coli* genome is A = T = 49.2%, G = C = 50.8%. In a random sequence of 4 639 221 nucleotides with these proportions, what is the expected number of occurrences of the sequence CTAG?

2.2 The *E. coli* genome contains a number of pairs of enzymes that catalyse the same reaction. How would this affect the use of knockout experiments (deletion or inactivation of individual genes) to try to discern function?

2.3 Which of the categories used to classify the functions of yeast proteins (see page 92) would be appropriate for classifying proteins from a prokaryotic genome?

2.4 Which occurred first, a man landing on the moon or the discovery of deep-sea hydrothermal vents? Guess first, then look it up.

2.5 Gardner syndrome is a condition in which large numbers of polyps develop in the lower gastrointestinal tract, leading inevitably to cancer if untreated. In every observed case, one of the parents is also a sufferer. What is the mode of the inheritance of this condition?

2.6 The gene for retinoblastoma is transmitted along with a gene for esterase D to which it is closely linked. However, either of the two alleles for esterase D can be transmitted with either allele for retinoblastoma. How do you know that retinoblastoma is not the direct effect of the esterase D phenotype?

2.7 If all somatic cells of an organism have the same DNA sequence, why is it necessary to have cDNA libraries from different tissues?

2.8 Suppose you are trying to identify a gene causing a human disease. You find a genetic marker 0.75 cM from the disease gene. To within approximately how many base pairs have you localized the gene you are looking for? Approximately how many genes is this region likely to contain?

2.9 Leber Hereditary Optic Neuropathy (LHON) is an inherited condition that can cause loss of central vision, resulting from mutations in mitochondrial DNA. You are asked to counsel a woman who has normal mitochondrial DNA and a man with LHON, who are contemplating marriage. What advice would you give them about the risk to their offspring of developing LHON?

2.10 Glucose-6-phosphate dehydrogenase deficiency is a single-gene recessive X-linked genetic defect affecting hundreds of millions of people. Clinical consequences include haemolytic anaemia and persistent neonatal jaundice. The gene has not been eliminated from the population because it confers resistance to malaria. In this case, general knowledge of metabolic pathways identified the protein causing the defect. Given the amino acid sequence of the protein, how would you determine the chromosomal location(s) of the corresponding gene?

2.11 Before DNA was recognized as the genetic material, the nature of a gene—in detailed biochemical terms—was obscure. In the 1940s, G. Beadle and E. Tatum observed that single mutations could knock out individual steps in biochemical pathways. On this basis they proposed the *one gene-one enzyme* hypothesis. On a photocopy of Fig. 2.1, draw lines linking genes in the figure to numbered steps in the sequence of reactions in the pathway. To what extent do the genes of the trp operon satisfy the one gene-one enzyme hypothesis and to what extent do they present exceptions?

2.12 The figure shows human chromosome 5 (left) and the matching chromosome from a chimpanzee. On a photocopy of this figure indicate which regions show an inversion of the banding pattern.

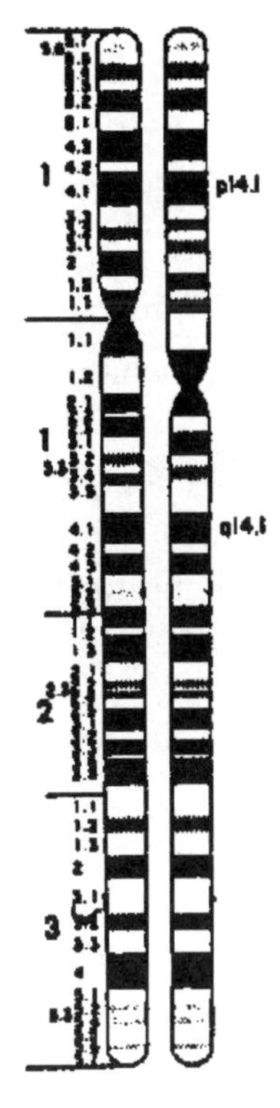

(From: Yunis, J. J. Sawyer, J. R. & Dunham, K. (1980), The striking resemblance of high-resolution G-banded chromosomes of man and chimpanzee, *Science*, **208**, 1145–1148.)

2.13 Describe in general terms how the FISH picture in Plate II would appear if the affected region of chromosome 20 were not deleted but translocated to another chromosome.

2.14 755 open reading frames entered the *E. coli* genome by horizontal transfer in the 14.4 million years since divergence from *Salmonella*. What is the average rate of horizontal transfer in kbp/year? To how many typical proteins (~300 amino acids) would this correspond? What per cent of known genes entered the *E. coli* genome via horizontal transfer?

2.15 To what extent is a living genome like a database? Which of the following properties are shared by living genomes and computer databases? Which are properties of living genomes but not databases? Which are properties of databases but not living genomes?

(a) Serve as repositories of information.

(b) Are self-interpreting.

(c) Different copies are not identical.

(d) Scientists can detect errors.

(e) Scientists can correct errors.

(f) There is planned and organized responsibility for assembling and disseminating the information.

Problems

2.1 What is the experimental evidence that shows that the genetic linkage map on any single chromosome is linearly ordered?

2.2 For *M. genitalium* and *H. influenzae*, what are the values of (a) gene density in genes/kbp, (b) average gene size in bp, (c) number of genes. Which factor contributes most highly to the reduction of genome size in *M. genitalium* relative to *H. influenzae*?

2.3 It is estimated that the human immune system can produce 10^{15} antibodies. Would it be feasible for such a large number of proteins each to be encoded entirely by a separate gene, the diversity arising from gene duplication and divergence? A typical gene for an IgG molecule is about 2000 bp long.

Weblems

2.1 On photocopies of Fig. 1.2, 1.3, and 1.4, indicate the positions of species for which full-genome sequences are known. (http://www.ebi.ac.uk/genomes/)

2.2 What are the differences between the standard genetic code and the vertebrate mitochondrial genetic code?

2.3 What is the chromosomal location of the human myoglobin gene?

2.4 What is the number of occurrences of the tetrapeptide CTAG in the *E. coli* genome? Is it over-represented or under-represented relative to the expectation for a random sequence of the same length and base composition as the *E. coli* genome? (See Exercise 2.1.)

2.5 Plot a histogram of the cumulative number of completed genome sequences in each year since 1995.

2.6 (a) How many predicted ORFs are there on Yeast chromosome X? (b) How many tRNA genes?

2.7 Which amino acid is entirely lacking in the proteins of *M. genitalium*? How does the genetic code of *M. genitalium* differ from the standard one?

2.8 In the human, 1 cM $\sim 10^6$ bp. In yeast, approximately how many base pairs correspond to 1 cM?

2.9 Sperm cells are active swimmers, and contain mitochondria. At fertilization the entire contents of a sperm cell enters the egg. How is it, therefore, that mitochondrial DNA is inherited from the mother only?

2.10 The box on pages 90–91 shows the duplications and divergences leading to the current human α and β globin gene clusters. (a) In which species, closely related to ancestors of humans, did these divergences take place? (b) In which species related to ancestors of humans did the developmental pattern of expression pattern ($\zeta_2\epsilon_2$ = embryonic; $\alpha_2\gamma_2$ = foetal; $\alpha_2\gamma_2$ = adult) emerge?

2.11 Are language groups more closely correlated with variations in human mitochondrial DNA or Y chromosome sequences? Suggest an explanation for the observed result.

2.12 The mutation causing sickle-cell anaemia is a single base change A→T, causing the change Glu→Val at position 6 of the β chain of haemoglobin. The base change occurs in the sequence 5′-GTGAG-3′(*normal*)→GTGTG (*mutant*). What restriction enzyme is used to distinguish between these sequences, to detect carriers? What is the specificity of this enzyme?

2.13 What mutation is the most common cause of phenylketonuria (PKU)?

2.14 Find three examples of mutations in the CFTR gene (associated with cystic fibrosis) that produce reduced but not entirely absent chloride channel function. What are the clinical symptoms of these mutations?

2.15 Find an example of a genetic disease that is: (a) Autosomal dominant (b) Autosomal recessive (other than cystic fibrosis) (c) X-linked dominant (d) X-linked recessive (e) Y-linked (f) The result of abnormal mitochondrial DNA (other than Leber's Hereditary Optic Neuropathy).

2.16 (a) Identify a state of the USA in which newborn infants are routinely tested for homocystinuria. (b) Identify a state of the USA in which newborn infants are *not* routinely tested for homocystinuria. (c) Identify a state of the USA in which newborn infants are routinely tested for biotinidiase. (d) Identify a state of the USA in which newborn infants are *not* routinely tested for biotinidiase. (e) What are the clinical consequences of failure to detect homocystinurea, or biotinidase deficiency?

2.17 (a) What is the normal function of the protein that is defective in Menke disease? (b) Is there a homologue of this gene in the *A. thaliana* genome? (c) If so, what is the function of this gene in *A. thaliana*?

2.18 *Duchenne muscular dystrophy* (DMD) is an X-linked inherited disease causing progressive muscle weakness. DMD sufferers usually lose the ability to walk by the age of 12, and life expectancy is no more than about 20–25 years. *Becker muscular dystrophy* (BMD) is a less severe condition involving the same gene. Both conditions are usually caused by deletions in a single gene, dystrophin. In DMD there is complete absence of functional protein; in BMD there is a truncated protein retaining some function. Some of the deletions in cases of BMD are longer than others that produce BMD. What distinguishes the two classes of deletions causing these two conditions?

2.19 What chromosome of the cow contains a region homologous to human chromosome region 8q21.12?

CHAPTER 3

Archives and information retrieval

Chapter contents

Introduction *118*
Database indexing and specification of search terms *118*
Follow-up questions *120*
Analysis of retrieved data *121*
The archives *121*
Nucleic acid sequence databases *122*
Genome databases *124*
Protein sequence databases *124*
Databases of structures *128*
Specialized, or 'boutique' databases *135*
Expression and proteomics databases *136*
Databases of metabolic pathways *138*
Bibliographic databases *139*
Surveys of molecular biology databases and servers *139*
Gateways to archives *140*
Access to databases in molecular biology *141*
ENTREZ *141*
The Sequence Retrieval System (SRS) *148*
The Protein Identification Resource (PIR) *149*
ExPASy—Expert Protein Analysis System *150*
Ensembl *151*
Where do we go from here? *152*
Recommended reading *152*
Exercises, Problems, and Weblems *153*

Learning goals

1. To understand the general types of data about the molecules and processes of life that are assembled in support of research and applications in biology, medicine, agriculture and technology.
2. To know the basic infrastructure of bioinformatics, in terms of the sites and responsibilities of the major archival projects.
3. To understand the basic concepts of information retrieval, including how to frame queries.
4. To have facility with general search engines on the Web, and specific sites for bioinformatics.
5. To know how to search for specific information about sequences, structures, metabolic pathways, relationships to disease, and how to launch analyses of the data recovered.

Introduction

This chapter introduces the information retrieval skills that will allow you to make effective use of the databanks. The goal is to give you familiarity with basic operations. It will then be easy to improve and develop your technique. Indeed, embedded in many databanks are tutorials which make it easy to explore their facilities.

Database indexing and specification of search terms

An index is a set of pointers to information in a database. In searching the entire World Wide Web, or a specialized database in molecular biology, you propose one or more search terms, and a program checks for them in its tables of indices. The model is that the entire database is composed of **entries**—discrete coherent parcels of information. The information retrieval software identifies entries with contents relevant to your interest. An example of the simplest paradigm is that you submit the term 'horse' and the program returns a list of entries that contain the term horse.

A full search of the Web would turn up information about many different aspects of horses—molecular biology, breeding, racing, poems about horses—most of which you don't want to see. For a successful search, it is not enough to mention what you *do* want—you must ensure that the desired responses don't get buried in extraneous rubbish. (Of course rubbish is merely whatever *other* people are interested in.)

To focus the results, information retrieval engines accept multiple query terms or keywords. A search for 'horse liver alcohol dehydrogenase' would produce responses specialized to this enzyme. The search would identify entries that contain all four keywords that you submitted: horse AND liver AND alcohol AND dehydrogenase. It would not return poems about horses among its top hits (except in the unlikely event that a poem contained all four keywords).

It is possible to ask for other logical combinations of indexing terms. For instance, if a search engine didn't know about transatlantic spelling differences, it would be useful to be able to search for 'hemoglobin OR haemoglobin'. (Note that a search for 'hemoglobin haemoglobin' would probably be interpreted as 'hemoglobin AND haemoglobin' which would pick up only documents written by international committees or orthographically-challenged expatriates.)

If you wanted to know about other dehydrogenases, you could ask for 'dehydrogenase NOT alcohol'. This would retrieve entries that contain the term dehydrogenase but did NOT contain the word alcohol. You would find entries about lactate dehydrogenase, malate dehydrogenase, etc. You would miss references to review articles that compared alcohol dehydrogenases to other dehydrogenases, or alignments of the sequences of many dehydrogenases including alcohol dehydrogenase. You might regret missing these.

Many database search engines will allow complex logical expressions such as '(haemoglobin OR hemoglobin) AND (dehydrogenase NOT alcohol)'. Construction of such expressions is an exercise in set theory, and it is helped by drawing Venn diagrams. Although the logic of a search is independent of the software used to query a database, different programs demand different syntax to express the same conditions. For example the query for dehydrogenase NOT alcohol might have to be entered as DEHYDROGENASE −ALCOHOL or DEHYDROGENASE !ALCOHOL.

Specialized databases, including those in molecular biology, impose a structure on the information, to separate different categories of information. This is essential. There are currently active biomedical scientists named E(lisabetta) Coli, (John D.) Yeast, (Patrice) Rat, and a large number of Rabbits, as well as several Crystals and Blots. If you wanted to find papers published by these investigators, it would be naive to perform a general search of a molecular biology database with any of their surnames. Many databases provide separate indexing and searching of different categories of information. They permit searching for papers of which E. Coli is an AUTHOR.

Some of the categories, such as taxonomy, have **controlled vocabularies**. Often these are presented to the user as pull-down menus. To do a search for 'globin NOT mammal', and pick out the relatively few entries about nonmammalian globins rather than the very many entries about globins, including human haemoglobins, that do not explicitly mention the term mammal, requires an information retrieval system that 'understands' the taxonomic hierarchy. Controlled vocabularies—limited, explicit, and carefully defined sets of terms—are also important in distributing queries among several databases.

A technical problem that frequently creates difficulty is how to enter terms containing nonstandard characters such as accent marks or umlauts, Greek letters, and, as already mentioned, differences between US and British spelling. A specialized database such as NCBI's ENTREZ can handle the US–British spelling differences with a synonym dictionary. Programs that index the entire Web usually do not. Ignore the accent marks and hope for the best.

Follow-up questions

When searching in databases, it is rare that you will find exactly what you want on the first round of probing. Usually you have to modify the query, on the basis of the results initially returned. Most information retrieval software permits consecutive, cumulative searches, with altered sets of search terms and/or logical relationships. Conversely, once you find what you looked for, you will often want to extend your search to find related material. If you find a gene sequence, you might want to know about homologous genes in other organisms. Or whether a three-dimensional structure of the corresponding protein is available. Or you might want to know about papers published about the gene.

For these subsidiary queries you need links between entries in the same or different databases. This is a special example of the question of how one 'browses' in electronic libraries—a difficult problem, the subject of current research.

To find homologous genes you would like links to other items in the *same* database (a database of gene sequences). To find structures, or bibliographical references, related to a gene, you would like links *between* different databases (from the database of gene sequences to a database of three-dimensional structures, or to a bibliographical database). As the number of databases grows, intercommunication among them has become a high-priority goal. Indeed, the interactivity of the databases in molecular biology is growing more and more effective, so that these operations are fairly easy now—formerly one had to do separate searches on isolated databases. This is a generalization of the original model of a database as a closed set of independent entries that can be selected only by their indexed contents.

To some extent database activities in bioinformatics can be classified into *archiving*—with the major goals of conservation and curation—and *interpreting*— the compilation of biological information in a form most useful to support research. Different archives specialize in different kinds of data—nucleic acid sequences, protein sequences, structures—for reasons in part historical and in part because of the different specialized curatorial skills required. Interpretative databases are free to combine information from any available sources. In most cases, archival and interpretative projects are carried out at the same institution and even by the same people.

Two aspects of the development of bioinformatics databases are apparent. One is the very great growth of individual database projects that recombine the archived data in different ways. The other is the combination of many individual databases into 'umbrella' sites. There is really no paradox—both are going on. (Genes also both multiply and combine.)

Most database unifications are merely extensions, with greater appearance of intimacy, of the collaborations or competitive efforts that in most cases formerly existed. We shall see, for instance, that protein sequence databases are coordinating their activities as UniProt; and that the protein structure databases are coordinating as the Worldwide Protein Data Bank. Interpro is an umbrella database that integrates the contents, features, and annotation of individual databases of protein families, domains, and functional sites, and contains links to others, including the Gene Ontology Consortium™ functional classification. It currently subsumes the PROSITE, Pfam, PRINTS, SMART and ProDom databases, and intends to assimilate others. (Resistance is futile.)

Analysis of retrieved data

Sometimes as a result of a search you will want to launch a program using the results retrieved as input. For instance, if you identify a protein sequence of interest, you might want to perform a PSI-BLAST search. This is not strictly a database-lookup problem, and formerly you would have to run a separate job, and feed the retrieved sequence to the application program by hand. However, like searches in multiple databases, information retrieval systems in molecular biology often provide facilities for initiating such processes. This makes for very much improved fluency in your sessions at the computer.

The archives

Although our knowledge of biological sequence and structure data is very far from complete, it is of quite respectable size, and growing extremely rapidly. Many scientists are working to generate the data, or to carry out research projects analysing the results. Archiving and distribution are carried out by particular databanking organizations.

Archiving of bioinformatics data was originally carried out by individual research groups motivated by an interest in the associated science. As the requirements for equipment and personnel grew—and the nature of the skills required changed, to include much more emphasis on computing—they have been made the responsibility of special national and even international projects, on a very large scale indeed. Anyone who has followed the entire history of these projects cannot help being impressed by their growth from small, low-profile and ill-funded projects carried out by a few dedicated individuals, to a multinational heavy industry subject to political takeovers and the scientific equivalent of leveraged buyouts.

Primary data collections related to biological macromolecules include:

- Nucleic acid sequences, including whole-genome projects
- Amino acid sequences of proteins
- Protein and nucleic acid structures
- Small-molecule crystal structures
- Protein functions
- Expression patterns of genes
- Metabolic pathways, and networks of interaction and control
- Publications

Nucleic acid sequence databases

The worldwide nucleic acid sequence archive is a triple partnership of The National Center for Biotechnology Information (USA), the EMBL Data Library (European Bioinformatics Institute, UK), and the DNA Data Bank of Japan (National Institute of Genetics, Japan). The groups exchange data daily. As a result the raw data are identical, although the format in which they are stored, and the nature of the annotation, vary among them. These databases curate, archive, and distribute DNA and RNA sequences collected from genome projects, scientific publications, and patent applications. To make sure that these fundamental data are freely available, scientific journals require deposition of new nucleotide sequences in a database as a condition for publication of an article. Similar conditions apply to amino acid sequences, and to nucleic acid and protein structures.

The nucleic acid sequence databases, as distributed, are collections of entries. Each entry has the form of a text file containing data and annotations for a single contiguous sequence. Many entries are assembled from several published papers reporting overlapping fragments of a complete sequence. Others are complete genomes.

Entries have a life cycle in the database. Because of the desire on the part of the user community for rapid access to data, new entries are made available before annotation is complete and checks are performed. Entries mature through the classes:

Unannotated → Preliminary → Unreviewed → Standard

Rarely, an entry 'dies'—a few have been removed when they were determined to be erroneous.

A sample DNA sequence entry from the EMBL data library, including annotations as well as sequence data, is the gene for bovine pancreatic trypsin inhibitor (the Box shows part of this entry, omitting most of the sequence itself).

The EMBL Data Library entry for the bovine pancreatic trypsin inhibitor gene

```
ID   BTBPTIG    standard; DNA; MAM; 3998 BP.
XX
AC   X03365; K00966;
XX
DT   18-NOV-1986 (Rel. 10, Created)
DT   20-MAY-1992 (Rel. 31, Last updated, Version 3)
XX
DE   Bovine pancreatic trypsin inhibitor (BPTI) gene
XX
KW   Alu-like repetitive sequence; protease inhibitor;
KW   trypsin inhibitor.
XX
OS   Bos taurus (cattle)
OC   Eukaryota; Animalia; Metazoa; Chordata; Vertebrata; Mammalia;
OC   Theria; Eutheria; Artiodactyla; Ruminantia; Pecora; Bovidae.
XX
RN   [1]
RP   1-3998
RA   Kingston I.B., Anderson S.;
RT   "Sequences encoding two trypsin inhibitors occur in strikingly
RT   similar genomic environments";
RL   Biochem. J. 233:443-450(1986).
XX
RN   [2]
RA   Anderson S., Kingston I.B.;
RT   "Isolation of a genomic clone for bovine pancreatic trypsin
RT   inhibitor by using a unique-sequence synthetic dna probe";
RL   Proc. Natl. Acad. Sci. U.S.A. 80:6838-6842(1983).
XX
DR   SWISS-PROT; P00974; BPT1_BOVIN.
XX
CC   Data kindly reviewed (08-DEC-1987) by Kingston I.B.
XX
FH   Key             Location/Qualifiers
FH
FT   misc_feature    795..800
FT                   /note="pot. polyA signal"
FT   misc_feature    835..839
FT                   /note="pot. polyA signal"
FT   repeat_region   837..847
FT                   /note="direct repeat"
FT   misc_feature    930..945
FT                   /note="sequence homologous to Alu-like
FT                   consensus seq."
FT   repeat_region   1035..1045
FT                   /note="direct repeat"
FT   misc_feature    2456..2461
FT                   /note="pot. splice signal"
FT   CDS             2470..2736
FT                   /note="put. precursor"
FT   misc_feature    2488..2489
FT                   /note="pot. intron/exon splice junction"
FT   misc_feature    2506..2507
FT                   /note="pot. intron/exon splice junction"
FT   CDS             2512..2685
FT                   /note="trypsin inhibitor (aa 1-58)"
FT   misc_feature    2698..2699
FT                   /note="pot. exon/intron splice junction"
FT   misc_feature    3690..3695
FT                   /note="pot. polyA signal"
FT   misc_feature    3729..3733
FT                   /note="pot. polyA signal"
XX
SQ   Sequence 3998 BP; 1053 A; 902 C; 892 G; 1151 T; 0 other;
     aattctgata atgcagagaa ctggtaagga gttctgattg ttctgcttga ttaaatgggt
     tgtaacagga tagtgtcttg tcctgatcct agcattcata tggtgtgtgt tctggggcaa
     gtcatctgca gtttcttcac ctgaacaggg ggaccaggtt acatgagttt cttaaaagat
     taccagtcat gagtatgaag agtttacact ttcctgatca atgacgtcca tttcccatca
                                   3720 nucleotides deleted ...
     gccaggtcaa actttggggt gtgttatttc cctgaatt
//
```

A **feature table** (lines beginning FT) is a component of the annotation of an entry that reports properties of specific regions, for instance coding sequences (CDS). Because these are designed to be readable by computer programs—for example, to translate a coding region to an amino acid sequence—they have a more carefully controlled format and a more restricted vocabulary. Development of controlled vocabularies and a shared dictionary and thesaurus for keywords and feature tables is also important in establishing links between different databases.

The feature table may indicate regions that:

- perform or affect function
- interact with other molecules
- affect replication
- are involved in recombination
- are a repeated unit
- have secondary or tertiary structure
- are revised or corrected

Genome databases

Although genome sequences form entries in the standard nucleic acid sequence archives, many species have special databases that bring together the genome sequence and its annotation with other data related to the species.

Protein sequence databases

In 2002, three protein sequence databases—The Protein Information Resource, at the National Biomedical Research Foundation of the Georgetown University Medical Center in Washington, DC, USA; and SWISS-PROT and TrEMBL, from the Swiss Institute of Bioinformatics in Geneva and the European Bioinformatics Institute in Hinxton, UK—coordinated their efforts, to form the *UniProt* consortium. The partners in this enterprise share the database but continue to offer separate information retrieval tools for access.

The PIR grew out of the very first sequence database, developed by Margaret O. Dayhoff—the pioneer of the field of bioinformatics. SWISS-PROT was developed at the Swiss Institute of Bioinformatics. TrEMBL contains the translations of genes identified within DNA sequences in the EMBL Data Library. TrEMBL entries are regarded as preliminary, and are converted—after curation and extended annotation—to full SWISS-PROT entries.

Today, almost all amino acid sequence information arises from translation of nucleic acid sequences. Information about ligands, disulphide bridges, subunit associations, post-translational modifications, glycosylation, effects of mRNA editing, etc., are not available from gene sequences. For instance, from genetic information alone one would not know that human insulin is a dimer linked by disulphide bridges. Protein sequence databanks collect this additional information from the literature and provide suitable annotations.

From UniProt, the entry for the amino acid sequence of the protein bovine pancreatic trypsin inhibitor, in SWISS-PROT format, is shown in the Box, pages 126–127. (The comparison of SWISS-PROT format with ENTREZ and PIR formats is the subject of Weblem 3.12.)

Databases associated with SWISS-PROT

Two related databases closely associated with SWISS-PROT are the ENZYME DB, and PROSITE, a set of motifs.

The ENZYME DB stores the following information about enzymes:

- EC Number: a numerical identifier assigned by the Enzyme Commission (authorized by the International Union of Biochemistry and Molecular Biology; see http://www.chem.qmw.ac.uk/iubmb/enzyme/)
- Recommended name
- Alternative names, if any
- Catalytic activity
- Cofactors, if any
- Pointers to SWISS-PROT and other data banks
- Pointers to disease associated with enzyme deficiency if any known

A sample entry in ENZYME DB

```
ID 1.14.17.3
DE PEPTIDYLGLYCINE MONOOXYGENASE.
AN PEPTIDYL ALPHA-AMIDATING ENZYME.
AN PEPTIDYLGLYCINE 2-HYDROXYLASE.
CA PEPTIDYLGLYCINE + ASCORBATE + O(2) = PEPTIDYL(2-HYDROXYGLYCINE) +
CA DEHYDROASCORBATE + H(2)O.
CF COPPER.
CC -!- PEPTIDYLGLYCINES WITH A NEUTRAL AMINO ACID RESIDUE IN THE
CC PENULTIMATE POSITION ARE THE BEST SUBSTRATES FOR THE ENZYME.
CC -!- THE ENZYME ALSO CATALYZES THE DISMUTATATION OF THE PRODUCT TO
CC GLYOXYLATE AND THE CORRESPONDING DESGLYCINE PEPTIDE AMIDE.
DR P10731, AMD_BOVIN ; P19021, AMD_HUMAN ; P14925, AMD_RAT ;
DR P08478, AMD1_XENLA; P12890, AMD2_XENLA;
```

The first two characters of each line identify the information that the line contains. For instance, ID = Identification, DE = Description = Official name, AN = Alternate name(s), CA = catalytic activity, CF = cofactor(s), CC = comments, DR = database reference (to SWISS-PROT).

PROSITE contains common patterns of residues of sets of proteins. Such a pattern (or motif, or signature, or fingerprint, or template) appears in a family of related proteins usually because of the requirements of binding sites that constrain the evolution of a protein family. Often they indicate distant relationships not otherwise detectable by comparing sequences. The consensus pattern for inorganic pyrophosphatase is: D-[SGN]-D-[PE]-[LIVM]-D-[LIVMGC]. The three conserved aspartates (D) bind divalent metal cations.

The PIR and associated databases

The PIR maintains several databases about proteins:

- PIR-PSD: the main protein sequence database
- iProClass: classification of proteins according to structure and function

Amino acid sequence entry for bovine pancreatic trypsin inhibitor

NiceProt View of Swiss-Prot: P00974

Entry information

Entry name	**BPT1_BOVIN**
Primary accession number	**P00974**
Secondary accession numbers	None
Entered in Swiss-Prot in	Release 01, July 1986
Sequence was last modified in	Release 10, March 1989
Annotations were last modified in	Release 44, June 2004

Name and origin of the protein

Protein name	**Pancreatic trypsin inhibitor [Precursor]**
Synonyms	**Basic protease inhibitor** **BPI** **BPTI** **Aprotinin**
Gene name	None
From	Bos taurus (Bovine) [TaxID: 9913]
Taxonomy	Eukaryota; Metazoa; Chordata; Craniata; Vertebrata; Euteleostomi; Mammalia; Eutheria; Cetartiodactyla; Ruminantia; Pecora; Bovidae; Bovinae; Bos.

References

[1] SEQUENCE FROM NUCLEIC ACID
MEDLINE=87283904; PubMed=2441071;
Creighton T.E., Charles I.G.;
"Sequences of the genes and polypeptide precursors for two bovine protease inhibitors";
J. Mol. Biol. 194:11–22(1987).
ADDITIONAL REFERENCES DELETED

Comments

- *FUNCTION*: Inhibits trypsin, kallikrein, chymotrypsin, and plasmin.
- *SUBCELLULAR LOCATION*: Secreted.
- *PHARMACEUTICAL*: Available under the name Trasylol (Mile). Used for inhibiting coagulation so as to reduce blood loss during bypass surgery.
- *SIMILARITY*: Contains 1 BPTI/Kunitz inhibitor domain.
- *DATABASE*: Name=Trasylol; Note=Clinical information on Trasylol; www="http://www.trasylol.com/".
  ADDITIONAL COMMENTS DELETED

Copyright

The Swiss-Prot entry is copyright. It is produced through a collaboration between the Swiss Institite of Bioinformatics and the EMBL outstation–the European Bioinformatics Institute. There are no restrictions on its use by non-profit institutions as long as its content is in no way modified and this statement is not removed. Usage by and for commercial entities requires a license agreement (See http://www.isb-sib.ch/announce/ or send an email to license@isb-sib.ch)

→

→

Cross-references

EMBL	M20934; AAD13685.1;-.
	ADDITIONAL CROSS-REFERENCES TO EMBL DELETED
PIR	S00277; TIBO.
PDB	1K09; 10-JUL-02.
	ADDITIONAL CROSS-REFERENCES TO PDB DELETED
InterPro	IPR002223; Kunitz_BPTI.
Pfam	PF00014; Kunitz_BPTI; 1.
	Pfam graphical view of domain structure.
PRINTS	PR00759; BASICPTASE.
ProDom	PD000222; Kunitz_BPTI; 1.
	[Domain structure/List of seq. sharing at least 1 domain]
SMART	SM00131; KU; 1.
PROSITE	PS00280; BPTI_KUNITZ_1;1.
	PS50279; BPTI_KUNITZ_1; 2.
	PROSITE graphical view of domain structure.
Implicit links to	HOVERGEN; BLOCKS; ProtoNet; ProtoMap; PRESAGE; DIP; ModBase; SMR; SWISS-2DPAGE; UniRef.

Keywords
Serine protease inhibitor; Signal; Pharmaceutical; 3D-structure.

Features

Key	From	To	Length	Description
SIGNAL	1	21	21	Potential.
PROPEP	22	35	14	
CHAIN	36	93	58	Pancreatic trypsin inhibitor.
PROPEP	94	100	7	
DOMAIN	40	90	51	BPTI/Kunitz inhibitor.
SITE	50	51	2	Reactive bond for trypsin.
DISULFID	40	90		
DISULFID	49	73		
DISULFID	65	86		
HELIX	38	41	4	
STRAND	53	59	7	
TURN	60	63	4	
STRAND	64	70	7	
STRAND	80	80	1	
HELIX	83	90	8	

Sequence information
Length: **100 AA** [This is the length of the unprocessed precursor]
Molecular weight: **10903 Da** [This is the MW of the unprocessed precursor]
CRC64: **6A778A4AD763FB19** [This is a checksum on the sequence]

```
        10         20         30         40         50         60
         |          |          |          |          |          |
MKMSRLCLSV ALLVLLGTLA ASTPGCDTSN QAKAQRPDFC LEPPYTGPCK ARIIRYFYNA

        70         80         90        100
         |          |          |          |
KAGLCQTFVY GGCRAKRNNF KSAEDCMRTC GGAIGPWENL
```

- ASDB: annotation and similarity database; each entry is linked to a list of similar sequences
- NRL_3D: a database of sequences and annotations of proteins of known structure deposited in the Protein Data Bank
- ALN: a database of protein sequence alignments
- RESID: a database of covalent protein structure modifications (recall that important structural features of proteins such as disulphide bridges are not inferrable from gene sequences, and will not appear in protein sequence databases derived solely by translation of genomic data)

The PIR has also created IESA: The Integrated Environment for Sequence Analysis, a site for information retrieval and launching of calculations.

The web server of PIR shows some of the richness of information retrieval tools available. It includes:

- FETCH DATABASE ENTRY
- PAIRWISE SEQUENCE ALIGNMENT
- PROT-FAM: classification by protein family of over 7000 multiple sequence alignments of protein families
- ATLAS: search text fields of databases, or scan sequences for short peptides
- ALERT: receive information about new database entries of interest by e-mail, automatically
- GATEWAY: pattern recognition, homology identification

Databases of structures

Structure databases archive, annotate and distribute sets of atomic coordinates. The major database for biological macromolecular structures is the Protein Data Bank (PDB). It contains structures of proteins, nucleic acids, and a few carbohydrates. Started by the late Walter Hamilton at Brookhaven National Laboratories, Long Island, New York, USA in 1971, the PDB is now managed by the Research Collaboratory for Structural Bioinformatics (RCSB), a distributed organization based at Rutgers University, in New Jersey; the San Diego Supercomputer Center, in California; and the National Institute of Standards and Technology, in Maryland, all in the USA. The parent web site of the Protein Data Bank is at http://www.rcsb.org.

The home page of the PDB contains links to the data files themselves, to expository and tutorial material including short news items and the PDB Newsletter, to facilities for deposition of new entries, and to specialized search software for retrieving structures.

Recently, the RCSB, the Molecular Structure Database and the European Bioinformatics Institute, and the Protein Data Bank Japan have formed the Worldwide Protein Data Bank (wwPDB), with the goal of producing a unified archive.

The box shows part of a Protein Data Bank* entry for a structure of *E. coli* thioredoxin.† The information contained includes:

- What protein is the subject of the entry, and what species it came from
- Who solved the structure, and references to publications describing the structure determination
- Experimental details about the structure determination, including information related to the general quality of the result such as resolution of an X-ray structure determination and stereochemical statistics
- The amino acid sequence
- What additional molecules appear in the structure, including cofactors, inhibitors, and water molecules
- Assignments of secondary structure: helix, sheet
- Disulphide bridges
- The atomic coordinates

Protein Data Bank entry 2TRX, *E. Coli* thioredoxin

```
HEADER    ELECTRON TRANSPORT                      19-MAR-90   2TRX
COMPND    THIOREDOXIN
SOURCE    (ESCHERICHIA $COLI)
AUTHOR    S.K.KATTI,D.M.LE*MASTER,H.EKLUND
REVDAT   2   15-JAN-93 2TRXA   1       HEADER COMPND
REVDAT   1   15-OCT-91 2TRX    0
JRNL        AUTH   S.K.KATTI,D.M.LE*MASTER,H.EKLUND
JRNL        TITL   CRYSTAL STRUCTURE OF THIOREDOXIN FROM ESCHERICHIA
JRNL        TITL 2 $COLI AT 1.68 ANGSTROMS RESOLUTION
JRNL        REF    J.MOL.BIOL.                  V. 212   167 1990
JRNL        REFN   ASTM JMOBAK  UK ISSN 0022-2836                 070
REMARK   1
REMARK   1 REFERENCE 1
REMARK   1  AUTH   A.HOLMGREN,B.-*O.SODERBERG,H.EKLUND,C.-*I.BRANDEN
REMARK   1  TITL   THREE-DIMENSIONAL STRUCTURE OF ESCHERICHIA COLI
REMARK   1  TITL 2 THIOREDOXIN-*S=2= TO 2.8 ANGSTROMS RESOLUTION
REMARK   1  REF    PROC.NAT.ACAD.SCI.USA        V.  72  2305 1975
REMARK   1  REFN   ASTM PNASA6  US ISSN 0027-8424                 040
REMARK   1 REFERENCE 2
REMARK   1  AUTH   B.-*O.SODERBERG,A.HOLMGREN,C.-*I.BRANDEN
REMARK   1  TITL   STRUCTURE OF OXIDIZED THIOREDOXIN TO 4.5 ANGSTROMS
REMARK   1  TITL 2 RESOLUTION
REMARK   1  REF    J.MOL.BIOL.                  V.  90   143 1974
REMARK   1  REFN   ASTM JMOBAK  UK ISSN 0022-2836                 070
REMARK   1 REFERENCE 3
REMARK   1  AUTH   A.HOLMGREN,B.-*O.SODERBERG
REMARK   1  TITL   CRYSTALLIZATION AND PRELIMINARY CRYSTALLOGRAPHIC
REMARK   1  TITL 2 DATA FOR THIOREDOXIN FROM ESCHERICHIA $COLI B
REMARK   1  REF    J.MOL.BIOL.                  V.  54   387 1970
REMARK   1  REFN   ASTM JMOBAK  UK ISSN 0022-2836                 070
REMARK   2
REMARK   2 RESOLUTION. 1.68 ANGSTROMS.
REMARK   3
REMARK   3 REFINEMENT. BY THE RESTRAINED LEAST-SQUARES PROCEDURE OF J.
REMARK   3  KONNERT AND W. HENDRICKSON AS MODIFIED BY B. FINZEL
REMARK   3  (PROGRAM *PROFFT*).  THE R VALUE IS 0.165 FOR 25969
REMARK   3  REFLECTIONS IN THE RESOLUTION RANGE 8.0 TO 1.68 ANGSTROMS
REMARK   3  WITH FOBS .GT. 3.0*SIGMA(FOBS)
REMARK   3
```

→

* Berman, H. M., Westbrook, J., Feng, Z., Gilliland, G., Bhat, T. N.,Weissig, H., Shindyalov, I. N. & Bourne, P. E. (2000), The Protein Data Bank Nucleic Acids Research, **28**, 235–242.

† Katti, S. K., LeMaster, D. M. & Eklund, H. (1990), Crystal structure of thioredoxin from *Escherichia coli* at 1.68 Å resolution, *J. Mol. Biol.*, **212**, 167–184.

Protein Data Bank entry 2TRX, ***E. Coli*** thioredoxin *(continued)*

```
REMARK   3  RMS DEVIATIONS FROM IDEAL VALUES (THE VALUES OF
REMARK   3      SIGMA, IN PARENTHESES, ARE THE INPUT ESTIMATED
REMARK   3      STANDARD DEVIATIONS THAT DETERMINE THE RELATIVE
REMARK   3      WEIGHTS OF THE CORRESPONDING RESTRAINTS)
REMARK   3    DISTANCE RESTRAINTS (ANGSTROMS)
REMARK   3      BOND DISTANCE                                   0.015(0.020)
REMARK   3      ANGLE DISTANCE                                  0.035(0.030)
REMARK   3      PLANAR 1-4 DISTANCE                             0.055(0.050)
REMARK   3    PLANE RESTRAINT (ANGSTROMS)                       0.021(0.020)
REMARK   3    CHIRAL-CENTER RESTRAINT (ANGSTROMS**3)            0.131(0.150)
REMARK   3    NON-BONDED CONTACT RESTRAINTS (ANGSTROMS)
REMARK   3      SINGLE TORSION CONTACT                          0.165(0.500)
REMARK   3      MULTIPLE TORSION CONTACT                        0.174(0.500)
REMARK   3      POSSIBLE HYDROGEN BOND                          0.180(0.500)
REMARK   3    CONFORMATIONAL TORSION ANGLE RESTRAINT (DEGREES)
REMARK   3      PLANAR (OMEGA)                                    4.0(3.0)
REMARK   3      STAGGERED                                        16.3(15.0)
REMARK   3      ORTHONORMAL                                      11.7(20.0)
REMARK   3    ISOTROPIC THERMAL FACTOR RESTRAINTS (ANGSTROMS**2)
REMARK   3      MAIN-CHAIN BOND                                  1.38(1.000)
REMARK   3      MAIN-CHAIN ANGLE                                 2.28(1.000)
REMARK   3      SIDE-CHAIN BOND                                  1.97(1.000)
REMARK   3      SIDE-CHAIN ANGLE                                 3.27(1.500)
REMARK   4
REMARK   4 THERE ARE TWO MOLECULES IN THE ASYMMETRIC UNIT.  THEY HAVE
REMARK   4 BEEN ASSIGNED CHAIN INDICATORS *A* AND *B*.  THEY HAVE BEEN
REMARK   4 REFINED INDEPENDENTLY WITHOUT IMPOSING NON-CRYSTALLOGRAPHIC
REMARK   4 SYMMETRY RESTRAINTS.
REMARK   5
REMARK   5 IN ADDITION TO THE METAL COORDINATION SPECIFIED ON CONECT
REMARK   5 RECORDS BELOW, THERE ARE BONDS TO OD1 AND OD2 OF ASP 10 IN
REMARK   5 A SYMMETRY-RELATED MOLECULE.  DUE TO SOME LIMITATIONS OF
REMARK   5 PROTEIN DATA BANK FORMAT, THESE BONDS CANNOT BE PRESENTED
REMARK   5 ON CONECT RECORDS.
REMARK   6
REMARK   6 CORRECTION. CORRECT CLASSIFICATION ON HEADER RECORD AND
REMARK   6  REMOVE E.C. CODE.  15-JAN-93.
SEQRES   1 A  108  SER ASP LYS ILE ILE HIS LEU THR ASP ASP SER PHE ASP
SEQRES   2 A  108  THR ASP VAL LEU LYS ALA ASP GLY ALA ILE LEU VAL ASP
SEQRES   3 A  108  PHE TRP ALA GLU TRP CYS GLY PRO CYS LYS MET ILE ALA
SEQRES   4 A  108  PRO ILE LEU ASP GLU ILE ALA ASP GLU TYR GLN GLY LYS
SEQRES   5 A  108  LEU THR VAL ALA LYS LEU ASN ILE ASP GLN ASN PRO GLY
SEQRES   6 A  108  THR ALA PRO LYS TYR GLY ILE ARG GLY ILE PRO THR LEU
SEQRES   7 A  108  LEU LEU PHE LYS ASN GLY GLU VAL ALA ALA THR LYS VAL
SEQRES   8 A  108  GLY ALA LEU SER LYS GLY GLN LEU LYS GLU PHE LEU ASP
SEQRES   9 A  108  ALA ASN LEU ALA
SEQRES   1 B  108  SER ASP LYS ILE ILE HIS LEU THR ASP ASP SER PHE ASP
SEQRES   2 B  108  THR ASP VAL LEU LYS ALA ASP GLY ALA ILE LEU VAL ASP
SEQRES   3 B  108  PHE TRP ALA GLU TRP CYS GLY PRO CYS LYS MET ILE ALA
SEQRES   4 B  108  PRO ILE LEU ASP GLU ILE ALA ASP GLU TYR GLN GLY LYS
SEQRES   5 B  108  LEU THR VAL ALA LYS LEU ASN ILE ASP GLN ASN PRO GLY
SEQRES   6 B  108  THR ALA PRO LYS TYR GLY ILE ARG GLY ILE PRO THR LEU
SEQRES   7 B  108  LEU LEU PHE LYS ASN GLY GLU VAL ALA ALA THR LYS VAL
SEQRES   8 B  108  GLY ALA LEU SER LYS GLY GLN LEU LYS GLU PHE LEU ASP
SEQRES   9 B  108  ALA ASN LEU ALA
FTNOTE   1
FTNOTE   1 RESIDUES PRO A 76 AND PRO B 76 ARE CIS PROLINES.
FTNOTE   2
FTNOTE   2 RESIDUES HIS A 6, LEU A 7, ILE A 23, ASP A 47, GLU A 48,
FTNOTE   2 LEU A 58, LEU A 80, HIS B 6, ASP B 47, LEU B 58, AND
FTNOTE   2 LEU B 80 HAVE BEEN MODELED AS TWO CONFORMERS.
FTNOTE   3
FTNOTE   3 RESIDUES 11 - 21 IN CHAIN B ARE DISORDERED.
HET     CU     109      1     COPPER ++ ION
HET     CU     109      1     COPPER ++ ION
HET    MPD     601      8     2-METHYL-2,4-PENTANEDIOL
HET    MPD     602      8     2-METHYL-2,4-PENTANEDIOL
HET    MPD     603      8     2-METHYL-2,4-PENTANEDIOL
HET    MPD     604      8     2-METHYL-2,4-PENTANEDIOL
HET    MPD     605      8     2-METHYL-2,4-PENTANEDIOL
HET    MPD     606      8     2-METHYL-2,4-PENTANEDIOL
HET    MPD     607      8     2-METHYL-2,4-PENTANEDIOL
HET    MPD     608      8     2-METHYL-2,4-PENTANEDIOL
FORMUL   3   CU    2(CU1 ++)
FORMUL   4  MPD    8(C6 H14 O2)
FORMUL   5  HOH   *140(H2 O1)
HELIX    1 A1A SER A   11  LEU A   17  1 DISORDERED IN MOLECULE B
HELIX    2 A2A CYS A   32  TYR A   49  1 BENT BY 30 DEGREES AT RES 39
HELIX    3 A3A ASN A   59  ASN A   63  1
HELIX    4 31A THR A   66  TYR A   70  5 DISTORTED H-BONDING C-TERMINS
HELIX    5 A4A SER A   95  LEU A  107  1
HELIX    6 A1B SER B   11  LEU B   17  1 DISORDERED IN MOLECULE B
HELIX    7 A2B CYS B   32  TYR B   49  1 BENT BY 30 DEGREES AT RES 39
HELIX    8 A3B ASN B   59  ASN B   63  1
HELIX    9 31B THR B   66  TYR B   70  5 DISTORTED H-BONDING C-TERMINS
HELIX   10 A4B SER B   95  LEU B  107  1
```

→

```
SHEET    1 B1A 5 LYS A   3  THR A   8  0
SHEET    2 B1A 5 LEU A  53  ASN A  59  1  O  VAL A  55   N  ILE A   5
SHEET    3 B1A 5 GLY A  21  TRP A  28  1  N  TRP A  28   O  LEU A  58
SHEET    4 B1A 5 PRO A  76  LYS A  82 -1  O  THR A  77   N  PHE A  27
SHEET    5 B1A 5 VAL A  86  GLY A  92 -1  N  GLY A  92   O  LYS A  82
SHEET    1 B1B 5 LYS B   3  THR B   8  0
SHEET    2 B1B 5 LEU B  53  ASN B  59  1  O  VAL B  55   N  ILE B   5
SHEET    3 B1B 5 GLY B  21  TRP B  28  1  N  TRP B  28   O  LEU B  58
SHEET    4 B1B 5 PRO B  76  LYS B  82 -1  O  THR B  77   N  PHE B  27
SHEET    5 B1B 5 VAL B  86  GLY B  92 -1  N  GLY B  92   O  LYS B  82
TURN     1 T1A THR A   8  SER A  11      III (TYPE I IN MOLECULE B)
TURN     2 T2A ALA A  29  CYS A  32      I
TURN     3 T3A TYR A  49  LYS A  52      II
TURN     4 T4A GLY A  74  THR A  77      VIB (INCLUDES CIS PRO 76)
TURN     5 T5A LYS A  82  GLU A  85      I'
TURN     6 T1B THR B   8  SER B  11      I (TYPE III IN MOLECULE A)
TURN     7 T2B ALA B  29  CYS B  32      I
TURN     8 T3B TYR B  49  LYS B  52      II
TURN     9 T4B GLY B  74  THR B  77      VIB (INCLUDES CIS PRO 76)
TURN    10 T5B LYS B  82  GLU B  85      I'
SSBOND   1 CYS A   32    CYS A   35
SSBOND   2 CYS B   32    CYS B   35
CRYST1   89.500   51.060   60.450  90.00 113.50  90.00 C 2           8
ORIGX1      1.000000  0.000000  0.000000        0.00000
ORIGX2      0.000000  1.000000  0.000000        0.00000
ORIGX3      0.000000  0.000000  1.000000        0.00000
SCALE1      0.011173  0.000000  0.004858        0.00000
SCALE2      0.000000  0.019585  0.000000        0.00000
SCALE3      0.000000  0.000000  0.018039        0.00000
ATOM      1  N   SER A   1      21.389  25.406  -4.628  1.00 23.22
ATOM      2  CA  SER A   1      21.628  26.691  -3.983  1.00 24.42
ATOM      3  C   SER A   1      20.937  26.944  -2.679  1.00 24.21
ATOM      4  O   SER A   1      21.072  28.079  -2.093  1.00 24.97
ATOM      5  CB  SER A   1      21.117  27.770  -5.002  1.00 28.27
ATOM      6  OG  SER A   1      22.276  27.925  -5.861  1.00 32.61
ATOM      7  N   ASP A   2      20.173  26.028  -2.163  1.00 21.39
ATOM      8  CA  ASP A   2      19.395  26.125  -0.949  1.00 21.57
ATOM      9  C   ASP A   2      20.264  26.214   0.297  1.00 20.89
ATOM     10  O   ASP A   2      19.760  26.575   1.371  1.00 21.49
ATOM     11  CB  ASP A   2      18.439  24.914  -0.856  1.00 22.14
ATOM     12  CG  ASP A   2      19.199  23.629  -0.576  1.00 23.23
ATOM     13  OD1 ASP A   2      20.107  23.371  -1.387  1.00 22.71
ATOM     14  OD2 ASP A   2      18.905  22.959   0.420  1.00 23.61

                      . . . protein atoms deleted

ATOM    844  N   ALA A 108      41.357  21.341   9.676  1.00 42.93
ATOM    845  CA  ALA A 108      42.151  20.619  10.674  1.00 46.31
ATOM    846  C   ALA A 108      42.632  19.312  10.013  1.00 48.21
ATOM    847  O   ALA A 108      41.703  18.483   9.767  1.00 49.54
ATOM    848  CB  ALA A 108      41.441  20.369  11.988  1.00 46.65
ATOM    849  OXT ALA A 108      43.857  19.249   9.766  1.00 49.19
TER     850      ALA A 108

  . . . second chain, and methane, pentane-diol molecules deleted

          HETATM 1749  O   HOH   401      30.339  33.478  16.727  1.00 17.61
          HETATM 1750  O   HOH   402      29.396  44.583   6.834  0.95 17.71

                 . . . 72 additional water molecules deleted
```

The PDB overlaps in scope with several other databases. The Cambridge Crystallographic Data Centre archives the structures of small molecules; oligonucleotides appear in both the CCDC and PDB. This information is extremely useful in studies of conformations of the component units of biological macromolecules, and for investigations of macromolecule-ligand interactions, including but not limited to applications to drug design. The Nucleic Acid Database (NDB) at Rutgers University, New Brunswick, New Jersey, USA complements the PDB. The BioMagResBank, at the Department of Biochemistry, University of Wisconsin, Madison, Wisconsin, USA, archives protein structures determined by Nuclear Magnetic Resonance.

The archives collect not only the results of structure determinations, but also the measurements on which they are based. The PDB keeps the new data from X-ray structure determinations, and the BioMagRes Bank those from NMR.

The PDB assigns a four-character identifier to each structure deposited. The first character is a number from 1–9. Do not expect mnemonic significance. In many cases several entries correspond to one protein—solved in different states of ligation, or in different crystal forms, or re-solved using better crystals or more accurate data collection techniques. For instance, there have been at least four generations of sperm whale myoglobin crystal structures.

It is easy to retrieve a structure if you know its identifier. From the RCSB home page, entering a PDB ID and selecting Explore gives a 1-page summary of the entry. Figure 3.1 shows the summary page for the thioredoxin structure, identifier 2TRX. Links from this page take you to:

- The publication in which the entry was described, via the bibliographic database PubMed

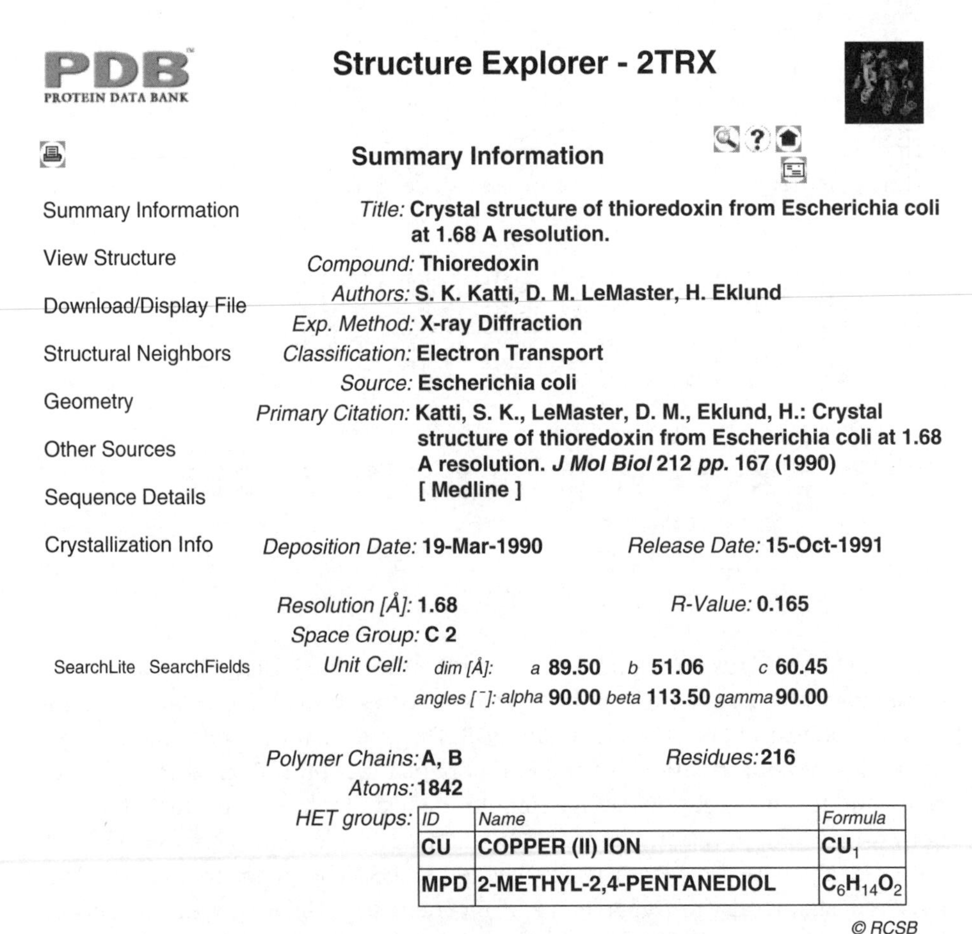

Fig. 3.1 The summary page for the PDB entry 2TRX, *E. coli* thioredoxin.

- Pictures of the structure (some of these may require that you install a viewing program on your computer)
- Access to the file containing the entry itself
- Lists of related structures, according to several different classifications of protein structures
- Stereochemical analysis—the distribution of bond lengths and angles, and conformational angles
- Sources of other information about this entry
- The sequence and secondary structure assignment
- Details about the crystal form and methods by which the crystals were produced

Fine, if you know the identifier. If not, how do you find it? A simple tool accessible from the PDB home page, called SearchLite, permits a search for keywords. Entering `coli and thioredoxin` returns 145 entries, including 2TRX and other crystal structures of the same molecule or mutants, but also several structures of Staphylococcal nuclease, because embedded in the nuclease structure entries is a reference to an article that contains the word thioredoxin in the title. The information returned would easily permit you to choose structures to look at or analyse, according to your particular interest in this family of molecules.

The PDB also offers more complex browsers. The Macromolecular Structure Database at the European Bioinformatics Institute (EBI) offers a useful list of facilities for searching and browsing the PDB including a search tool called OCA. OCA is a browser database for protein structure and function, integrating information from numerous databanks. Developed originally by J. Prilusky, OCA is supported by the EBI and is available there and at numerous mirror sites. (The name OCA, in addition to being the Spanish word for goose, has the same relationship to PDB as A. C. Clarke's computer HAL in the movie 2001 has to IBM.)

Another useful information source available at the EBI is the database of Probable Quaternary Structures (PQS) of the biologically active forms of proteins. Often the asymmetric unit of the crystal structure, as deposited in the PDB entry, contains only part of the active unit, or alternatively multiple copies of the active unit. In many cases it is not obvious how to go from the deposited entry to the active form, and this information is available in PQS.

Indicators of structure quality

X-ray crystal structure analysis produces estimates of the positions of the atoms in a molecule and of their effective sizes, known as **B-factors**. An important feature of the experimental data (the absolute values of the Fourier coefficients of the electron density) is that all atoms contribute to all observations. It is difficult to estimate errors in individual atomic positions.

Crystal structure determinations are at the mercy of the degree of order in different parts of the molecule. (Order is the extent to which different unit cells of the crystal are exact copies of one another.)

The degree of order governs the available **resolution** of the experimental data. Resolution is an index of potential quality of an X-ray structure determination. It measures the ratio of the number of parameters to be determined to the number of observations. In structure determinations of small organic molecules or of minerals, this ratio is usually generous: ~10. But for a typical protein crystal:

	Low resolution			...		**High**
Resolution in Å	4.0	3.5	3.0	2.5	2.0	1.5
Ratio of observations to parameters	0.3	0.4	0.6	1.1	2.2	3.8

(Resolution measures the fineness of the details that can be distinguished, hence the lower the number, the higher the resolution.)

In addition to disorder, errors in crystal structures reflect both errors in data and errors in solving the structure. A comparison of four independently-solved structures of interleukin-1β showed an average variation in atomic position of 0.84 Å, higher than the expected experimental error.

Many crystallographers deposit their experimental data along with the solved structures. This permits detailed checks on the results. But in many cases the experimental data are not available. How can one then assess the quality of a structure? B-factors provide important clues; high B-factors in an entire region suggest that the region has not been well-determined. This usually reflects imperfect order in the crystal. Programs can flag stereochemical outliers—exceptions to regularities common to well-determined protein structures. The entries corresponding to the PDB entries in www.cmbi.kun.nl/gv/pdbreport describe diagnostic analysis and identification of problems and outliers.

But although outliers are relatively easy to *detect*, it is difficult to decide whether they are correct but unusual features of the structure, or the result of errors in building the model, or the inevitable result of crystal disorder. Proper assessment requires access to the experimental data; and fixing real errors may well require the attention of an experienced crystallographer. The conclusion seems inescapable that structure factors should be archived and available.

Nuclear Magnetic Resonance (NMR)

NMR is the second major technique for determining macromolecular structure. It produces structures that are generally correct in topology but not as precise as a good X-ray structure determination and therefore less useful for the study of fine structural details. Crystallographers report a single structure, or only a small number. NMR spectroscopists usually produce a family of ~10–20 related structures or even more, calculated from the same experimental data. Comparison across such an ensemble indicates precision; regions in which the local variation in structure is small are well defined by the data. This is a rough equivalent of the crystallographer's B-factor.

Web resources: Protein and nucleic acid structures

Home page of Protein Data Bank:
http://www.rcsb.org

Home page of EBI Macromolecular Structure Database:
http://msd.ebi.ac.uk/

Home page of BioMagResBank:
http://www.bmrb.wisc.edu/

Searching the Protein Data Bank:

Home page of SCOP (Structural Classification of Proteins):
http://scop.mrc-lmb.cam.ac.uk/scop/

List of browsers: http://pdb-browsers.ebi.ac.uk/browse_it.shtml

OCA: http://oca.ebi.ac.uk/oca-bin/ocamain

Database of Protein Quaternary Structure:
http://pqs.ebi.ac.uk/

Reports of structure quality:
http://www.cmbi.kun.nl/gv/pdbreport

Classifications of protein structures

Several web sites offer hierarchical classifications of the entire Protein Data Bank according to the folding patterns of the proteins (see Chapter 5):

- SCOP: Structural Classification of Proteins
- CATH: Class/Architecture/Topology/Homologous superfamily?
- DALI: based on extraction of similar structures from distance matrices
- CE: a database of structural alignments

These sites are useful general entry points to protein structural data. For instance, SCOP offers facilities for searching on keywords to identify structures, navigation up and down the hierarchy, generation of pictures, access to the annotation records in the PDB entries, and links to related databases.

Specialized, or 'boutique' databases

Many individuals or groups select, annotate, and recombine data focused on particular topics, and include links affording streamlined access to information about subjects of interest.

For instance, the protein kinase resource is a specialized compilation that includes sequences, structures, functional information, laboratory procedures, lists of interested scientists, tools for analysis, a bulletin board, and links.

The HIV protease database archives structures of Human Immunodeficiency Virus 1 proteinases, Human Immunodeficiency Virus 2 proteinases, and Simian Immunodeficiency Virus Proteinases, and their complexes; and provides tools for their analysis and links to other sites with AIDS-related information. This database contains some crystal structures not deposited in the PDB.

In the field of immunology:

- IMGT, the international ImMunoGeneTics database, is a high-quality integrated database specializing in Immunoglobulins, T-cell receptors and Major Histocompatibility Complex (MHC) molecules of all vertebrate species. The IMGT server provides a common access to all Immunogenetics data. At present, it includes two databases: IMGT/LIGM-DB, a comprehensive database of immunoglobulin and T-cell receptor gene sequences from human and other vertebrates, with translation for fully annotated sequences; and IMGT/HLA-DB, a database of the human MHC referred to as HLA (Human Leucocyte Antigens).
- KABAT—Database of Sequences of Proteins of Immunological Interest—North-Western University (USA)
- MHCPEP—Major Histocompatibility Complex Binding Peptides Database—WEHI (Melbourne, Australia)

Web resources: Databases for specific protein families

Protein kinases:
http://www.sdsc.edu/kinases/

HIV proteases:
http://www-fbsc.ncifcrf.gov/HIVdb/

Immunology:
IGMT: http://imgt.cines.fr
KABAT: http://immuno.bme.nwu.edu/
MHCPEP: http://wehih.wehi.edu.au/mhcpep/

Expression and proteomics databases

Recall the central dogma: DNA makes RNA makes protein. Genomic databases contain DNA sequences. Expression databases record measurements of *mRNA* levels, usually via ESTs (expressed sequence tags-short terminal sequences of cDNA synthesized from mRNA) describing patterns of gene transcription. Proteomics databases record measurements on *proteins*, describing patterns of gene translation.

Comparisons of expression patterns give clues to: (1) the function and mechanism of action of gene products, (2) how organisms coordinate their control over metabolic processes in different conditions—for instance yeast under aerobic or anaerobic conditions, (3) the variations in mobilization of genes at different stages of the cell cycle, or of the development of an organism, (4) mechanisms of antibiotic resistance in bacteria, and consequent suggestion of targets for drug development (5) the response to challenge by a parasite, (6) the response to medications of different types and dosages, to guide effective therapy.

There are many databases of ESTs. In most, the entries contain fields indicating tissue of origin and/or subcellular location, state of development, conditions of growth, and quantitation of expression level. Within GenBank the dbEST collection currently contains almost 23 million entries, from 719 species, led by:

Species with largest number of entries in dbEST

Species	Number of entries
Homo sapiens (human)	5 654 825
Mus musculus + *domesticus* (mouse)	4 235 142
Ciona intestinalis (primitive chordate)	684 280
Rattus sp. (rat)	636 658
Triticum aestivum (wheat)	559 149
Danio rerio (zebrafish)	532 545
Gallus gallus (chicken)	494 605
Bos taurus (cattle)	465 743
Zea mays (maize)	415 211
Xenopus tropicalis	392 901
Xenopus laevis (African clawed frog)	385 714
Drosophila melanogaster (fruit fly)	382 439
Hordeum vulgare + subsp. *vulgare* (barley)	356 856
Glycine max (soybean)	334 668
Sus scrofa (pig)	328 573
Arabidopsis thaliana (thale cress)	322 641
Caenorhabditis elegans (nematode)	298 805
Oryza sativa (rice)	284 007

Some EST collections are specialized to particular tissues (e.g. muscle, teeth) or to species. In many cases there is an effort to link expression patterns to other knowledge of the organism. For instance, the Jackson Lab Gene Expression Information Resource Project for Mouse Development coordinates data on gene expression and developmental anatomy.

Many databases provide connections between ESTs in different species, for instance, linking human and mouse homologues, or relationships between human disease genes and yeast proteins. Other EST collections are specialized to a type of protein, for instance, cytokines. A large effort is focussed on cancer: integrating information on mutations, chromosomal rearrangements, and changes in expression patterns, to identify genetic changes during tumour formation and progression.

Although of course there is a close relationship between patterns of transcription and patterns of translation, direct measurements of protein contents of cells and tissues—proteomics—provides additional valuable information. Because of differential rates of translation of different mRNAs, measurements of proteins directly give a more accurate description of patterns of gene expression than measurements of transcription. Post-translational modifications can be detected *only* by examining the proteins.

Proteome analysis involves separation, identification, and determination of the quantitative amounts of proteins in a sample (see Chapter 6). Proteome databases store images of gels, and their interpretation in terms of protein patterns. For each protein, an entry typically records (see Weblem 3.21):

- identification of protein
- relative amount
- function
- mechanism of action
- expression pattern
- subcellular localization
- related proteins
- post-translational modifications
- interactions with other proteins
- links to other databases

Bioinformatics is contributing to the development of these databases, and also to the development of algorithms for comparing and analysing the patterns they contain.

Databases of metabolic pathways

The Kyoto Encyclopedia of Genes and Genomes (KEGG) collects individual genomes, gene products and their functions, but its special strengths lie in its integration of biochemical and genetic information. KEGG focuses on interactions: molecular assemblies, and metabolic and regulatory networks. It has been developed under the direction of M. Kanehisa.

KEGG organizes five types of data into a comprehensive system:

1. Catalogues of chemical compounds in living cells
2. Gene catalogues
3. Genome maps
4. Pathway maps
5. Orthologue tables

The catalogues of chemical compounds and genes—items 1 and 2—contain information about particular molecules or sequences. Item 3, genome maps, integrates the genes themselves according to their appearance on chromosomes. In some cases knowing that a gene appears in an operon can provide clues to its function.

Item 4, the pathway maps, describe potential networks of molecular activities, both metabolic and regulatory. A metabolic pathway in KEGG is an idealization corresponding to a large number of possible metabolic cascades. It can generate a real metabolic pathway of a particular organism, by matching the proteins of that organism to enzymes within the reference pathways.

One enzyme in one organism would be referred to in KEGG in its orthologue tables, item 5, which link the enzyme to related ones in other organisms. This permits analysis of relationships between the metabolic pathways of different organisms.

KEGG derives its power from the very dense network of links among these categories of information, and additional links to many other databases to which the system maintains access. Two examples of the kinds of questions that can be treated by KEGG are:

- It has been suggested that simple metabolic pathways evolve into more complex ones by gene duplication and subsequent divergence. Searching the pathway catalogue for sets of enzymes that share a folding pattern will reveal clusters of paralogues.
- KEGG can take the set of enzymes from some organism and check whether they can be integrated into known metabolic pathways. A gap in a pathway suggests a missing enzyme or an unexpected alternative pathway.

Bibliographic databases

MEDLINE (based at the US National Library of Medicine) integrates the biomedical literature, including very many papers dealing with subjects in molecular biology not overtly clinical in content. It is included in PubMed, a bibliographical database offering abstracts of scientific articles, integrated with other information retrieval tools of the National Center for Biotechnology Information (NCBI) within the National Library of Medicine (http://www.ncbi.nlm.nih.gov/PubMed/).

One very effective feature of PubMed is the option to retrieve *related articles*. This is a very quick way to 'get into' the literature of a topic. Combined with the use of a general search engine for web sites that do not correspond to articles published in journals, fairly comprehensive information is readily available about most subjects. Here's a tip: if you are trying to start to learn about an unfamiliar subject, try adding the keyword *tutorial* to your search in a general search engine, or the keyword *review* to your search in PubMed.

Almost all scientific journals now place their tables of contents, and in many cases their entire issues, on web sites. The US National Institutes of Health have established a centralized web-based library of scientific articles, called PubMed Central (http://www.pubmedcentral.nih.gov/). In collaboration with scientific journals, the NCBI is organizing the electronic distribution of the full texts of published articles.

A new organization, the Public Library of Science, has the goal of making the scientific (including medical) literature publicly and freely accessible. A non-profit organization, the Public Library of Science has received support from foundations for its efforts in distributing literature published by others, and to start its own publications, which will permit exploration of different relationships—including but not limited to economic ones—between authors, publishers and readers.

Surveys of molecular biology databases and servers

It is difficult to explore any topic in molecular biology on the web without quickly bumping into a list of this nature. Lists of web resources in molecular biology are very common. They contain, to a large extent, the same information, but vary widely in their 'look and feel' aspects. The real problem is that unless they are curated they tend to degenerate into lists of dead links. (A draft of this section contained a reference to a web site that contained a reasonable survey. Returning to it two months later, the name of the site had changed, and over half of the sites listed had disappeared.)

This book does not contain a long annotated list of relevant and recommended sites, for the following reasons: (1) You don't want a long list, you need a short one. (2) The Web is too volatile for such a list to stay useful for very long. *It is much more effective to use a general search engine to find what you want at the moment you want it.* Each year the January issue of the journal Nucleic Acids Research contains a set of articles on databases in molecular biology. This is an invaluable reference.

Moreover, the content of the databases is expanding all the time. If you try the searches described in examples in this chapter you will obtain more 'hits' than the results printed here. (Indeed, I have not hesitated to use older sets of results if, because they contain more variety than the latest results, they seem more informative. The problem of suppressing extensive redundancy in responses to websearches is a challenge for research in the field of information retrieval.)

My advice is: spend some time browsing; it won't take you long to find a site that appears reasonably stable and has a style compatible with your methods of work. Alternatively, here's a site that is comprehensive and shows signs of a commitment to keeping it up to date: `http://www.expasy.org/alinks.html`. It is a suitable site for starting a browsing session.

Gateways to archives

Databases of nucleic acid and protein sequences maintain facilities for a very wide variety of information retrieval and analysis operations. Categories of these operations include:

1. **Retrieval of sequences from the database** Sequences can be 'called up' either on the basis of features of the annotations, or by patterns found within the sequences themselves.
2. **Sequence comparison** This is not a facility, this is a heavy industry! It was introduced in Chapter 1 and will be discussed in detail in Chapter 4. It includes the very important searches for relatives.
3. **Translation of DNA sequences to protein sequences**
4. **Simple types of structure analysis and prediction** For example, statistical methods for predicting the secondary structure of proteins from sequences alone, including hydrophobicity profiles—from which the transmembrane proteins can generally be identified (see page 193).
5. **Pattern recognition** It is possible to search for all sequences containing a pattern or combination of patterns, expressed as probabilities for finding certain sets of residues at consecutive positions. In DNA sequences, these may be

recognition sites for enzymes such as those responsible for splicing interrupted genes. In proteins, short and localized patterns sometimes identify molecules that share a common function even if there is no obvious overall relationship between their sequences. PROSITE is a collection of these protein 'signature' patterns.

6. **Molecular graphics** is necessary to provide intelligible depictions of very complicated systems. Typical applications of molecular graphics include:
   - Mapping residues believed to be involved in function, onto the three-dimensional framework of a protein. Often this will isolate an active site.
   - Classifying and comparing the folding patterns of proteins.
   - Analysing changes between closely-related structures, or between two conformational states of a single molecule,
   - Studying the interaction of a small molecule with a protein, in order to attempt to assign function, or for drug development,
   - Interactive fitting of a model to the noisy and fuzzy image of the molecule that arises initially from the measurements in solving protein structures by X-ray crystallography.
   - Design and modelling of new structures.

Access to databases in molecular biology

How to learn web skills

It would be difficult to learn to ride a bicycle by reading a book describing the sets of movements required, much less one about the theory of the gyroscope. Similarly, the place to learn web skills is at a terminal, running a browser. True enough, but there is always a certain initial period of difficulty and imbalance. Here the goal is only to provide some temporary assistance to get you started. Then, off you go!

This section contains introductions to some of the major databanks and information retrieval systems in molecular biology. In each case we show relatively simple searches and applications. When appropriate, unique features of each system will be emphasized.

ENTREZ

The National Center for Biotechnology Information, a component of the United States National Library of Medicine, maintains databases and avenues of access to them. ENTREZ offers access via the following database divisions:

- Protein
- Peptide
- Nucleotide
- Structure
- Genome

- Popset—information about populations
- OMIM—Online Mendelian Inheritance in Man

Links between various databases are a strong point of NCBI's system. The starting point for retrieval of sequences and structures is called ENTREZ: http://www.ncbi.nlm.nih.gov/Entrez/.

Let us pick a molecule—human neutrophil elastase—and search for relevant entries in the different sections of ENTREZ.

Search in ENTREZ protein database

Go to http://www.ncbi.nlm.nih.gov/Entrez/. Select `Protein: sequence database`, enter the search terms HUMAN ELASTASE and click on GO.

The Box shows fifteen answers returned by the program. (In a browser, you will also find links to the sequence databank entries.) The top hit is ELASTASE 1 PRECURSOR [HOMO SAPIENS]; other responses include elastases from other species, inhibitors from human and from leech, and tyrosyl-tRNA synthetase. (Why should a leech protein and tRNA synthetase show up in a search for human elastase? See Weblem 3.9.) Later we shall see how to tune the query to eliminate these extraneous responses.

ENTREZ responses to *human elastase* in PROTEIN database

1. elastase 1 precursor [Homo sapiens]
   gi—4731318—gb—AAD28441.1—AF120493_1[4731318]
2. ALPHA-1-ANTITRYPSIN PRECURSOR (ALPHA-1 PROTEASE INHIBITOR) (ALPHA-1-ANTIPROTEINASE)
   gi—1703025—sp—P01009—A1AT_HUMAN[1703025]
3. elastase [Mus musculus]
   gi—7657060—ref—NP_056594.1—[7657060]
4. proteinase 3 [Mus musculus]
   gi—6755184—ref—NP_035308.1—[6755184]
5. ANTIMICROBIAL PEPTIDE ENAP-2
   gi—7674025—sp—P56928—ENA2_HORSE[7674025]
6. AMBP PROTEIN PRECURSOR [CONTAINS: ALPHA-1-MICROGLOBULIN (PROTEIN HC)
   (COMPLEX-FORMING GLYCOPROTEIN HETEROGENEOUS IN CHARGE);
   INTER-ALPHA-TRYPSIN INHIBITOR LIGHT CHAIN (ITI-LC) (BIKUNIN) (HI-30)]
   gi—122801—sp—P02760—AMBP_HUMAN[122801]
7. ELAFIN PRECURSOR (ELASTASE-SPECIFIC INHIBITOR) (ESI) (SKIN-DERIVED ANTILEUKOPROTEINASE) (SKALP)
   gi—119262—sp—P19957—ELAF_HUMAN[119262]
8. ANTILEUKOPROTEINASE
   gi—113637—sp—P22298—ALK1_PIG[113637]

→

→

9. ANTILEUKOPROTEINASE 1 PRECURSOR (ALP) (HUSI-1) (SEMINAL PROTEINASE
INHIBITOR) (SECRETORY LEUKOCYTE PROTEASE INHIBITOR) (BLPI) (MUCUS PROTEINASE INHIBITOR) (MPI)
gi—113636—sp—P03973—ALK1_HUMAN[113636]
10. ALPHA-2-MACROGLOBULIN PRECURSOR (ALPHA-2-M)
gi—112911—sp—P01023—A2MG_HUMAN[112911]
11. tyrosyl-tRNA synthetase [Homo sapiens]
gi—4507947—ref—NP_003671.1—[4507947]
12. pancreatic elastase IIB [Homo sapiens]
gi—7705648—ref—NP_056933.1—[7705648]
13. protease inhibitor 3, skin-derived (SKALP) [Homo sapiens]
gi—4505787—ref—NP_002629.1—[4505787]
14. pancreatic elastase I (allele HEL1-36)—human (fragment)
gi—7513237—pir—S70441[7513237]
15. guamerin—Korean leech

The format of the responses is as follows. In each case, the first line gives the name and synonyms of the molecule, and the species of origin. Note that Greek letters are spelt out. The last line gives references to the source databanks: gi = GenInfo Identifier, (see page 25), gb = GenBank accession number, sp = Swiss-Prot, pir = Protein Identification Resource, ref = the Reference Sequence project of NCBI. The entries retrieved include elastases from human and other species, and also inhibitors of elastase.

Opening the entry corresponding to the first hit retrieves the file shown in the next Box. The first lines are mostly database housekeeping—accession numbers, molecule name, date of deposition, etc. Then descriptive material such as the source, this case human, with the full taxonomic classification, credit to the scientists who deposited the entry, and literature references. Finally the particular scientific information: the location of the gene, and its product (CDS = coding sequence), and the sequence itself (see Exercise 3.2).

Searches in ENTREZ nucleotide database

We next look again for HUMAN ELASTASE, this time in the Nucleotide database. Let us try to tune the search, to eliminate the responses that refer to elastase inhibitors.

1. Select NUCLEOTIDE at the ENTREZ site.
2. Click on LIMITS, select ORGANISM from the pulldown menu, type HOMO SAPIENS in the search box.
3. Next select SUBSTANCE NAME from the pulldown menu, and then type AND ELASTASE in the search box.

Top result of search for human elastase in ENTREZ Protein database

```
LOCUS       AF120493_1    258 aa                    PRI       03-AUG-2000
DEFINITION  elastase 1 precursor [Homo sapiens].
ACCESSION   AAD28441
PID         g4731318
VERSION     AAD28441.1  GI:4731318
DBSOURCE    locus AF120493 accession AF120493.1
KEYWORDS    .
SOURCE      human.
  ORGANISM  Homo sapiens
            Eukaryota; Metazoa; Chordata; Craniata; Vertebrata; Euteleostomi;
            Mammalia; Eutheria; Primates; Catarrhini; Hominidae; Homo.
REFERENCE   1  (residues 1 to 258)
  AUTHORS   Talas,U., Dunlop,J., Khalaf,S., Leigh,I.M. and Kelsell,D.P.
  TITLE     Human elastase 1: evidence for expression in the skin and the
            identification of a frequent frameshift polymorphism
  JOURNAL   J. Invest. Dermatol. 114 (1), 165-170 (2000)
  MEDLINE   20087075
   PUBMED   10620133
REFERENCE   2  (residues 1 to 258)
  AUTHORS   Talas,U., Dunlop,J., Leigh,I.M. and Kelsell,D.P.
  TITLE     Direct Submission
  JOURNAL   Submitted (15-JAN-1999) Centre for Cutaneous Research, Queen Mary
            and Westfield College, 2 Newark Street, London E1 2AT, UK
COMMENT     Method: conceptual translation supplied by author.
FEATURES             Location/Qualifiers
     source          1..258
                     /organism="Homo sapiens"
                     /db_xref="taxon:9606"
                     /chromosome="12"
                     /map="12q13"
                     /cell_type="keratinocyte"
     Protein         1..258
                     /product="elastase 1 precursor"
     CDS             1..258
                     /gene="ELA1"
                     /coded_by="AF120493.1:42..818"
ORIGIN
        1 mlvlyghstq dlpetnarvv ggteagrnsw psqislqyrs ggsryhtcgg tlirqnwvmt
       61 aahcvdyqkt frvvagdhnl sqndgteqyv svqkivvhpy wnsdnvaagy diallrlaqs
      121 vtlnsyvqlg vlpqegaila nnspcyitgw gktktngqla qtlqqaylps vdyaicssss
      181 ywgstvkntm vcaggdgvrs gcqgdsggpl hclvngkysl hgvtsfvssr gcnvsrkptv
      241 ftqvsayisw innviasn
//
```

4. Finally select TEXT WORD from the pulldown menu, and then type NOT INHIBITOR in the search box. Now click on GO.

If you click on Details, you will find:

HOMO SAPIENS[ORGANISM] AND ELASTASE[SUBSTANCE NAME] NOT INHIBITOR [TEXT WORD]

The search returns over 400 hits, including many individual clones. The top hit (see Box) is: HOMO SAPIENS ELASTASE 1 PRECURSOR (ELA1) MRNA, COMPLETE CDS. The term 'complete cds' means complete coding sequence.

Compare this file with the result of searching in the Protein database (see Exercise 3.5).

Top result of search for human elastase in ENTREZ Nucleotide database

```
LOCUS       AF120493      952 bp    mRNA            PRI       03-AUG-2000
DEFINITION  Homo sapiens elastase 1 precursor (ELA1) mRNA, complete cds.
ACCESSION   AF120493
VERSION     AF120493.1  GI:4731317
KEYWORDS    .
SOURCE      human.
  ORGANISM  Homo sapiens
            Eukaryota; Metazoa; Chordata; Craniata; Vertebrata; Euteleostomi;
            Mammalia; Eutheria; Primates; Catarrhini; Hominidae; Homo.
REFERENCE   1  (bases 1 to 952)
  AUTHORS   Talas,U., Dunlop,J., Khalaf,S., Leigh,I.M. and Kelsell,D.P.
  TITLE     Human elastase 1: evidence for expression in the skin and the
            identification of a frequent frameshift polymorphism
  JOURNAL   J. Invest. Dermatol. 114 (1), 165-170 (2000)
  MEDLINE   20087075
   PUBMED   10620133
REFERENCE   2  (bases 1 to 952)
  AUTHORS   Talas,U., Dunlop,J., Leigh,I.M. and Kelsell,D.P.
  TITLE     Direct Submission
  JOURNAL   Submitted (15-JAN-1999) Centre for Cutaneous Research, Queen Mary
            and Westfield College, 2 Newark Street, London E1 2AT, UK
FEATURES             Location/Qualifiers
     source          1..952
                     /organism="Homo sapiens"
                     /db_xref="taxon:9606"
                     /chromosome="12"
                     /map="12q13"
                     /cell_type="keratinocyte"
     gene            1..952
                     /gene="ELA1"
     CDS             42..818
                     /gene="ELA1"
                     /codon_start=1
                     /product="elastase 1 precursor"
                     /protein_id="AAD28441.1"
                     /db_xref="GI:4731318"
                     /translation="MLVLYGHSTQDLPETNARVVGGTEAGRNSWPSQISLQYRSGGSR
                     YHTCGGTLIRQNWVMTAAHCVDYQKTFRVVAGDHNLSQNDGTEQYVSVQKIVVHPYWN
                     SDNVAAGYDIALLRLAQSVTLNSYVQLGVLPQEGAILANNSPCYITGWGKTKTNGQLA
                     QTLQQAYLPSVDYAICSSSSYWGSTVKNTMVCAGGDGVRSGCQGDSGGPLHCLVNGKY
                     SLHGVTSFVSSRGCNVSRKPTVFTQVSAYISWINNVIASN"
BASE COUNT      226 a    261 c    250 g    215 t
ORIGIN
        1 ttggtccaag caagaaggca gtggtctact ccatcggcaa catgctggtc ctttatggac
       61 acagcaccca ggaccttccg gaaaccaatg cccgcgtagt cggagggact gaggccggga
      121 ggaattcctg gccctctcag atttccctcc agtaccggtc tggaggttcc cggtatcaca
      181 cctgtggagg gacccttatc agacagaact gggtgatgac agctgctcac tgcgtggatt
      241 accagaagac tttccgcgtg gtggctggag accataacct gagccagaat gatggcactg
      301 agcagtacgt gagtgtgcag aagatcgtgg tgcatccata ctggaacagc gataacgtgg
      361 ctgccggcta tgacatcgcc ctgctgcgcc tggcccagag cgttaccctc aatagctatg
      421 tccagctggg tgttctgccc caggagggag ccatcctggc taacaacagt ccctgctaca
      481 tcacaggctg gggcaagacc aagaccaatg ggcagctggc ccagaccctg cagcaggctt
      541 acctgccctc tgtggactat gccatctgct ccagctcctc ctactggggc tccactgtga
      601 agaacaccat ggtgtgtgct ggtggagatg gagttcgctc tggatgccag ggtgactctg
      661 gggggccccct ccattgcttg gtgaatggca agtattctct ccatggagtg accagctttg
      721 tgtccagccg ggctgtaat gtctccagga agcctacagt cttcacccag gtctctgctt
      781 acatctcctg gataaataat gtcatcgcct ccaactgaac atttcctga gtccaacgac
      841 cttcccaaaa tggttcttag atctgcaata ggacttgcga tcaaaaagta aaacacattc
      901 tgaaagacta ttgagccatt gatagaaaag caaataaaac tagatataca tt
//
```

Searches in ENTREZ genome database

A search for HUMAN ELASTASE returns:

1. NC_000967 CAENORHABDITIS ELEGANS CHROMOSOME III[64] LCL—WORM_CHR_III
2. NC_001099 HOMO SAPIENS CHROMOSOME 19[19] REF—NC_001099—HSAP-19
3. NC_001065 HOMO SAPIENS CHROMOSOME 14[14] REF—NC_001065—HSAP-14
4. NC_001044 HOMO SAPIENS CHROMOSOME 11[11] REF—NC_001044—HSAP-11
5. NC_001008 HOMO SAPIENS CHROMOSOME 6[6] REF—NC_001008—HSAP-6

Why should a *C. elegans* protein appear in a search for human elastase? The entry NC_000967 is chromosome III of *C. elegans* in its entirety. Comments on one of the genes detected include:

```
gene="T07A5.1" /note="weak similarity with elastase (PIR accesssion number A406659)"
```

Many other genes in *C. elegans* are annotated with similarities to human proteins. However, although *C. elegans* does contain an elastase, this is *not* flagged as similar to human elastase, although it is a homologue.

Searches in ENTREZ structure database

Is the three-dimensional structure of human elastase known? Select the STRUCTURE database, from the choices to the left of the query box, and rerun the search. The program returns at least five answers:

1JK3	CRYSTAL STRUCTURE OF HUMAN MMP-12 (MACROPHAGE ELASTASE) AT TRUE ATOMIC RESOLUTION
1HAZ	SNAPSHOTS OF SERINE PROTEASE CATALYSIS: (C) ACYL-ENZYME INTERMEDIATE BETWEEN PORCINE PANCREATIC ELASTASE AND HUMAN BETA-CASOMORPHIN-7 JUMPED TO PH 9 FOR 1 MINUTE
1HAX	SNAPSHOTS OF SERINE PROTEASE CATALYSIS: (A) ACYL-ENZYME INTERMEDIATE BETWEEN PORCINE PANCREATIC ELASTASE AND HUMAN BETA-CASOMORPHIN-7 AT PH 5
1B0F	CRYSTAL STRUCTURE OF HUMAN NEUTROPHIL ELASTASE WITH MDL 101, 146
1QIX	PORCINE PANCREATIC ELASTASE COMPLEXED WITH HUMAN BETA- CASOMORPHIN-7

The designations 1JK3, 1HAZ, 1HAZ, 1BOF, and 1QIX are entry codes from the Protein Data Bank.

OOPS!—we may not realize it, but we have missed many useful entries. There are many elastase structures solved in complex with inhibitors, which we have asked the system to reject. Deleting NOT INHIBITORS and rerunning the query returns several more structures.

Searches in the bibliographic database PubMed

Perhaps it is time to look at what people have had to say about our molecule. Of course the literature on elastase is huge. A search in PubMed for HUMAN ELASTASE returns over 7500 entries. To prune the results, let us try to find citations to articles describing the role of elastase in disease. A search for HUMAN ELASTASE DISEASE returns over 1600 entries. What about specific elastase **mutants** related to human disease? A search for HUMAN ELASTASE DISEASE MUTATION returns more than 40 articles, in reverse chronological order. Here are 10 of them:

Hermans MH, Touw IP. Significance of neutrophil elastase mutations versus G-CSF receptor mutations for leukemic progression of congenital neutropenia. Blood. 2001 Apr 1;97(7):2185–6. No abstract available.

Li FQ, Horwitz M. Characterization of mutant neutrophil elastase in severe congenital neutropenia. J Biol Chem. 2001 Apr 27;276(17):14230–41.

Ye S. Polymorphism in matrix metalloproteinase gene promoters: implication in regulation of gene expression and susceptibility of various diseases. Matrix Biol. 2000 Dec;19(7):623–9. Review.

Dale DC, Person RE, Bolyard AA, Aprikyan AG, Bos C, Bonilla MA, Boxer LA, Kannourakis G, Zeidler C, Welte K, Benson KF, Horwitz M. Mutations in the gene encoding neutrophil elastase in congenital and cyclic neutropenia. Blood. 2000 Oct 1;96(7):2317–22.

McGettrick AJ, Knott V, Willis A, Handford PA. Molecular effects of calcium binding mutations in Marfan syndrome depend on domain context. Hum Mol Genet. 2000 Aug 12;9(13):1987–94.

Rashid MH, Rumbaugh K, Free in PMC, Passador L, Davies DG, Hamood AN, Iglewski BH, Kornberg A. Polyphosphate kinase is essential for biofilm development, quorum sensing, and virulence of Pseudomonas aeruginosa. Proc Natl Acad Sci U S A. 2000 Aug 15;97(17):9636–41.

Jormsjo S, Ye S, Moritz J, Walter DH, Dimmeler S, Zeiher AM, Henney A, Hamsten A, Eriksson P. Allele-specific regulation of matrix metalloproteinase-12 gene activity is associated with coronary artery luminal dimensions in diabetic patients with manifest coronary artery disease. Circ Res. 2000 May 12;86(9):998–1003.

Talas U, Dunlop J, Khalaf S, Leigh IM, Kelsell DP. Human elastase 1: evidence for expression in the skin and the identification of a frequent frameshift polymorphism. J Invest Dermatol. 2000 Jan;114(1):165–70.

Horwitz M, Benson KF, Person RE, Aprikyan AG, Dale DC. Mutations in ELA2, encoding neutrophil elastase, define a 21-day biological clock in cyclic haematopoiesis. Nat Genet. 1999 Dec;23(4):433–6.

Griffin MD, Torres VE, Grande JP, Kumar R. Vascular expression of polycystin. J Am Soc Nephrol. 1997 Apr;8(4):616–26.

There are references to a relation between mutations in neutrophil elastase and neutropenia—a low level of a type of white blood cells called neutrophils. To pursue this, we can look for elastase in the database of human genetic disease:

Online Mendelian Inheritance in Man (OMIM™)

OMIM is a database of human genes and genetic disorders. It was originally compiled by V. A. McKusick, M. Smith and colleagues and published on paper. The National Center for Biotechnology Information (NCBI) of the US National Library of Medicine has developed it into a database accessible from the Web, and introduced links to other archives of related information, including sequence databanks and the medical literature. OMIM is now well integrated with the NCBI information retrieval system ENTREZ. A related database, the OMIM Morbid Map, treats genetic diseases and their chromosomal locations.

The response to ELASTASE in a search of OMIM describes the results linking mutations in the gene to cyclic neutropenia.

The collection of results on elastase that we have assembled would support research on the system; for instance, we could map elastase mutants onto the structure of the molecule to see whether we could derive clues to the cause of cyclic neutropenia.

The Sequence Retrieval System (SRS)

SRS, originally developed by T. Etzold, is an integrated system for information retrieval from many different sequence databases, and for feeding the sequences retrieved into analytic tools such as sequence comparison and alignment programs.

SRS can search a total of 141 databases of protein and nucleotide sequences, metabolic pathways, 3D structures and functions, genomes, and disease and phenotype information (see Box). These include many small databases such as the Prosite and Blocks databases of protein structural motifs, transcription factor databases, and databases specialized to certain pathogens.

Some categories of databases searchable from SRS

Nucleotide Sequence	Literature
Uniprot	Mapping
Protein Function	Protein Structure
Enzymes	Metabolic Pathways
Mutation, SNP	Gene Ontology

In addition to the number and variety of databases to which it offers access, SRS offers tight links among the databases, and fluency in launching applications. A search in a single database component can be extended to a search in the complete network; that is, entries in all databases pertaining to a given protein can be found easily. Similarity searches and alignments can be launched directly, without saving the responses in an intermediate file.

In an SRS session, you begin by selecting one or more of the databases in which to search. The databases are grouped by category: nucleotide sequence-related, protein-related, etc. Then you can enter a set of query terms. As with ENTREZ, you may search for them either in all fields, or assign terms to categories. The program will respond with a set of entries containing your terms. As follow-on queries one might:

1. Examine one of the sequences identified by linking to the file retrieved.
2. Select one or more of the sequences identified and search other databases for related entries.
3. Launch an application, such as a secondary structure prediction or a multiple sequence alignment.

Other options on the search results page allow you to create and download reports on the selected matches. This might be simply a listing of the sequences, or the result of a more complex analysis of the results. Applying the multiple sequence alignment program CLUSTAL-W to the results produces an alignment such as appears in Plate III.

The Protein Identification Resource (PIR)

The PIR is an effective combination of a carefully curated database, information retrieval access software, and a workbench for investigations of sequences. The PIR also produces the Integrated Environment for Sequence Analysis (IESA). Think of this as an analysis package sitting on top of a retrieval system. Its functionality includes browsing, searching and similarity analysis, and links to other databases. Users may:

- Browse by annotations.
- Search selected text fields for different annotations, such as Superfamily, Family, Title, Species, Taxonomy group, Keywords and Domains.
- Analyse sequences using BLAST or FASTA Searches, Pattern Match, Multiple alignment.
- Global and Domain Search, and Annotation-sorted Search.
- View Statistics for Superfamily, Family, Title, Species, Taxonomy group, Keywords, Domains, Features.
- View Links to other databases, including PDB, COG, KEGG, WIT, and BRENDA.
- Select Specialized Sequence Groups such as Human, Mouse, Yeast and *E. coli* genomes.

The URLs for search of PIR by Text terms are:
In the US: http://www-nbrf.georgetown.edu/pirwww/search/textpsd.html
In Europe: http://www.mips.gsf.de

One feature of the PIR International system is the search for a specific peptide. Looking at the alignment of mammalian elastases in Plate III, we note at positions 220–228 a conserved motif: most of the sequences contain CNGDSGGPLN.

In the PIR, we can select PATTERN/PEPTIDE MATCH and search for exact matches for the subsequence CNGDSGGPLN giving 63 results.

Returning to the alignment table (Plate III), variations in the pattern appear in some molecules. The more general search for C[RNQF]GDSG[GS]PL[HNV], in which [XYZ] means a position containing either X or Y or Z, would pull out all the mammalian elastases in the alignment, plus a total of 82 sequences in all. Even these are not all the elastase homologues in the databank, as one could find by running a PSI-BLAST search for any of the sequences, or, remaining strictly within PIR, by looking up elastase in the PROT-FAM database. The pattern matches 20 families, all serine proteinases.

We are well on the way to generating a complete list of homologues.

ExPASy—Expert Protein Analysis System

ExPASy is the information retrieval and analysis system of the Swiss Institute of Bioinformatics, which (in collaboration with the European Institute of Bioinformatics) also produces the protein sequence databases SWISS-PROT and TrEMBL. TrEMBL contains translations of nucleotide sequences from the EMBL Data Library not yet fully integrated into SWISS-PROT.

Opening the main web page of ExPASy (`http://www.expasy.org`) and selecting SWISS-PROT and TrEMBL gives access to a set of information retrieval tools, including a link to SRS. There is also the option of searching SWISS-PROT directly. If we select FULL TEXT SEARCH and probe SWISS-PROT with the single term ELASTASE, we find ELNE_HUMAN, the real goal of our search, and around 150 other hits, including many inhibitors. One elastase homologue found is from the blood fluke: CERC_SCHMA. Both sequences are precursors; in the following alignment of these two sequences, upper case letters indicate the mature enzyme:

```
CERC_SCHMA  --msnrwrfvvvvtlftycltfervstwlIRSGEPVQHPAEFPFIAFLTTER-TMCTGSL 57
ELNE_HUMAN  mtlgrrlaclflacvlpalllggtalaseIVGGR-RARPHAWPFMVSLQLRGGHFCGATL 59
              :..*    :.:. ::.  *     . : * .*.     :*  :**:. *   .  :* .:*

CERC_SCHMA  VSTRAVLTAGHCVCSPLPVIRVSFLTLRNGDQQGIHHQPSGVKVAPGYMPSCMSARQRRP 117
ELNE_HUMAN  IAPNFVMSAAHCVAN——VNVRAVRVVLGAHNLSRREP——TRQVFAVQRIFENGYDP     111
            ::.. *::*.***..  :.*  : : * ::  :::*    .  :  . :  .   *

CERC_SCHMA  IAQTLSGFDIAIVMLAQMVNLQSGIRVISLPQPSDIPPPGTGVFIVGYGRDDNDRDPSRK 177
ELNE_HUMAN  VNLLN--DIVILQLNGSATINANVQVAQLPAQGRRLGNGVQCLAMGWGLLGRNRG——    164
            :        **.*: *    ..:::.::*  .**   .      *. : :*:*  ..:*.

CERC_SCHMA  NGGILKKGRATIMECRHATNGNPICVKAGQNFGQLPAPGDSGGPLLPS-LQGPVLGVVSH 236
ELNE_HUMAN  IASVLQELNVTVVTS-LCRRSNVCTLVRGRQAG—VCFGDSGSPLVCNGLIHGIASFVRG  221
            ..:*::  ..*:: .    . ..*  :   *:: * .  ****.**: . *     : ..*

CERC_SCHMA  GVTLPNLPDIIVEYASVARMLDFVRSNI----------------   264
ELNE_HUMAN  GCASGLYPDAFAPVAQFVNWIDSIIQRSEDNPCPHPRDPDPASRTH 267
            * :   ** :. *.... :* : ..
```

The structure of human neutrophil elastase is known from X-ray crystallography, but that of the blood fluke elastase is not.

One of the unique facilities of the ExPASy server is the link to SWISS-MODEL, an automatic web server for building homology models. Opening SWISS-MODEL and choosing FIRST APPROACH MODE (the simplest), we can simply enter the

SWISS-PROT code CERC_SCHMA, and launch the application. Model building is not a trivial operation, so the job is done off-line and the results sent by e-mail.

We shall discuss SWISS-MODEL further in Chapter 5.

Ensembl

Ensembl (http://www.ensembl.org) is intended to be the universal information source for the human genome. The goals are to collect and annotate all available information about human DNA sequences, link it to the master genome sequence, and make it accessible to the many scientists who will approach the data with many different points of view and different requirements. To this end, in addition to collecting and organizing the information, very serious effort has gone into developing computational infrastructure. Suitable conventions of nomenclature are established: it is not trivial to devise a scheme for maintaining stable identifiers in the face of data that will be undergoing not only growth but revision. The most visible result of these efforts is the web site, very rich in facilities both for browsing and for focussing in on details.

Ensembl is a joint project of the European Bioinformatics Institute and The Sanger Centre; participants include E. Birney, M. Clamp, T. Cox and T. J. P. Hubbard. However, Ensembl is organized as an open project, encouraging outside contributions. All but the most naive of readers must recognize the great demands this will place on quality control procedures.

Data collected in Ensembl includes genes, SNPs, repeats, and homologies. Genes may either be known experimentally, or deduced from the sequence. Because the experimental support for annotation of the human genome is so variable, Ensembl presents the supporting evidence for identification of every gene. Very extensive linking to other databases containing related information, such as Online Mendelian Inheritance in Man (OMIM), or expression databases, extend the accessible information.

Ensembl is structured around the human genome sequence. Users may identify regions via several types of lookups or searches:

- BLAST searches on a sequence or fragment
- Browsing—starting at the chromosome level then zooming in
- Gene name
- Relation to diseases, via OMIM
- ENSEMBL ID if the user knows it
- General text search

A text search in Ensembl for BRCA1 produced the page displayed, showing the region around the BRCA1 locus. The upper frame shows a megabase, mapped to the q21.2 and q21.31 bands of chromosome 17. It reports markers, and assigned genes. The bottom frame shows a more detailed view. Note the control panels between the two frames that permit navigation and 'zooming'. The bottom frame shows a 0.1 megabase region, reporting many more details, including the detailed structure of the BRCA1 gene, and the SNPs observed.

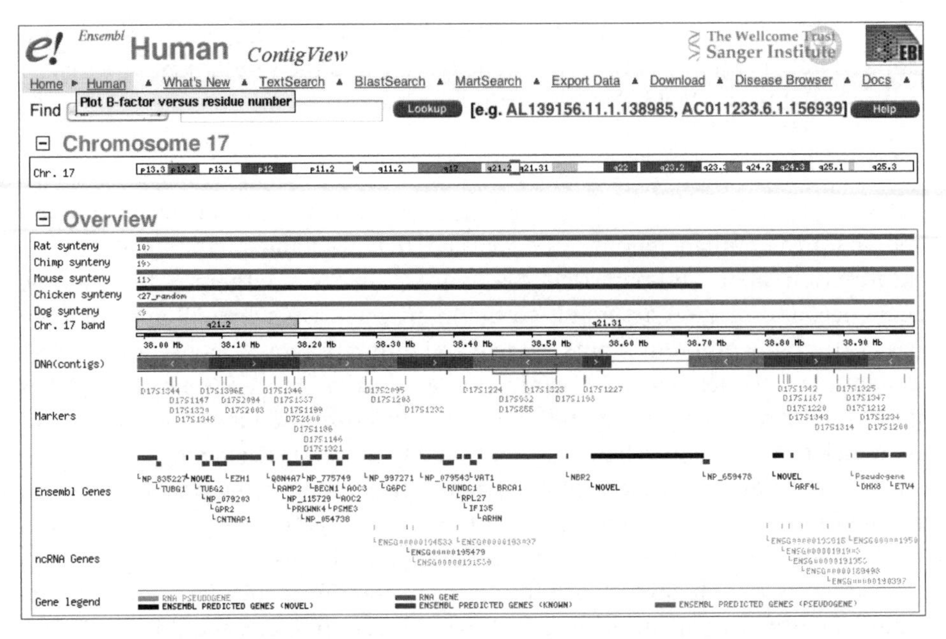

Fig. 3.4 A portion of the ENSEMBL page showing the region of the human genome surrounding BRAC1.

Where do we go from here?

We have visited only a few of the many databanks in molecular biology accessible on the Web. In the short term, readers will explore these sites and others, and become familiar with not only with the contents of the Web but its dynamics—the appearance and disappearance of sites and links. There are various biological metaphors for the Web—as an ecosystem that is evolving, or that is growing polluted by dead sites and links to dead sites. Unfortunately, there is no effective mechanism for decay and recycling as in the organic world!

Databanks are developing more effective avenues of intercommunication, to the point where ever more intimate links shade into apparent coalescence. The time is not far off when there will be one molecular biology databank, with many avenues of access. Scientists will be able to configure their own access to selected slices of the information, creating 'virtual databases' tailored to their own needs.

Recommended Reading

Each year the January issues of the journal *Nucleic Acid Research* contains a set of articles on databases in molecular biology. This is an invaluable reference.

Bishop, M. J., *Genetics databases* (London: Academic, 1999). [A compendium of databases, access and analysis.]

Zdobnov, E. M., Lopez, R., Apwelier, R. & Etzold, T. (2002), The EBI SRS server—new features, *Bioinformatics*, **18**, 1149–1150.

Exercises, Problems, and Weblems

Exercises

3.1 A database of vehicles has entries for the following: bicycle, tricycle, motorcycle, car. It stores only the following information about each entry: (1) how many wheels (a number), and (2) source of propulsion = human or engine. For every possible pair of vehicles, devise a logical combination of query terms referring to either the exact value or the range in the number of wheels, and to the source of propulsion, that will return the two selected vehicles and no others.

3.2 The Box on page 144 showed the NCBI protein entry for human elastase 1 precursor. On a photocopy of this page, indicate which items are (a) purely database housekeeping, (b) peripheral data such as literature references, (c) the results of experimental measurements, (d) information inferred from experimental measurements.

3.3 Write a PERL script to extract the amino acid sequence from an entry in the ENTREZ protein sequence database as shown in the Box, page 144, and convert it to FASTA format.

3.4 Compare the file retrieved by a search in NCBI for human elastase under Protein (page 144) and Nucleotide (page 145). On photocopies of these two pages, mark with a highlighter all items that the two files have in common.

Weblems

3.1 Retrieve the complete SWISS-PROT entry for bovine pancreatic trypsin inhibitor (*not* pancreatic secretory trypsin inhibitor) and the complete PIR entry for this protein. What information does each have that the other does not?

3.2 Find a list of official and unofficial mirror sites of the Protein Data Bank. Which is closest to you?

3.3 Find all structures of sperm whale myoglobin in the Protein Data Bank and draw a histogram of their dates of deposition.

3.4 Find protein structures determined by Peter Hudson, alone or with colleagues.

3.5 Design a search string for use with the Protein Data Bank tool SearchLite that would return *E. coli* thioredoxin structures but *not* Staphylococcal nuclease structures.

3.6 For what fraction of structures determined by X-ray crystallography deposited in the Protein Data Bank have structure factor files also been deposited?

3.7 Protein Data Bank entry 8XIA contains the structure of one monomer of D-Xylose isomerase from *Streptomyces rubiginosus*. What is the probable quaternary structure? How was the geometry of the assembly corresponding to the probable quaternary structure derived from the coordinates in the entry?

3.8 Find structural neighbours of Protein Data Bank entry 2TRX (*E. coli* thioredoxin), according to SCOP, CATH, FSSP, and CE. Which, if any, structures do *all* these classifications consider structural neighbours of 2TRX? Which structures are considered structural neighbours in some but not all classifications?

3.9 Why did an ENTREZ search in the protein category for HUMAN ELASTASE return a tRNA synthetase?

3.10 The Box on page 154 contains the amino acid sequence of human elastase 1 precursor. What sequence differences are there between this and the mature protein?

3.11 What is the relation between the elastase sequences recovered from searching the NCBI and the PIR?

3.12 Using SWISS-PROT directly, or SRS, recover the SWISS-PROT entry for human elastase. What information does this file contain that does not appear in (a) the corresponding entry in ENTREZ (protein) and (b) the corresponding entry in PIR?

3.13 What homologues of human neutrophil elastase can be identified by PSI-BLAST?

3.14 Search for structures of elastases using the Protein Data Bank search facilities. Compare the results with those from ENTREZ, described in the text.

3.15 Which gene in *C. elegans* encodes a protein similar in sequence to human elastase?

3.16 What is the chromosomal location of the human gene for glucose-6-phosphate dehydrogenase?

3.17 Pseudogenes in eukaryotes can be classified into those that arose by gene duplication and divergence, and those reinserted into the genome from mRNA by a retrovirus, called *processed* pseudogenes. Processed pseudogenes can be identified by the absence of introns. Which if any of the pseudogenes in the human globin gene clusters are processed pseudogenes?

3.18 Preliminary genetic analysis on the way to isolating the gene associated with cystic fibrosis bracketed it between the MET oncogene and RFLP D7S8. It was then estimated that this region contained 1–2 million bp, and might contain 100–200 genes. (a) How many base pairs long did this region actually turn out to be? (b) How many expressed genes is this region now believed to contain?

3.19 The gene for Berardinelli-Seip syndrome was initially localized between two markers on chromosome band 11q13—D11S4191 and D11S987. How many base pairs are there in the interval between these two markers?

3.20 Is there a database available on the Web that specifically collects structural and thermodynamic information on protein-nucleic acid interactions?

3.21 The Yeast proteome database contains an entry for cdc6, the protein that regulates initiation of DNA replication. (a) On what chromosome is the gene for yeast cdc6? (b) What post-translational modification does this protein undergo to reach its mature active state? (c) What are the closest known relatives of this protein in other species? (d) With what other proteins is yeast cdc6 known to interact? (e) What is the effect of distamycin A on the activity of yeast cdc6? (f) What is the effect of actinomycin D on the the activity of yeast cdc6?

CHAPTER 4

Alignments and phylogenetic trees

Chapter contents

Introduction to sequence alignment *158*

The dotplot *160*

Dotplots and sequence alignments *165*

Measures of sequence similarity *171*

Scoring schemes *171*

Computing the alignment of two sequences *175*

Variations and generalizations *175*

Approximate methods for quick screening of databases *176*

The dynamic programming algorithm for optimal pairwise sequence alignment *176*

Significance of alignments *182*

Multiple sequence alignment *186*

Applications of multiple sequence alignments to database searching *188*

Profiles *189*

PSI-BLAST *191*

Hidden Markov Models *193*

Phylogeny *198*

Phylogenetic trees *203*

Clustering methods *205*

Cladistic methods *206*

The problem of varying rates of evolution *207*

Computational considerations *208*

Recommended reading 209

Exercises, Problems, and Weblems 210

Learning goals

1. To understand the concept of sequence alignment: the assignment of residue-residue correspondences.
2. To know how to construct and interpret dotplots, and the relationship between dotplots and alignments.
3. To be able to define and distinguish the Hamming distance and Levenshtein distance as measures of dissimilarity of character strings.
4. To understand the basis of scoring schemes for string alignment, including substitution matrices and gap penalties.
5. To appreciate the difference between global alignments and local alignments, and to understand the use of approximate methods for quick screening of databases.
6. To understand the significance of Z-scores, and to know how to interpret *P*-values and *E*-values returned by database searches.
7. To be able to interpret multiple alignments of amino acid sequences, and to make inferences from multiple sequence alignments about protein structures.
8. To be able to define and distinguish the concepts of homology, similarity, clustering, and phylogeny.
9. To become expert in the use of PSI-BLAST and related programs.
10. To appreciate the use of profile methods and Hidden Markov Models in database searching.
11. To understand the contents and significance of phylogenetic trees, and the methods available for deriving them, including maximum parsimony and maximum likelihood; to know the role and use of an outgroup in derivation of a phylogenetic tree.

Introduction to sequence alignment

Given two or more sequences, we initially wish to:

- measure their similarity
- determine the residue-residue correspondences
- observe patterns of conservation and variability
- infer evolutionary relationships

If we can do this, we will be in a good position to go fishing in databanks for related sequences. A major application is to the annotation of genomes, involving assignment of structure and function to as many genes as possible.

How can we define a quantitative measure of sequence similarity? To compare the nucleotides or amino acids that appear at corresponding positions in two or more sequences, we must first assign those correspondences. *Sequence alignment is the identification of residue-residue correspondences.* It is *the* basic tool of bioinformatics.

Any assignment of correspondences that preserves the order of the residues within the sequences is an alignment. Gaps may be introduced.

Given two text strings: first string = `a b c d e`

second string = `a c d e f`

a reasonable alignment would be:

```
a b c d e -
a - c d e f
```

We must define criteria so that an algorithm can choose the *best* alignment. For the sequences `gctgaacg` and `ctataatc`:

An uninformative alignment:

```
- - - - - - - g c t g a a c g
c t a t a a t c - - - - - - -
```

An alignment without gaps:

```
g c t g a a c g
c t a t a a t c
```

An alignment with gaps:

```
g c t g a - a - - c g
- - c t - a t a a t c
```

And another:

```
g c t g - a a - c g
- c t a t a a t c -
```

Most readers would consider the last of these alignments the best of the four. To decide whether it is the best of *all* possibilities, we need a way to examine all possible alignments systematically. Then we need to compute a score reflecting the quality of each possible alignment, and to identify the alignment with the optimal score. The optimal alignment may not be unique: several different alignments may give the same best score. Moreover, even minor variations in the scoring scheme may change the ranking of alignments, causing a different one to emerge as the best.

These examples illustrate **pairwise sequence alignments**. However, usually we can find large families of similar sequences, by identifying homologues in different species. A mutual alignment of more than two sequences is called a **multiple sequence alignment**. Multiple sequence alignments are much more informative than pairwise sequence alignments, in terms of revealing patterns of conservation.

The dotplot

The dotplot is a simple picture that gives an overview of pairwise sequence similarity. Less obvious is its close relationship to alignments.

The dotplot is a table or matrix. The rows correspond to the residues of one sequence and the columns to the residues of the other sequence. In its simplest form, the positions in the dotplot are left blank if the residues are different, and filled if they match. Stretches of similar residues show up as diagonals in the upper left–lower right (Northwest–Southeast) direction.

Example 4.1

Dotplot showing identities between short name (DOROTHYHODGKIN) and full name (DOROTHYCROWFOOTHODGKIN) of a famous protein crystallographer.

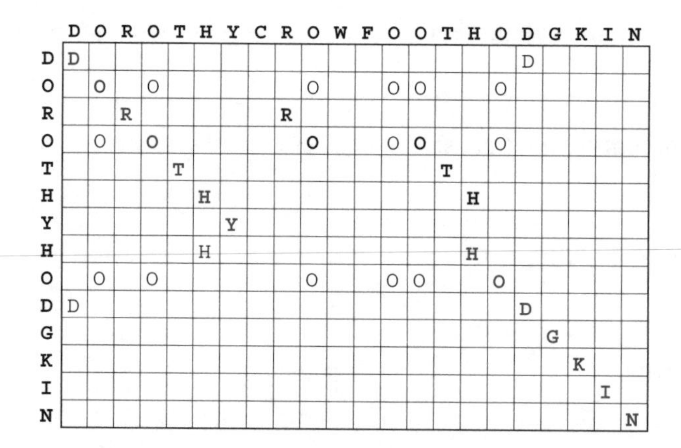

	D	O	R	O	T	H	Y	C	R	O	W	F	O	O	T	H	O	D	G	K	I	N
D	D																	D				
O		O		O						O			O	O			O					
R			R						R													
O		O		O						O			O	O			O					
T					T										T							
H						H										H						
Y							Y															
H						H										H						
O		O		O						O			O	O			O					
D	D																	D				
G																			G			
K																				K		
I																					I	
N																						N

Letters corresponding to *isolated* matches are shown in non-bold type. The longest matching regions, shown in red, are the first and last names DOROTHY and HODGKIN. Shorter matching regions, such as the OTH of dorOTHy and crowfoOTHodgkin, or the RO of doROthy and cROwfoot, are noise.

Example 4.2

Dotplot showing identities between a repetitive sequence (ABRACADABRA-CADABRA) and itself. The repeats appear on several subsidiary diagonals parallel to the main diagonal.

	A	B	R	A	C	A	D	A	B	R	A	C	A	D	A	B	R	A
A	A			A		A		A			A		A		A			A
B		B							B							B		
R			R							R							R	
A	A			A		A		A			A		A		A			A
C					C							C						
A	A			A		A		A			A		A		A			A
D							D							D				
A	A			A		A		A			A		A		A			A
B		B							B							B		
R			R							R							R	
A	A			A		A		A			A		A		A			A
C					C							C						
A	A			A		A		A			A		A		A			A
D							D							D				
A	A			A		A		A			A		A		A			A
B		B							B							B		
R			R							R							R	
A	A			A		A		A			A		A		A			A

Example 4.3

Dotplot showing identities between the palindromic sequence MAX I STAY AWAY AT SIX AM and itself. The palindrome reveals itself as a stretch of matches *perpendicular* to the main diagonal.

	M	A	X	I	S	T	A	Y	A	W	A	Y	A	T	S	I	X	A	M
M	M																		M
A		A					A		A		A		A					A	
X			X														X		
I				I												I			
S					S										S				
T						T								T					
A		A					A		A		A		A					A	
Y								Y				Y							
A		A					A		A		A		A					A	
W										W									
A		A					A		A		A		A					A	
Y								Y				Y							
A		A					A		A		A		A					A	
T						T								T					
S					S										S				
I				I												I			
X			X														X		
A		A					A		A		A		A					A	
M	M																		M

This is not just word play—regions in DNA recognized by transcription regulators or restriction enzymes have sequences related to palindromes, crossing from one strand to the other:

→

Example 4.3 *(continued)*

EcoRI recognition site: GAATTC

CTTAAG

Within each strand a region is followed by its reverse complement (see Exercise 4.9 and Problem 4.8). Longer regions of DNA or RNA containing inverted repeats of this form can form stem-loop structures. In addition, some transposable elements in plants contain true (approximate) palindromic sequences—inverted repeats of noncomplemented sequences, on the same strand. The following example appears in the Wheat dwarf virus genome: `ttttcgtgagtgcgcggaggctttt`.

The dotplot gives a quick pictorial statement of the relationship between two sequences. Obvious features of similarity stand out. For example, a dotplot relating the mitochondrial ATPase-6 genes from a lamprey (*Petromyzon marinus*) and dogfish shark (*Scyliorhinus canicula*) shows that the similarity of the sequences is weakest near the beginning. This gene codes for a subunit of the ATPase complex. In the human, mutations in this gene cause Leigh syndrome, a neurological disorder of infants produced by the effects of impaired oxidative metabolism on the brain during development.

A disadvantage of the dotplot is that its 'reach' into the realm of distantly-related sequences is poor. In analysing sequences, one should always look at a dotplot to be sure of not missing anything obvious, but be prepared to apply more subtle tools.

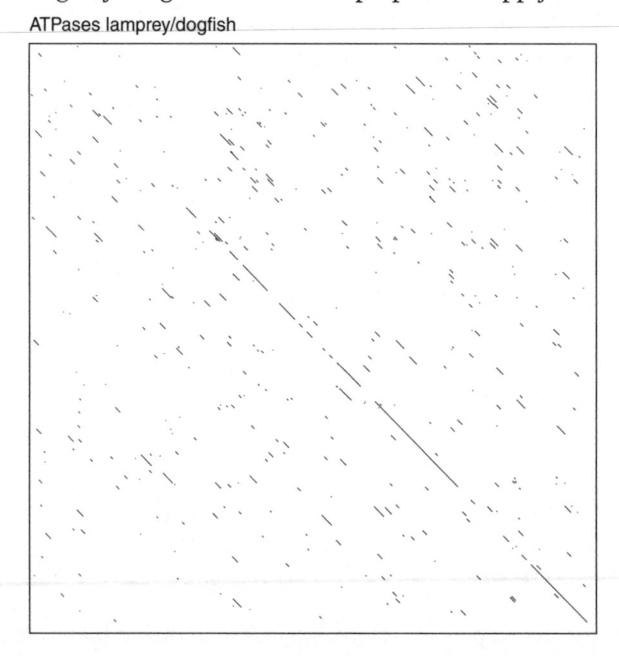

Often regions of similarity are displaced, to appear on parallel but not collinear diagonals. This indicates that insertions or deletions have occurred in the segments between the similar regions. A dotplot relating the PAX-6 protein of mouse and the eyeless protein of *Drosophila melanogaster* shows three extended regions of similarity with different lengths of sequence between them. Two of the regions are near the beginning of the sequences and one is near the middle. The section between the second and third regions of similarity is longer in the mouse sequence than in the *Drosophila* sequence.

Filtering the results can reduce the noise in a dotplot. In the comparison of the ATPase sequences, dots were not shown unless they were at the centre of a consecutive region or 15 residues containing at least 6 matches. The PERL program for dotplots (see Box) allows the user to set values for a **window** (length of region of consecutive residues) and a **threshold** (number of matches required within the window).

Web resources: Dotplots

E. L. Sondhammer's program Dotter computes and displays dotplots. It allows the user to control the calculation and alter the appearance of the display by adjusting parameters interactively.
http://www.cgb.ki.se/cgb/groups/sonnhammer/Dotter.html
To use the full set of features of Dotter it is necessary to install it locally.

A web site that offers interactive dotplotting is:
http://www.isrec.isb-sib.ch/java/dotlet/Dotlet.html

PERL Example 4.1 A program to draw dotplots

The program shown reads:

1. A general title for the job, printed at the top of the output drawing (first line of input).
2. Parameters specifying the filtering parameters *window* and *threshold* (second line of input). A dot will appear in the dotplot if it is in the centre of a stretch of residues of length *window* such that the number of matches is ≥ *threshold*.
3. The two sequences, each beginning with a title line and ending with a *.

The program draws a dotplot similar to those shown in the text. The output is in a graphical language called PostScript™, which can be displayed or printed on many devices.

```
#!/usr/bin/perl
#dotplot.pl -- reads two sequences and prints dotplot

# read input

$/ = "";
$_ = <DATA>; $_ =~ s/#(.*)\n/\n/g;
$_ =~ /^(.*)\n\s*(\d+)\s+(\d+)\s*\n(.*)\n([A-Za-z\n]*)\*\s*\n(.*)\n([A-Za-z\n]*)\*/
$title = $1; $nwind = $2; $thresh = $3;
$seqt1 = $4; $seq1 = $5; $seqt2 = $6; $seq2 = $7;
$seq1 =~ s/\n//g; $seq2 =~ s/\n//g; $n = length($seq1); $m = length($seq2);

# postscript header

print <<EOF;
%!PS-Adobe-
/s /stroke load def /l /lineto load def /m /moveto load def /r /rlineto load def
/n /newpath load def /c /closepath load def /f /fill load def
1.75 setlinewidth 30 30 translate /Helvetica findfont 20 scalefont setfont
EOF

#print matrix

$dx = 500.0/$n;  $mdx = -$dx; $dy = 500.0/$m;
if ($dy < $dx) {$dx = $dy;} $dy = $dx; $xmx = $n*$dx; $ymx = $m*$dx;
print "0 510 m ($title NWIND = $nwind) show\n";
printf "0 0 m 0 %9.2f l %9.2f %9.2f l %9.2f 0 l c s\n", $ymx,$xmx,$ymx,$xmx;

for ($k = $nwind - $m + 1; $k < $n - $nwind; $k++) {
    $i = $k; $j = 1; if ($k < 1) {$i = 1; $j = 2 - $k;}
    while ($i <= $n - $nwind && $j <= $m - $nwind) {
        $_ = (substr($seq1,$i -1,$nwind) ^ substr($seq2,$j -1,$nwind));
        $mismatch = ($_ =~ s/[^\x0]//g);
        if ($mismatch < $thresh) {
            $xl = ($i - 1)*$dx; $yb = ($m - $j)*$dy;
            printf "n %9.2f %9.2f m %9.2f 0 r 0 %9.2f r %9.2f 0 r c f\n",
                   $xl,$yb,$dx,$dy,$mdx;
        }
        $i++; $j++;
    }
}
print "showpage\n";

__END__
ATPases lamprey / dogfish                          #TITLE
15 6                                               #WINDOW, THRESHOLD
Petromyzon marinus mitochondrion                   #SEQUENCE 1
atgacactagatatctttgaccaatttacctccccaaca
atatttgggcttccactagcctgattagctatactagcccctagctta
```

```
atattagtttcacaaacaccaaaatttatcaaatctcgttatcacacacta
cttacacccatcttaacatctattgccaaacaactctttcttccaataaac
caacaagggcataaatgagccttaatttgtatagcctctataatatttatc
ttaataattaatcttttaggattattaccatatacttatacaccaactacc
caattatcaataaacataggattagcagtgccactatgactagctactgtc
ctcattgggttacaaaaaaaaccaacagaagccctagcccacttattacca
gaaggtaccccagcagcactcattcccatattaattatcattgaaactatt
agtctttttatccgacctatcgccctaggagtccgactaaccgctaattta
acagctggtcacttacttatacaactagtttctataacaacctttgtaata
attcctgtcatttcaatttcaattattacctcactacttcttctatta
ctaacaattctggagttagctgttgctgtaatccaggcatatgtatttatt
ctacttttaactctttatctgcaagaaaacgttt*
Scyliorhinus canicula mitochondrion        #SEQUENCE 2
atgattataagctttttgatcaattcctaagtccctcctttctagga
atcccactaattgccctagctatttcaattccatgattaatatttccaacaccaacc
aatcgttgacttaataatcgattattaactcttcaagcatgatttattaaccgatttatt
tatcaactaatacaacccataaatttaggaggacataaatgagctatcttatttacagcc
ctaatattattttaattaccatcaatcttctaggtctccttccatatacttttacgcct
acaactcaactttctcttaatatagcctttgccctgcccttatggcttacaactgtatta
attggtatatttaatcaaccaaccattgccctagggcacttattacctgaaggtacccca
accccttagtaccagtactaatcattatcgaaaccatcagtttatttattcgaccatta
gccttaggagtccgattaacagccaacttaacagctggacatctccttatacaattaatc
gcaactgcggcctttgtccttttaactataataccaaccgtggccttactaacctcccta
gtcctgttcctattgactattttagaagtggctgtagctataattcaagcatacgtattt
gtccttcttttaagcttatatctacaagaaaacgtataa*
```

Dotplots and sequence alignments

The dotplot captures in a single picture not only the overall similarity of two sequences, but also the complete set and relative quality of different possible alignments. Any path through the dotplot from upper left to lower right, moving at each point only East, South or Southeast, corresponds to a possible alignment. If two sequences are closely related, the alignment can be read directly off the dotplot.

Figure 4.1 shows an example based on the Dorothy Hodgkin dotplot. If the direction of the 'move' between successive cells is diagonal, two pairs of

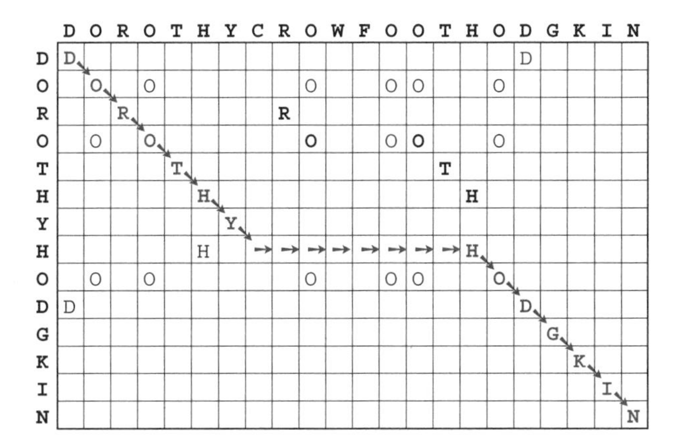

Fig. 4.1 Any path through the dotplot from upper left to lower right passes through a succession of cells, *each of which picks out a pair of positions, one from the row and one from the column, that correspond in the alignment; or that indicates a gap in one of the sequences.* The path need not pass through filled-in points only. However, the more filled-in points on the path, the more matching residues in the alignment.

successive residues appear in the alignment without an insertion between them. If the direction of the move is horizontal, a gap is introduced in the sequence indexing the rows. If the direction of the move is vertical (none in this example), a gap would be introduced in the sequence indexing the columns. Note that no moves can be directed up or to the left, as this would correspond to aligning several residues of one sequence with only one residue of the other. The path indicated by the arrows corresponds to the obvious alignment:

```
DOROTHY--------HODGKIN
DOROTHYCROWFOOTHODGKIN
```

Another way to think of a path through the dotplot is as an *edit script*; that is, the prescription of a series of operations that transforms the sequence that indexes the columns—the 'horizontal' sequence—into the sequence that indexes the rows—the 'vertical' sequence. Each move tells us to perform an operation—a substitution, an insertion, or a deletion. When the end of the path is reached, the effect will be to change one sequence into the other. In general, several different sequences of edit operations may convert one string into the other in the same number of steps, but they may induce different alignments.

It should be emphasized that although a sequence of edit operations derived from an optimal alignment *may* correspond to an actual evolutionary pathway, it is impossible to *prove* that it does. The larger the edit distance, the larger the number of reasonable evolutionary pathways between two sequences.

Example 4.4 Dotplots and alignments

Let us compare the appearance of dotplots between pairs of proteins with increasingly more distant relationships. Figure 4.2 shows the dotplot comparisons of the sulphydryl proteinase papain from papaya, with four homologues—the close relative, kiwi fruit actinidin, and more distant relatives, human procathepsin L, human cathepsin B, and *Staphyloccus aureus* Staphopain. The sequence alignments are also shown. As the sequences progressively diverge, it becomes more and more difficult to spot the correct alignment in the dotplot. The alignments shown were derived from comparisons of the structures.

→

→

```
ALIGNMENT OF 9pap     and 2act
SCORE = 5324  NPOS = 219 NIDENT = 102  %IDENT = 46.58

IPEYVDWRQKGAVTPVKNQGSCGSCWAFSAVVTIEGIIKIRTGNLNQYSEQELLDCDR--
 | |||||  |||   | || || ||||||  | ||| ||  | |   ||||| || |
LPSYVDWRSAGAVVDIKSQGECGGCWAFSAIATVEGINKITSGSLISLSEQELIDCGRTQ

RSYGCNGGYPWSALQ-LVAQYGIHYRNTYPYEGVQRYCRSREKGPYAAKTDGVRQVQPYN
   || |||     |      ||     |||      |            |    |   |
NTRGCDGGYITDGFQFIINDGGINTEENYPYTAQDGDCDVALQDQKYVTIDTYENVPYNN

QGALLYSIANQPVSVVLQAAGKDFQLYRGGIFVGPCGNKVDHAVAAVGYGP----NYILI
  ||      ||||| | |||  |  |  ||| ||||  ||||   ||||      |
EWALQTAVTYQPVSVALDAAGDAFKQYASGIFTGPCGTAVDHAIVIVGYGTEGGVDYWIV

KNSWGTGWGENGYIRIKRGTGNSYGVCGLYTSSFYPVKN
|||| | ||| || || |  |   | ||  |   ||||
KNSWDTTWGEEGYMRILRNVGGA-GTCGIATMPSYPVKY
```

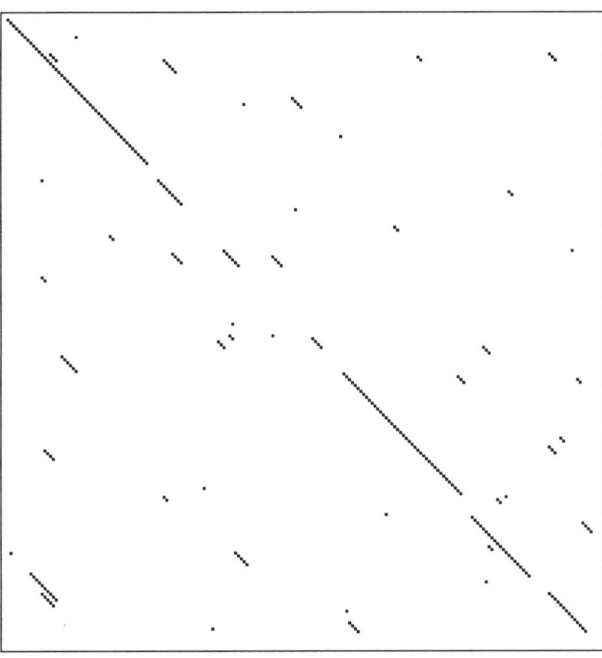

Fig. 4.2a Alignment of papaya papain and kiwi fruit actinidin, with the corresponding dotplot.

→

Example 4.4 (*continued*)

```
ALIGNMENT OF 9pap     and 1cjl

SCORE = 3214 NPOS = 220 NIDENT  = 81  %IDENT = 36.82

IPEYVDWRQKGAVTPVKNQGSCGSCWAFSAVVTIEGIIKIRTGNLNQYSEQELLDCD--R
     ||| || |||||||| ||| |||||    ||     || |   ||| | ||
V----DWREKGYVTPVKNQGQCGSSWAFSATGALEGQMFRKTGRLISLSEQNLVDCSGPE

RSYGCNGGYPWSALQLVAQY-GIHYRNTYPYEGVQRYCRSREKGPYAAKTDGVRQVQPYN
   |||||    | | |    |      ||||     |    |    |   |
GNEGCNGGLMDYAFQYVQDNGGLDSEESYPYEATEESCKYNPKYS-VANDAGFVDIPKQE

QGALLYSIANQPVSVVLQAAGKDFQLYRGGIFVGP--CGNKVDHAVAAVGYG---PNYIL
           | ||    |    |  |  ||    |       || |  ||||     | |
KALMKAVATVGPISVAIDAGHESFLFYKEGIYFEPDCSSEDMDHGVLVVGYGFESNKYWL

IKNSWGTGWGENGYIRIKRGTGNSYGVCGLYTSSFYPVKN
 |||||  ||  ||         |    ||      ||
VKNSWGEEWGMGGYVKMAKDRRN-H--CGIASAASYPTV-
```

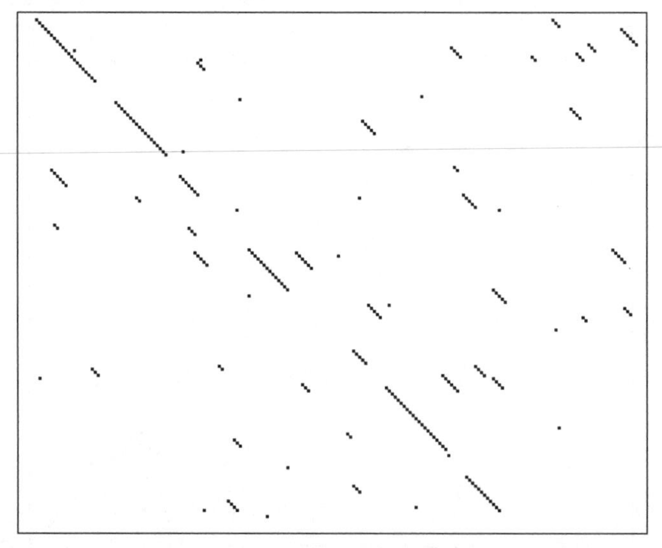

Fig. 4.2b Alignment of papaya papain and human procathepsin L, with the corresponding dotplot. This dotplot shows that there are several similar regions, but it would be difficult to generate a complete sequence alignment from the dotplot.

```
ALIGNMENT OF 9pap     and 1huc

SCORE   =  2073  NPOS = 251  NIDENT  = 66  %IDENT = 26.29

IPEYVD-WRQKGAVTPVKNQGSCGSCWAFSAVVTIEGIIKIRTGNLNQYSEQELLD-C-D
       | |         ||||||||||  ||  |   | | |      |   ||  |
--DAREQWPQCPTIKEIRDQGSCGSCWAFGAVEAISDRICIHTNVSVEVSAEDLLTCCGS

RRSYGCNGGYP------WSALQLVAQYGI--HYRN-TY-----P--YEGVQRYCRSREKG
    |||||||      |    ||       |     |          |    |
MCGDGCNGGYPAEAWNFWTRKGLVSGGLYESHVGCRPYSIPPCEHHVNGSRPPCTGEGDT

PYAAK------TDGVRQVQPYNQGALLYSIANQPVSV-V-----LQ---AAGKDFQLYRG
|   |           |   |       |                        || ||
PKCSKICEPGYSPTYKQDKHYGYNSYSVSNSEKDIMAEIYKNGPVEGAFSVYSDFLLYKS

GIFVGPCGNKV-DHAVAAV--GY--GPNYILIKNSWGTGWGENGYIRIKRGTGNSYGVCG
|      |     ||      |   |  | |  ||| | || ||   | ||     |
GVYQHVTGEMMGGHAIRILGWGVENGTPYWLVANSWNTDWGDNGFFKILRGQ-DHCGIES

LYTSSFYPVKN
       |
EVVAGI-PRTD
```

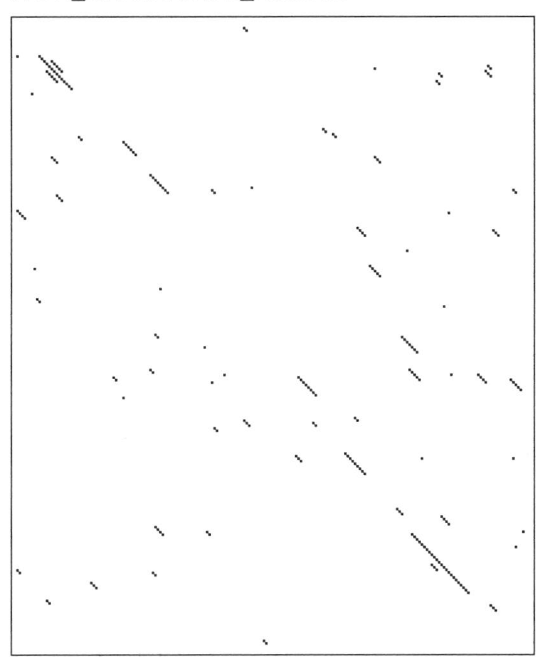

Fig. 4.2c Alignment of papaya papain and human liver cathepsin B, with the corresponding dotplot. Note, in *both* the sequence alignment and the dotplot, the higher similarity at the beginning and end of the sequences than in the middle region.

Example 4.4 (*continued*)

```
ALIGNMENT OF 9pap     and 1cv8

SCORE = -290  NPOS = 219  NIDENT  = 25 %IDENT = 11.42

IPEYVDWRQKGAVTPVKNQGSCGSCWAFSAVVTIEGIIKIRTGNLNQYSEQELLDCDRRS
                                                    |     |
----------------------------------------------EQYVNKLENFKIRE

YGCNGGYPWSALQLVAQYGIHYRNTYPYEGVQRYCRSREKG-PYAAKTDGVRQVQPY---
   | |                 | |  | | |       |          |
TQGNNGWCAGYTMSALLNATYNTNKYHAEAVMRFLHPNLQGQQFQFTGLTPREMIYFGQT

--NQGALLYSIANQPVSVVLQAAGKDFQLYRGGIFVGPCGNKVDHAVAAVGYGPNYILIK
      ||           |    |      |       |    || | ||
QGRSPQLLNRMTTYNEVDNLTKNNKGIAIL-GSRVESRNGMHAGHAMAVVGNAKLNNGQE

NSWGTGWGENGYIRIKRGTGNSYGVCGLYTSSFYPVKN-
         ||               |
VIIIWNPWDNGFMTQDAKNNVIPVSNGDHYQWYSSIYGY
```

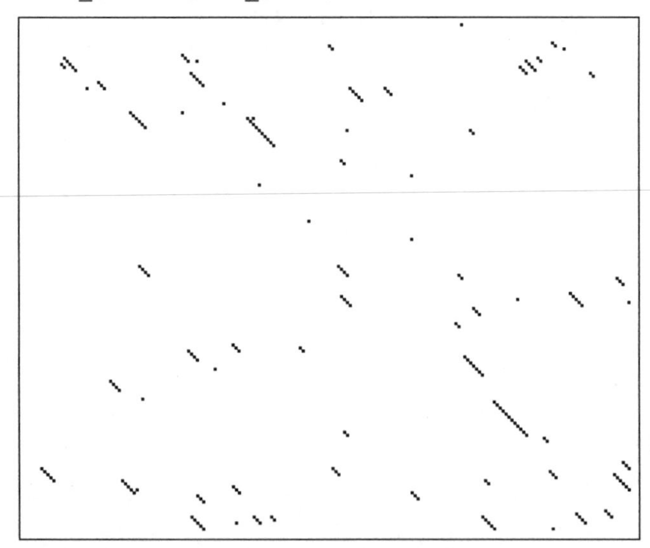

Fig. 4.2d Alignment of papaya papain and *S. aureus* staphopain, with the corresponding dotplot. The alignment of these two sequences is not derivable from this dotplot.

Measures of sequence similarity

To go beyond 'alignment by eyeball' via dotplots, we must define quantitative measures of sequence similarity and difference.

Given two character strings, two measures of the distance between them are:

(1) The **Hamming distance**, defined between two strings of equal length, is the number of positions with mismatching characters.

(2) The **Levenshtein**, or **edit distance**, between two strings of not necessarily equal length, is the minimal number of 'edit operations' required to change one string into the other, where an edit operation is a deletion, insertion or alteration of a single character in either sequence. A given sequence of edit operations induces a unique alignment, but not vice versa.

For example:

```
agtc      Hamming distance = 2
cgta

ag-tcc    Levenshtein distance = 3
cgctca
```

For applications to molecular biology, recognize that certain changes are more likely to occur naturally than others. For example, amino acid substitutions tend to be conservative: the replacement of one amino acid by another with similar size or physicochemical properties is more likely to have occurred than its replacement by another amino acid with dissimilar properties. Or, the deletion of a succession of contiguous bases or amino acids is a more probable event than the independent deletion of the same number of bases or amino acids at noncontiguous positions in the sequences. Therefore we wish to assign variable weights to different edit operations. A computer program can then determine not just minimal edit distances but optimal alignments. It can score each path through the dotplot, by adding up the scores of the individual steps. For each substitution, it adds the score of the mutation, depending on the pair of residues involved. For horizontal and vertical moves, it adds a suitable gap penalty.

Scoring schemes

A scoring system must account for residue substitutions, and insertions or deletions. (An insertion, from one sequence's point of view, is a deletion as seen by the other!) Deletions, or gaps in a sequence, will have scores that depend on their lengths.

Hamming and Levenshtein distances measure the *dissimilarity* of two sequences: similar sequences give small distances and dissimilar sequences give large distances. It is common in molecular biology to define scores as measures of sequence *similarity*. Then similar sequences give high scores and dissimilar sequences give low scores. These are equivalent formulations. Algorithms for optimal alignment can seek either to minimize a dissimilarity measure, or to maximize a scoring function.

Example 4.5

Transition mutations (*purine*↔*purine* and *pyrimidine* ↔ *pyrimidine;* that is, a↔g and t↔c) are more common than *transversions* (*purine* ↔ *pyrimidine;* that is, (a or g) ↔ (t or c)). Suggest a substitution matrix that reflects this.

One possibility is:

	a	t	g	c
a	20	10	5	5
t	10	20	5	5
g	5	5	20	10
c	5	5	10	20

For nucleic acid sequences, it is common to use a simple scheme for substitutions: +1 for a match, −1 for a mismatch, or a more complicated scheme based on the higher frequency of transition mutations than transversion mutations.

For proteins, a variety of scoring schemes have been proposed. We might group the amino acids into classes of similar physicochemical type, and score +1 for a match within residue class, and −1 for residues in different classes. We might try to devise a more precise substitution score from a combination of properties of the amino acids. Alternatively, we might try to let the proteins teach us an appropriate scoring scheme. M. O. Dayhoff did this first, by collecting statistics on substitution frequencies in the protein sequences then known. Her results were used for many years to score alignments. They have been superseded by newer matrices based on the very much larger set of sequences that has subsequently become available.

Derivation of substitution matrices

As sequences diverge, mutations accumulate. To measure the relative probability of any particular substitution, for instance Serine → Threonine, we can count the number of Serine → Threonine changes in pairs of aligned homologous sequences. We could use the relative frequencies of such changes to form a scoring matrix for substitutions. A likely change will score higher than a rare one. But, what if there have been multiple substitutions at certain sites? This will bias the statistics. We can avoid this problem by restricting our samples to sequences that are sufficiently similar that we can assume that no position has changed more than once.

A measure of sequence divergence is the **PAM**: 1 PAM = 1 Percent Accepted Mutation. Thus, two sequences 1 PAM apart have 99% identical residues. For pairs of sequences within the 1 PAM level of divergence, it is likely that there has been no more than one change at any position. Collecting statistics from pairs of sequences as closely related as this, and correcting for different amino acid abundances, produces the **1 PAM substitution matrix**. To produce a matrix appropriate to more widely divergent sequences, we can take powers of this matrix. The PAM250 level, corresponding to ~20% overall sequence identity, is the lowest sequence similarity for which we can generally hope to produce a correct alignment by sequence analysis alone. It is therefore the appropriate level to choose for practical work (see Box, Substitution matrices for scoring amino acid sequence similarity). (Several authors have derived substitution matrices appropriate in different ranges of overall sequence similarity.)

The occurrence of reversions, either directly or via one or more other changes, produces an apparent slowdown in mutation rates as sequences progressively diverge. The relation between PAM score and % sequence identity is:

PAM	0	30	80	110	200	250
% identity	100	75	50	60	25	20

The PAM250 matrix of M. O. Dayhoff is shown in the Box. It expresses scores as **log-odds** values:

Score of mutation $i \leftrightarrow j$

$$= \log_{10} \frac{\text{observed i} \leftrightarrow \text{j mutation rate}}{\text{mutation rate expected from amino acid frequencies}}$$

The numbers are multiplied by 10, simply to avoid decimal points. The matrix entries reflect the probabilities of mutational events. A value of +2—for instance, C ↔ S—implies that in related sequences the mutation would be expected to occur 1.6 times more frequently than random. The calculation is as follows: The matrix entry 2 corresponds to the actual value 0.2 because of the scaling. The value 0.2 is $\log_{10}$ of the relative expectation value of the mutation. Because $\log_{10}$ (1.6) = 0.2, the expectation value is 1.6.

The probability of two independent mutational events is the product of their probabilities. By using logs, we have scores that we can add up rather than multiply, a computational convenience.

The BLOSUM matrices

S. Henikoff and J. G. Henikoff developed the family of BLOSUM matrices for scoring substitutions in amino acid sequence comparisons. Their goal was to replace the Dayhoff matrix with one that would perform best in identifying distant relationships, making use of the much larger amount of data that had become available since Dayhoff's work.

The BLOSUM matrices are based on the BLOCKS database of aligned protein sequences; hence the name BLOcks SUbstitution Matrix. From regions of closely-related proteins alignable without gaps, Henikoff and Henikoff calculated the ratio of the number of observed pairs of amino acids at any position, to the number of pairs expected from the overall amino acid frequencies. Like the Dayhoff matrix, the results are expressed as log-odds. In order to avoid overweighting closely-related sequences, the Henikoffs replaced groups of proteins that have sequence identities higher than a threshold by either a single representative or a weighted average. The threshold 62% produces the commonly-used BLOSUM62 substitution matrix (see Box). This is offered by all programs as an option and as the default by most. BLOSUM matrices, and other recently-derived substitution parameters, have superseded the Dayhoff matrix in applications.

Substitution matrices used for scoring amino acid sequence similarity

Rows and columns are in alphabetical order of the THREE-letter amino acid names. Only the lower triangles of the matrices are shown, as the substitution probabilities are taken as symmetric. (Not because we are sure that the rate of any substitution is the same as the rate of its reverse, but because we cannot determine the differences between the two rates.)

The Dayhoff PAM250 matrix:

	A	R	N	D	C	Q	E	G	H	I	L	K	M	F	P	S	T	W	Y	V
Ala (A)	2																			
Arg (R)	−2	6																		
Asn (N)	0	0	2																	
Asp (D)	0	−1	2	4																
Cys (C)	−2	−4	−4	−5	12															
Gln (Q)	0	1	1	2	−5	4														
Glu (E)	0	−1	1	3	−5	2	4													
Gly (G)	1	−3	0	1	−3	−1	0	5												
His (H)	−1	2	2	1	−3	3	1	−2	6											
Ile (I)	−1	−2	−2	−2	−2	−2	−2	−3	−2	−5										
Leu (L)	−2	−3	−3	−4	−6	−2	−3	−4	−2	2	6									
Lys (K)	−1	3	1	0	−5	1	0	−2	0	−2	−3	5								
Met (M)	−1	0	−2	−3	−5	−1	−2	−3	−2	2	4	0	6							
Phe (F)	−3	−4	−3	−6	−4	−5	−5	−5	−2	1	2	−5	0	9						
Pro (P)	1	0	0	−1	−3	0	−1	0	0	−2	−3	−1	−2	−5	6					
Ser (S)	1	0	1	0	0	−1	0	1	−1	−1	−3	0	−2	−3	1	2				
Thr (T)	1	−2	0	0	−2	−1	0	0	−1	0	−2	0	−1	−3	0	1	3			
Trp (W)	−6	2	−4	−7	−8	−5	−7	−7	−3	−5	−2	−3	−4	0	−6	−2	−5	17		
Tyr (Y)	−3	−4	−2	−4	0	−4	−4	−5	0	−1	−1	−4	−2	7	−5	−3	−3	0	10	
Val (V)	0	−2	−2	−2	−2	−2	−2	−1	−2	4	2	−2	2	−1	−1	−1	0	−6	−2	4

The BLOSUM62 matrix:

	A	R	N	D	C	Q	E	G	H	I	L	K	M	F	P	S	T	W	Y	V
Ala (A)	4																			
Arg (R)	−1	5																		
Asn (N)	−2	0	6																	
Asp (D)	−2	−2	1	6																
Cys (C)	0	−3	−3	−3	9															
Gln (Q)	−1	1	0	0	−3	5														
Glu (E)	−1	0	0	2	−4	2	5													
Gly (G)	0	−2	0	−1	−3	−2	−2	6												
His (H)	−2	0	1	−1	−3	0	0	−2	8											
Ile (I)	−1	−3	−3	−3	−1	−3	−3	−4	−3	4										
Leu (L)	−1	−2	−3	−4	−1	−2	−3	−4	−3	2	4									
Lys (K)	−1	2	0	−1	−3	1	1	−2	−1	−3	−2	5								
Met (M)	−1	−1	−2	−3	−1	0	−2	−3	−2	1	2	−1	5							
Phe (F)	−2	−3	−3	−3	−2	−3	−3	−3	−1	0	0	−3	0	6						
Pro (P)	−1	−2	−2	−1	−3	−1	−1	−2	−2	−3	−3	−1	−2	−4	7					
Ser (S)	1	−1	1	0	−1	0	0	0	−1	−2	−2	0	−1	−2	−1	4				
Thr (T)	0	−1	0	−1	−1	−1	−1	−2	−2	−1	−1	−1	−1	−2	−1	1	5			
Trp (W)	−3	−3	−4	−4	−2	−2	−3	−2	−2	−3	−2	−3	−1	1	−4	−3	−2	11		
Tyr (Y)	−2	−2	−2	−3	−2	−1	−2	−3	2	−1	−1	−2	−1	3	−3	−2	−2	2	7	
Val (V)	0	−3	−3	−3	−1	−2	−2	−3	−3	3	1	−2	1	−1	−2	−2	0	−3	−1	4

Scoring insertions/deletions, or 'gap weighting'

To form a complete scoring scheme for alignments, we need, in addition to the substitution matrix, a way of scoring gaps. How important are insertions and deletions, relative to substitutions? Distinguish gap initiation:

```
aaagaaa
aaa-aaa
```

from gap extension:

```
aaaggggaaa
aaa----aaa
```

For aligning DNA sequences, the alignment program CLUSTAL-W recommends use of the identity matrix for substitution (+1 for a match, 0 for a mismatch) and gap penalties 10 for gap initiation and 0.1 for gap extension by one residue. For aligning protein sequences, the recommendations are to use the BLOSUM62 matrix for substitutions, and gap penalties 11 for gap initiation and 1 for gap extension by one residue.

Computing the alignment of two sequences

Now that we have a scoring scheme, we can apply it to finding optimal alignments—we seek the alignment that maximizes the score. A famous algorithm to determine the global optimal alignments of two sequences is based on a mathematical technique called dynamic programming. (Details are described at the end of this section.) This algorithm has been extremely important in molecular biology. Two of its noteworthy features are:

- The good news is that the method is guaranteed to give a *global* optimum. It will find the *best* alignment score, given the choice of parameters—substitution matrix (M) and gap penalty—with no approximation.
- The bad news is that many alignments may give the same optimal score. And none of these need correspond to the biologically correct alignment. For instance, in comparing the α- and β-chains of chicken haemoglobin, W. Fitch and T. Smith found 17 alignments all of which give the same optimal score, one of which is correct (on the basis of the structures, the court of last resort). There are 1317 alignments with scores within 5% of the optimum.

Another item of bad news is technical: The time required to align two sequences of lengths n and m is proportional to $n \times m$, because this is the size of the edit matrix that must be filled in. This means that the dynamic-programming method is not convenient to use for searching in an entire sequence database for a match to a probe sequence, and even less convenient for 'all-against-all' alignments. The database search problem is in effect the problem of matching a probe sequence to a region of a very long sequence, the length of the entire database.

Variations and generalizations

Variations of the dynamical-programming method apply to two related alignment questions (see also Box, page 26):

- ***global alignment:*** find the best alignment of one entire sequence with another entire sequence.
- ***local alignment:*** find the best alignment of some segment of one sequence against some segment of another sequence. (This includes probing a database with a single sequence, regarding a database as a single very long sequence.)

The global alignment algorithm was first applied to biological sequence alignment by S. B. Needleman and C. D. Wunsch. T. Smith and M. Waterman modified it to identify local matches.

Approximate methods for quick screening of databases

It is routine to screen genes from a new genome against the databases, for similarity to other sequences. Approximate methods can detect close relationships well and quickly but are inferior to the exact ones in picking up very distant relationships. In practice, they give satisfactory performance in the many cases in which the probe sequence is fairly similar to one or more sequences in the databank, and they are therefore worth trying first.

A typical approximation approach would take a small integer *k*, and determine all instances of each *k*-tuple of residues in the probe sequence that occur in any sequence in the database. A candidate sequence is a sequence in the databank containing a large number of matching *k*-tuples, with equivalent spacing in probe and candidate sequences. For a selected set of candidate sequences, approximate optimal alignment calculations are then carried out, with the time- and space-saving restriction that the paths through the matrix considered are restricted to bands around the diagonals containing the many matching *k*-tuples. There are several variations on this theme.

The dynamic programming algorithm for optimal pairwise sequence alignment*

A chart implicitly containing all possible alignments can be constructed as a matrix similar to that used in drawing the dotplot. The residues of one sequence index the rows, the residues of the other sequence index the columns. Any path through the matrix from upper left to lower right corresponds to an alignment (see Fig. 4.1). The task is to find the path that has the lowest cost, and the difficulty is that there are a very large number of paths to consider.

As an illustration, suppose you wanted to drive from Malmö in Southern Sweden to Tromsø in Northern Norway (see Fig. 4.3). Your route will consist of a number of segments, taking you through a succession of intermediate cities. There are many choices of different combinations of segments to produce a complete, continuous path.

The computational approach to finding the optimal path begins by assigning a numerical measure of the 'cost' to each of the possible individual segments of the journey. This 'cost' is not simply the financial outlay, but a more general estimate

* Optional section. Readers in doubt may consider the remarks in Lesk, A. M. (1988), TATA for now . . ., *Trends Biochem. Sci.*, **13**, 410.

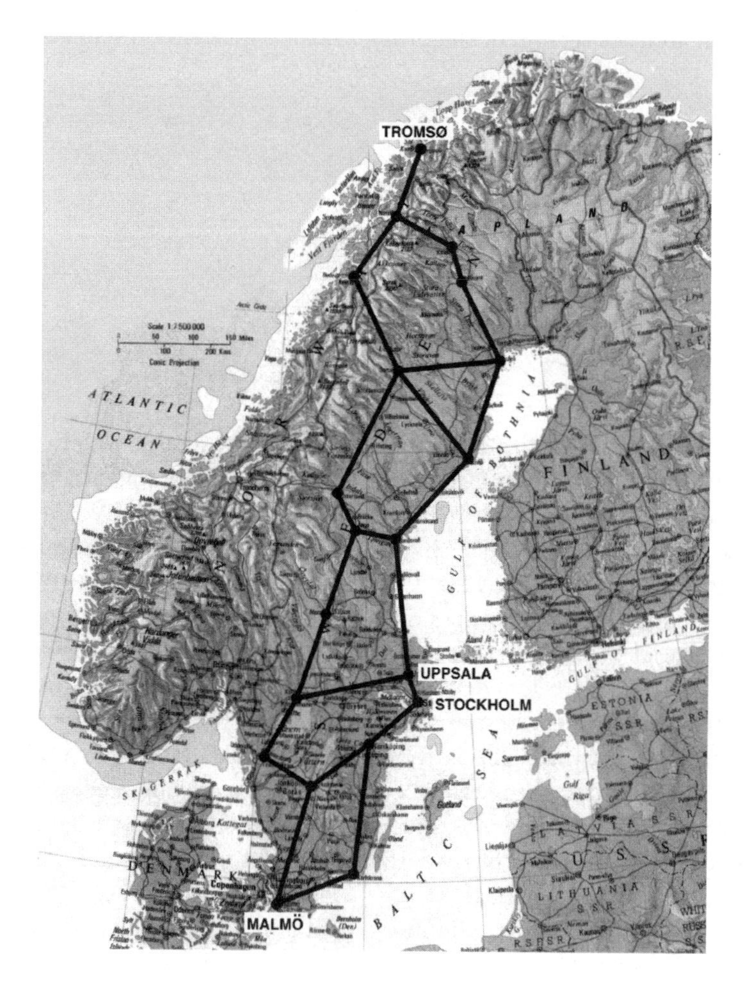

Fig. 4.3 Possible routes from Malmö to Tromsø. How can you determine an optimal route? (© Collins Bartholomew 1980. Reproduced by permission of HarperCollins Publishers.)

of your relative preferences for different portions of the route. The distance travelled will clearly be an important component of the cost, but other factors such as the quality of the roads and the opportunities for sightseeing also contribute. For any route selected, the overall cost of the trip is the sum of the costs of the individual segments. Clearly it is inefficient to repeat any leg of the journey, or to visit any city twice, so you will agree that every intermediate stop will be north of the previous one. This formalism is expressed in terms of minimizing a cost rather than maximizing a score; for our purposes the two approaches are equivalent. An algorithm can explore the possible combinations to determine an optimal overall route.

Here is an abstract version of the problem:

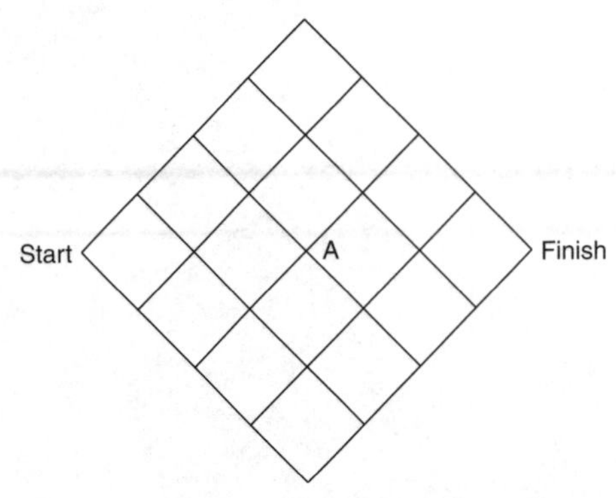

To grasp the essential idea of dynamic programming, first consider: How many paths from Start to Finish pass through A? There are 6 paths from Start to A. (Write them all down.) Therefore, by symmetry there are 6 paths from A to finish, and a total of 36 paths from Start to Finish passing through A. (Why?) Assuming that we have assigned costs to the individual steps, do we have to check all 36 paths to find the path of minimum cost that goes from Start to Finish, passing through A? No—here is the crucial observation: *The choice of the best path from A to Finish is independent of the choice of path from the Start to A.* If we determine the best of the 6 paths from Start to A, and we determine the best of the 6 paths from A to Finish, the best path from Start to Finish passing through A is: the best path from Start to A *followed by* the best path from A to finish. No more than 12 of the paths through A need be considered.

Even greater simplification is possible by systematically resubdividing the problem. The dynamic programming method for finding the optimal path through the matrix is based on this idea.

A statement of the optimal alignment problem and the dynamical programming solution is as follows: Given two character strings, possibly of unequal length: $A = a_1a_2 \ldots a_n$ and $B = b_1b_2 \ldots b_m$, where each a_i and b_j is a member of an alphabet set $\mathcal{A}$, consider sequences of edit operations that convert A and B to a common sequence. Individual edit operations include:

Substitution of b_j for a_i — represented (a_i, b_j).
Deletion of a_i from sequence A — represented (a_i, ϕ).
Deletion of b_j from sequence B — represented as (ϕ, b_j).

If we extend the alphabet set to include the null character ϕ: $\mathcal{A}^+ = \mathcal{A} \cup \{\phi\}$, a sequence of edit operations is a set of ordered pairs (x, y), with $x, y \in \mathcal{A}^+$.

A *cost function d* is defined on edit operations:

$d(a_i, b_j)$ = cost of a mutation in an alignment in which position i of sequence A corresponds to position j of sequence B, and the mutation substitutes $a_i \leftrightarrow b_j$.

$d(a_i, \phi)$ or $d(\phi, b_j)$ = cost of a deletion or insertion.

Define the minimum weighted distance between sequences A and B as

$$D(A, B) = \min_{A \to B} \Sigma d(x, y)$$

where $x, y \in A^+$ and the minimum is taken over all sequences of edit operations that convert A and B into a common sequence.

The problem is to find $D(A, B)$ and one or more of the alignments that correspond to it.

An algorithm that solves this problem in $\mathcal{O}(mn)$ time creates a matrix $\mathcal{D}(i, j)$, $i = 0, \ldots n; j = 0, \ldots m$, such that $\mathcal{D}(i, j)$ is the minimal distance between the strings that consist of the first i characters of A and the first j characters of B. Then $\mathcal{D}(n, m)$ will be the required minimal distance $D(A, B)$.

The algorithm computes $\mathcal{D}(i, j)$ by recursion. The value of $\mathcal{D}(i, j)$ corresponds to the conversion of the initial subsequences $A_i = a_1a_2 \ldots a_i$ and $B_j = b_1b_2 \ldots b_j$ into a common sequence by L edit operations S_k, $k = 1, \ldots L$, which can be considered to be applied in increasing order of position in the strings. Consider *undoing* the last of these edit operations. The resulting truncated sequence of edit operations, S_k, $k = 1, \ldots L - 1$, is a sequence of edit operations for converting a substring of A_i and a substring of B_j into a common result. What is more, it must be an *optimal* sequence of edit operations for these substrings, for if some other sequence S'_k were a lower-cost sequence of operations for these substrings, then S'_k followed by S_L would be a lower-cost sequence of operations than S_k for converting A_i to B_j. Therefore there should be a recursive method for calculating the $\mathcal{D}(i, j)$.

Recognize the correspondence between individual edit operations and steps between adjacent squares in the matrix (see Fig. 4.1):

$(i - 1, j - 1) \rightarrow (i, j)$	corresponds to the substitution $a_i \rightarrow b_j$.
$(i - 1, j) \rightarrow (i, j)$	corresponds to the deletion of a_i from A.
$(i, j - 1) \rightarrow (i, j)$	corresponds to the insertion of b_j into A at position i.

Sequences of edit operations correspond to stepwise paths through the matrix

$$(i_0, j_0) = (0, 0) \rightarrow (i_1, j_1) \rightarrow \cdots (n, m)$$

where $0 \leq i_{k+1} - i_k \leq 1$, (for $0 \leq k \leq n - 1$), $0 \leq j_{k+1} - j_k \leq 1$ (for $0 \leq k \leq m - 1$). Considering the possible sequences of edit operations and the corresponding paths through the matrix, the predecessor of an optimal string of edit operations leading from (0,0) to (i, j), where $i, j > 0$, must be an optimal sequence of edit operations leading to one of the cells $(i - 1, j)$, $(i - 1, j - 1)$, or $(i, j - 1)$; and, correspondingly, $\mathcal{D}(i, j)$ must depend only on the values of $\mathcal{D}(i - 1, j)$, $\mathcal{D}(i - 1, j - 1)$, and $\mathcal{D}(i, j - 1)$, (together of course with the parameterization specified by the cost function d).

The algorithm is then as follows:

Compute the $(m + 1) \times (n + 1)$ matrix $\mathcal{D}$ by applying:

(1) the initialization conditions on the top row and left column:

$$D(i, 0) = \sum_{k=0}^{i} d(a_k, \phi)$$

$$D(0, j) = \sum_{k=0}^{j} d(\phi, b_k)$$

These values impose the gap penalty on unmatched residues at the beginning of either sequence.

And then

(2) the recurrence relations:

$$D(i, j) = min\{D(i-1, j) + d(a_i, \phi, D(i-1, j-1) + d(a_i, b_j), D(i, j-1) + d(\phi, b_j)\}$$

for $i = 1, \ldots n; j=1, \ldots m$. This means: consider all three possible steps to $\mathcal{D}(i, j)$:

Operation	Cumulative cost
insert a gap in sequence A	$\mathcal{D}(i-1,j)+d(a_i,\phi)$
substitute $a_i \leftrightarrow b_j$	$\mathcal{D}(i-1,j-1)+d(a_i,b_j)$
insert a gap in sequence B	$\mathcal{D}(i,j-1)+d(\phi,b_j)$

From these, choose the minimal value. For each cell record not only the value $\mathcal{D}(i, j)$ but a pointer back to (one or more of) the cell(s) $(i-1, j)$, $(i-1, j-1)$ or $(i, j-1)$ selected by the minimization operation. Note that more than one predecessor may give the same value.

When the calculations are complete, $\mathcal{D}(n, m)$ is the optimal distance $D(A, B)$. An alignment corresponding to the sequence of edit operations recorded by the pointers can be recovered by tracing a path back through the matrix from (n, m) to $(0, 0)$. This alignment corresponding to the minimal distance $D(A, B) = \mathcal{D}(n, m)$ may well not be unique.

Example 4.6 Pairwise Sequence Alignment

Align the strings A = `ggaatgg` and B = `atg`, according to the simple scoring scheme: match = 0, mismatch = 20, insertion or deletion = 25.

Here is the state of play after the top row and leftmost column have been initialized (italic), and the element in the second row and second column has been entered as **20** (boldface):

	ϕ	a	t	g
ϕ	*0*	*25*	*50*	*75*
g	*25*	**20**		
g	*50*			
a	*75*			
a	*100*			
t	*125*			
g	*150*			
g	*175*			

The value of **20** was chosen as the minimum of 25 + 25 (horizontal move, or insert gap into string `atg`), 0 + 20 (substitution a↔g), and 25 + 25 (vertical move, or insert gap into string `ggaatgg`). Because the substitution (the diagonal move) provided the minimal value, the cell containing 0 in the upper left hand corner of the matrix is the predecessor of the cell in which we have just entered the 20. (If two or even three of the possible moves produce the same value, the resulting cell has multiple predecessors.)

→

Here is the matrix after completion of the calculation:

	ϕ		a		t		g
ϕ	0	←	25	←	50	←	75
	↑	↖		↖		↖	
g	25		20	←	45		50
	↑	↖	**↑**	↖		↖	
g	50		45		40		45
	↑	**↖**		↖	↑	↖	
a	75		50		65		60
	↑	**↖**	**↑**	↖		↖	↑
a	100		75		70		85
	↑		↑	**↖**		↖	
t	125		100		75		90
	↑		↑		**↑**	**↖**	
g	150		125		100		75
	↑		↑		↑	**↖**	**↑**
g	175		150		125		100

It includes the traceback information in the form of arrows pointing from each cell to its predecessor(s). For some applications we may need only the value of $D(A, B)$ but not an alignment; if so, it is unnecessary to save the pointers. (This has implications for the space required to perform the calculation.) Boldface arrows delineate the paths of optimal alignment retracing a trail of predecessors from lower right, back to upper left. In some cases, one cell may show two predecessors. These correspond to alternative alignments with the same score.

There are two cells at which the traceback path branches. This gives a total of four optimal alignments with equal score:

```
ggaatgg    ggaatgg    ggaatgg    ggaatgg

---atg-    ---at-g    --a-tg-    --a-t-g
```

With a gap-weighting scheme assigning a smaller penalty to gap extension than to gap initiation the first two of these would score better than the others. However, more sophisticated gap-weighting schemes require more complicated recurrence formulas for filling the matrix.

This algorithm determines the optimal *global* alignment of two sequences. It is inappropriate for detection of local regions of high similarity within two sequences, or for probing a long sequence with a short fragment, because it imposes gap penalties *outside* the similar regions. The method of T. Smith and M. Waterman solves this problem. Their modifications of the basic dynamic programming algorithm finds optimal local alignments; that is, it selects the substrings from both sequences that are most similar to each other. Their changes affect:

(1) **Initialization of the matrix**—setting the values of the top row and left column. In the Smith–Waterman method the top row and left column are

Example 4.6 *(continued)*

set to 0. As a result, either sequence can slide along the other before alignment starts, without incurring any gap penalty against the residues it leaves behind.

(2) **Filling in the matrix** In the example of global alignment, at each step a choice is forced among match, insertion or deletion, even if none of these choices is attractive and even if a succession of unattractive choices degrades the score along a path containing a well-fitting local region. The Smith–Waterman method adds the fourth option: end the region being aligned.

(3) **Scoring and traceback** The score of a global alignment is the number in the matrix element at the lower right. In the Smith–Waterman method it is the optimal value encountered, wherever in the matrix it appears. For global alignment, traceback to determine the actual alignment starts at the lower-right cell. In the Smith–Waterman method it starts at the cell containing the optimal value and continues back only as far as the region of local similarity continues.

The Smith–Waterman method would report a unique local optimum for our example:

```
ggaatgg
  atg
```

Note that no gaps appear outside the region matched.

(Example adapted from: Tyler, E. C., Horton, M. R. & Krause, P. R. (1991), A review of algorithms for molecular sequence comparison, *Comp. Biomed. Res*, **24**, 72–96.)

Significance of alignments

Suppose alignment reveals an intriguing similarity between two sequences. Is the similarity significant or could it have arisen by chance? (We raised this question in Chapter 1.) For some simple phenomena—tossing a coin or rolling dice—it is possible to calculate exactly the expected distribution of results, and the likelihood of any particular result. For sequences it is not trivial to define the population from which the alignment is selected. For instance, to take random strings of nucleotides or amino acids as controls ignores the bias arising from nonrandom composition.

A practical approach to the problem is as follows: If the score of the alignment observed is no better than might be expected from a *random permutation* of the sequence, then it is likely to have arisen by chance. We may randomize one of the sequences, many times, realign each result to the second sequence (held fixed), and collect the distribution of resulting scores. Figure 4.4 shows a typical

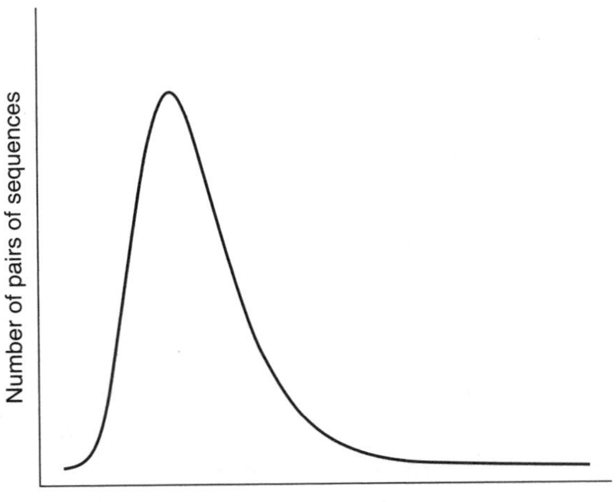

Fig. 4.4 Optimal local alignment scores for pairs of random amino acid sequences of the same length follow an extreme-value distribution. Note the long tail at the right. This means that a score several standard deviations above the mean has a higher probability of arising by chance (that is, it is *less* significant) than if the scores followed a normal distribution. This graph shows the probability distribution function. The formula for the corresponding cumulative distribution function, that is, for any score x, the probability of observing a score $\geq x$ is:

$$P(\text{Score} \geq x) = 1 - \exp(-Ke^{-\lambda x}),$$

where K and λ are parameters related to the position of the maximum and the width of the distribution.

result. For database searches, use the population of results returned from the entire database as the population with which to measure the statistics.

Clearly if the randomized sequences score as well as the original one, the alignment is unlikely to be significant. We can measure the mean and standard deviation of the scores of the alignments of randomized sequences, and ask whether the score of the original sequence is unusually high. The Z-score reflects the extent to which the original result is an outlier from the population:

$$Z\text{-score} = \frac{\text{score} - \text{mean}}{\text{standard deviation}}$$

A Z-score of 0 means that the observed similarity is no better than the average of random permutations of the sequence, and might well have arisen by chance. Other values used as measures of significance are P = the probability that the observed match could have happened by chance, and, for database searching, E = the number of matches as good as the observed one that would be expected to appear by chance in a database of the size probed (see Box, page 184).

Many 'rules of thumb' are expressed in terms of per cent identical residues in the optimal alignment. If two proteins have over 45% identical residues in their optimal alignment, the proteins will have very similar structures, and are very likely to have a common or at least a related function. If they have over 25% identical residues, they are likely to have a similar general folding pattern. On the other hand, observations of a lower degree of sequence similarity cannot rule out

How to play with matches but not get burned

Pairwise alignments and database searches often show tenuous but tantalizing sequence similarities. How can we decide whether we are seeing a true relationship? Statistics cannot answer biological questions directly, but can tell us the likelihood that a similarity as good as the one observed would appear, just by chance, among unrelated sequences. To do this we want to compare our result with alignments of the same sequences to a large population. This 'control' population should be similar in general features to our aligned sequences, but should contain few sequences related to them. Only if the observed match stands out from the population can we regard it as significant.

To what population of sequences should we compare our alignment? For pairwise alignments, we can pick one of the two sequences, make many scrambled copies of it using a random-number generator, and align each permuted copy to the second sequence. For probing a database, the entire database provides a comparison population.

Alignments of our sequence to each member of the control population generates a large set of scores. How does the score of our original alignment rate? Several statistical parameters have been used to evaluate the significance of alignments:

- The Z-score is a measure of how unusual our original match is, in terms of the mean and standard deviation of the population scores. If the original alignment has score S,

$$\text{Z-score of S} = \frac{\text{S} - \text{mean}}{\text{standard deviation}}$$

  A Z-score of 0 means that the observed similarity is no better than the average of the control population, and might well have arisen by chance. The higher the Z-score, the greater the probability that the observed alignment has not arisen simply by chance. Experience suggests that Z-scores ≥ 5 are significant.

- Many programs report P = the probability that the alignment is better than random. The relationship between Z and P depends on the distribution of the scores from the control population, which do *not* follow the normal distribution.

  A rough guide to interpreting P values:

$P \leq 10^{-100}$	exact match
P in range 10^{-100}–10^{-50}	sequences very nearly identical, e.g. alleles or SNPs
P in range 10^{-50}–10^{-10}	closely-related sequences, homology certain
P in range 10^{-5}–10^{-1}	distant relatives, usually
$P > 10^{-1}$	match probably insignificant

→

→

- For database searches, some programs (including PSI-BLAST) report *E*-values. The *E*-value of an alignment is the expected number of sequences that give the same *Z*-score or better if the database is probed with a random sequence. *E* is found by multiplying the value of *P* by the size of the database probed. Note that *E* but not *P* depends on the size of the database. Values of *P* are between 0 and 1.0. Values of *E* are between 0 and the number of sequences in the database searched.

  A rough guide to interpreting *E* values:

$E \leq 0.02$	sequences probably homologous
E between 0.02 and 1	homology can't be ruled out
$E > 1$	you'd have to expect this good a match just by chance

Statistics are a useful guide, but not a substitute for thinking carefully about the results, and further analysis of ones that look promising!

homology. R. F. Doolittle defined the region of 18%–25% sequence identity as the 'twilight zone' in which the suggestion of homology is tantalizing but dangerous. Below the twilight zone is a region where pairwise sequence alignments tell very little. Lack of significant sequence similarity does not preclude similarity of structure.

Although the twilight zone is a treacherous region, we are not entirely helpless. In deciding whether there is a genuine relationship, the 'texture' of the alignment is important—are the similar residues isolated and scattered throughout the sequence; or are there 'icebergs'—local regions of high similarity (another term of Doolittle's), which may correspond to a shared active site? We may need to rely on other information, about shared ligands or function. Of course if the structures are known, we could examine them directly.

Some illustrative examples:

- Sperm whale myoglobin and lupin leghaemoglobin have 15% identical residues in optimal alignment. This is even below Doolittle's definition of the twilight zone. But we also know that both molecules have similar three-dimensional structures, both contain a haem group, and both bind oxygen. They are indeed distantly-related homologues.
- The sequences of the N- and C-terminal halves of rhodanese have 11% identical residues in optimal alignment. If these appeared in independent proteins, one could not conclude from the sequences alone that they were related. However, their appearance in the same protein suggests that they arose via gene duplication and divergence. The striking similarity of their structures confirms their relationship.

- As a cautionary note, consider the proteinases chymotrypsin and subtilisin. They have 12% identical residues in optimal alignment. These enzymes have a common function, and a common Ser — His — Asp catalytic triad. However, they have dissimilar folding patterns, and are not related. Their common function and mechanism is an example of convergent evolution. This case serves as a warning against special pleading for relationships between proteins with dissimilar sequences on the basis of similarities of function and mechanism!

Multiple sequence alignment

'One amino acid sequence plays coy; a pair of homologous sequences whisper; many aligned sequences shout out loud.' In nature, even a single sequence contains all the information necessary to dictate the fold of the protein. How does a multiple sequence alignment make the information more intelligible? Alignment tables expose patterns of amino acid conservation, from which distant relationships may be more reliably detected. Structure prediction tools also give more reliable results when based on multiple sequence alignments than on single sequences.

Visual examination of multiple sequence alignment tables is one of the most profitable activities that a molecular biologist can undertake away from the lab bench. Don't even THINK about not displaying them with different colours for amino acids of different physicochemical type. A reasonable colour scheme (not the only one) is:

Colour	Residue Type	Amino Acids
Yellow	small nonpolar	Gly, Ala, Ser, Thr
Green	hydrophobic	Cys, Val, Ile, Leu, Pro, Phe, Tyr, Met, Trp
Magenta	polar	Asn, Gln, His
Red	negatively charged	Asp, Glu
Blue	postively charged	Lys, Arg

To be informative a multiple alignment should contain a distribution of closely- and distantly-related sequences. If all the sequences are very closely related, the information they contain is largely redundant, and few inferences can be drawn. If all the sequences are very distantly related, it will be difficult to construct an accurate alignment (unless all the structures are available), and in such cases the quality of the results, and the inferences they might suggest, are questionable. Ideally, one has a complete range of similarities, including distant relatives linked through chains of close relationships.

Case Study 4.1: Structural inferences from multiple sequence alignment of thioredoxins

Thioredoxins are enzymes found in all cells. They participate in a broad range of biological processes, including cell proliferation, blood clotting, seed germination, insulin degradation, repair of oxidative damage, and enzyme regulation. The common mechanism of these activities is the reduction of protein disulphide bonds.

Plate IV shows a multiple sequence alignment of 16 thioredoxins. The structure of *E. coli* thioredoxin contains a central five-stranded β-sheet flanked on either side by α-helices; these helices and strands are indicated by the symbols α and β. Other thioredoxins are expected to share most but not all of the secondary structure of the *E. coli* enzyme. The plate also shows a summary of the alignment as a *sequence logo*, in which letters of different sizes indicate different proportions of amino acids. T. Schneider and M. Stephens designed sequence logos; this example was produced using the web server http://weblogo.berkeley.edu.

Structural and functional features of thioredoxins that we might hope to identify from the multiple sequence alignment include (see Fig. 4.5 and Plate IV):

- **The most highly conserved regions probably correspond to the active site.** The disulphide bridge between residues 32 and 35 in *E. coli* thioredoxin is part of a WCGPC[K or R] motif conserved in the family. Other regions conserved in the sequences, including the PT at residues 75–77 and the GA at residues 92–93, are involved in substrate binding.
- **Regions rich in insertions and deletions probably correspond to surface loops. A position containing a conserved Gly or Pro probably corresponds to a turn.** Turns correlated with insertions and deletions occur at positions 9, 20, 60 and 95. The conserved glycine at position 92 in

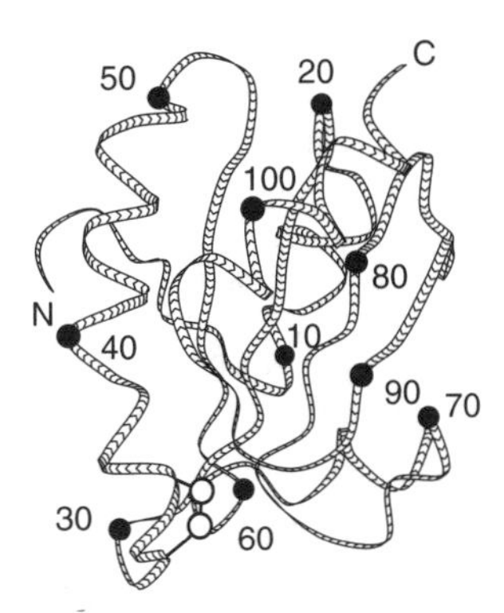

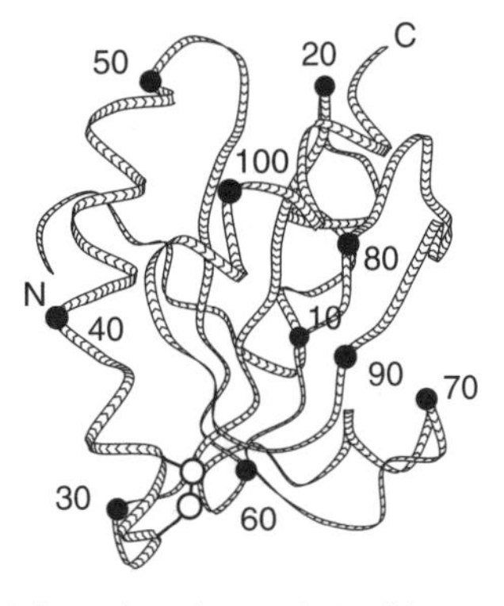

Fig. 4.5 The structure of *E. coli* thioredoxin [2TRX] (see also Plate IV). Residue numbers correspond to those in the multiple sequence alignment table. The N- and C-termini are also marked. Spheres indicate positions of the Cα atoms of every tenth residue. The reactive disulphide bridge between Cys32 and Cys35 appears between the numerals 30 and 60.

→

Case Study 4.1 *(continued)*

E. coli thioredoxin is indeed part of a turn. It is in an unusual mainchain conformation, one that is easily accessible only to glycine (see Chapter 5). The conserved proline at position 76 in *E. coli* thioredoxin is also associated with a turn. It is in another unusual mainchain conformation, this one easily accessible only to proline.

- A conserved pattern of hydrophobicity with spacing 2 (that is, every other residue)—with the intervening residues more variable and including hydrophilic residues—suggests a β-strand on the surface. This pattern is observable in the β-strand between residues 50 and 60.
- A conserved pattern of hydrophobicity with spacing ~4 suggests a helix. This pattern is observable in the region of helix between residues 40 and 49.

Thioredoxins are members of a superfamily including many more-distantly-related homologues. These include glutaredoxin (hydrogen donor for ribonucleotide reduction in DNA synthesis), protein disulphide isomerase (which catalyses exchange of mismatched disulphide bridges in protein folding), phosducin (a regulator of G-protein signalling pathways), and glutathione S-transferases (chemical defence proteins). Implicit in the multiple sequence alignment table of the thioredoxins themselves are patterns that should be applicable to identifying these more distant relatives.

Applications of multiple sequence alignments to database searching

Searching in databases for homologues of known proteins is a central theme of bioinformatics. Indeed it brooked no delay; we introduced it in Chapter 1 with the application of PSI-BLAST. We reconsider database searching here, with the goal of trying to understand how we can best use available information to build effective procedures. The goals are high sensitivity—picking up even very distant relationships—and high selectivity—minimizing the number of sequences reported that are not true homologues. Here we discuss how to apply multiple sequences. In Chapter 5 we shall discuss how to apply structural information in addition.

We recognize a familiar face by reacting to its integral appearance rather than to individual features. Similarly, multiple sequence alignments contain subtle patterns that characterize families of proteins.

During the last decade, great progress has been made in devising methods for applying multiple sequence alignments of known proteins to identify related sequences in database searches. The results are central to contemporary applications of bioinformatics, including the interpretion of genomes. Three important methods are: Profiles, PSI-BLAST and Hidden Markov Models.

Profiles

Profiles express the patterns inherent in a multiple sequence alignment of a set of homologous sequences. They have several applications:

- They permit greater accuracy in alignments of distantly-related sequences.
- Sets of residues that are highly conserved are likely to be part of the active site, and give clues to function.
- The conservation patterns facilitate identification of other homologous sequences.
- Patterns from the sequences are useful in classifying subfamilies within a set of homologues.
- Sets of residues that show little conservation, and are subject to insertion and deletion, are likely to be in surface loops. This information has been applied to vaccine design, because such regions are likely to elicit antibodies that will cross-react well with the native structure.
- Most structure-prediction methods are more reliable if based on a multiple sequence alignment than on a single sequence. Homology modelling, for instance, depends crucially on correct sequence alignments.

To use profile patterns to identify homologues, the basic idea is to match the query sequences from the database against the sequences in the alignment table, giving higher weight to positions that are conserved than to those that are variable. If a region is absolutely conserved, such as the WCGPC motif in thioredoxins, the procedure should all but insist on finding it. But being too compulsive risks missing interesting distant relatives; some leeway should be allowed.

What is needed is a quantitative measure of conservation. For each position in the table of aligned sequences, take an inventory of the distribution of amino acids. For instance, for positions 25–30 of the thioredoxin alignment:

Residue number	Number of each amino acid																			
	A	C	D	E	F	G	H	I	K	L	M	N	P	Q	R	S	T	V	W	Y
25	1									2								13		
26			16																	
27					16															
28																7	1		5	3
29	16																			
30			1	4									2			1	7	1		

Given a query sequence representing a potential thioredoxin homologue, we want to evaluate its similarity to the query sequence, in such a way that agreement with the known sequences at the absolutely conserved positions—for instance 26, 27 and 29—contributes a very high score, and disagreement at these positions contributes a very low score. For moderately conserved positions, such as 28, we want a modest positive contribution to the score if the query sequence has an S or a W at this position, and a smaller contribution if it has T or Y. The general idea is to score each residue from the query sequence based on the amino acid distribution at that position in the multiple sequence alignment table.

A tempting but overly simple approach would be to use the inventories as scores directly. For example, if the residues in a query sequence that correspond to positions 25–30 in thioredoxin contains the sequence VDFSAE, this fragment would score 13 + 16 + 16 + 7 + 16 + 4 = 72. This is almost the greatest value possible. The alternative query sequence ACGWAP would score 1 + 0 + 0 + 5 + 16 + 2 = 24, a much lower value. Of course for each query sequence we have to test all possible alignments with the multiple-alignment table, and take the largest total score. The highest-scoring sequences best fit the patterns implicit in the table.

This simple approach would work if our table contained a large and unbiased sample of thioredoxin sequences. But only in this case would the simple inventory give a correct picture of the *potential* distributions of residues at each position. If our sample were small, the pattern derived would be unlikely to reflect the complete repertoire. Or, if the sample contained a large subset of similar sequences, these would be over-represented in the inventories. For instance, we can see in Plate IV that vertebrate thioredoxins form a very closely-related set. If we included twenty more vertebrate thioredoxins in the alignment, the profile would recognize only vertebrate thioredoxins effectively.

Substitution matrices suggest how to make the inventory 'fuzzy' and thereby more general.

The observed amino acid distribution at any residue position is a 20-membered array: $(a_1, a_2, a_3, \ldots a_{20})$, where a_i is the number of amino acids of type i observed at that position. (For position 25 of the thioredoxins, $a_1 = 1$ because 1 alanine is observed, and $a_{18} = 13$, representing the valines.) Then in the simplest scheme, the score of an alanine at position 25 is just 1; the score of a Val is 13; in general, the score of an amino acid of type i is a_i. In this scheme the rows of the inventory itself provide the arrays a needed for scoring each position.

A better scoring scheme would evaluate any amino acid according to its chance of being substituted for one of the observed amino acids. If $D(i, j)$ is the amino acid substitution matrix—BLOSUM62, perhaps—then amino acid i could score $a_1D(i, 1) + a_2D(i, 2) \ldots a_{20}D(i, 20)$. This scheme distributes the score among observed amino acids, weighted according to the substitution probability. An amino acid in the query sequence could score high *either* if it appears frequently in the inventory at this position, or if it has a high probability of arising by mutation from residue types that are common at this position. This approach is more effective in detection of distant relatives from a limited set of known sequences. In this case, the scoring vector for amino acids is the product of the substitution matrix and the rows of the inventory array. An even better approach is to use as the amino acid distribution a combination of the observed inventory and a general background level of amino acid composition.

The result is a set of probability scores for each amino acid (or gap) at each position of the alignment, called a **position-specific scoring matrix**. An alternative method of deriving a position-specific scoring matrix, based on three-dimensional structures, is described in Chapter 5.

Given a query sequence, and the position-specific scoring matrices derived from a profile, the calculations required to find the optimal score over all alignments of the query sequence with the profile are extensions of the dynamic programming methods of aligning two sequences.

A weakness of simple profiles is that the multiple sequence alignment must be provided in advance, and is taken as fixed. PSI-BLAST and Hidden Markov Models gain power by integrating the alignment step with the collection of statistics.

PSI-BLAST

PSI-BLAST is a program that searches a databank for sequences similar to a query sequence. It is a development of the earlier program BLAST = Basic Local Sequence Alignment Tool. The BLAST program and its variants (see Box) check each entry in the databank *independently* against a query sequence. PSI-BLAST begins with such a one-at-a-time search. It then derives pattern information from a multiple sequence alignment of the initial hits, and reprobes the database using the pattern. Then it repeats the process, fine-tuning the pattern in successive cycles.

BLAST programs come in several flavours

Program	Type of query sequence	Search in database of
BLASTP	amino acid sequence	protein sequences
BLASTX	translated nucleotide sequence	protein sequences
TBLASTN	amino acid sequence	translated nucleotide sequences
TBLASTX	translated nucleotide sequence	translated nucleotide sequences
PSI-BLAST	amino acid sequence	protein sequence database

These programs compare amino acid sequences with amino acid sequences, using by default BLOSUM62 matrix. Searches involving nucleotide sequences, either as query sequence or in the database searched, are carried out by translating nucleotide sequences to amino acid sequence in all six possible reading frames. Another program in this family, BLASTN, compares nucleic acid query sequences with nucleic acid databanks directly.

The problem that BLAST was originally designed to solve is that full-blown dynamic programming methods are rather slow for complete searches in a large databank. Often the databank contains close matches to the query sequence. Less sensitive but faster programs are quite capable of identifying the close matches, and if that is what you want, fine. For example, if you want to search for homologues of a mouse protein in the human genome, the similarity is likely to be high and an approximate method likely to find it. But if you want to search for homologues of a human protein in *C. elegans* or yeast, the relationship may be more tenuous; and more sophisticated, slower methods may be required. (It may come as a surprise, but computer time requirements are still a consideration. For although computing is becoming less expensive, the sizes of the databanks and the number of searches desired, on a worldwide basis, are growing. The net effect is that the pressure on computing resources is increasing.)

The method used by BLAST goes back, in a sense, to the dotplot approach, checking for well-matching local regions. For each entry in the database, it checks for short contiguous regions that match a short contiguous region in the query sequence, using a substitution scoring matrix but allowing no gaps. An approach in which candidate regions of *fixed length* are identified initially can be made very fast by the use of lookup tables.

Once BLAST identifies a well-fitting region, it tries to extend it. In some versions gaps are allowed. The output of BLAST is the set of local segment matches. In an example from Chapter 1:

```
  My.care.is.loss.of.care,.by.old.care.done,
    |||||||||    |||||||||||||    |||||| ||
Your.care.is.gain.of.care,.by.new.care.won
```

even a very simple algorithm could pick up all matching regions of four contiguous residues and then combine and extend them (see Problem 4.5).

PSI-BLAST, using iterated pattern search (see Box), is much more powerful than simple pairwise BLAST in picking up distant relationships. PSI-BLAST correctly identifies three times as many homologues as BLAST in the region below 30% sequence identity. It is therefore a very useful method for analysing whole genomes.

The only methods based entirely on sequence analysis that do better than PSI-BLAST are Hidden Markov Models. These are described in the next section. To achieve significantly better performance it is necessary to make explicit use of structural information. This is discussed in the next chapter.

A flowchart for PSI-BLAST

1. Probe each sequence in the chosen database independently for local regions of similarity to the query sequence, using a BLAST-type search but allowing gaps.
2. Collect significant hits. Construct a multiple sequence alignment table between the query sequence and the significant local matches.
3. Form a profile from the multiple sequence alignment.
4. Reprobe the database with the profile, still looking only for local matches.
5. Decide which hits are statistically significant and retain these only.
6. Go back to step 2, until a cycle produces no change. This accounts for the 'Iterated' in the program title (Position Sensitive Iterated, PSI).

Hidden Markov Models

A Hidden Markov Model (HMM for short) is a computational structure for describing the subtle patterns that define families of homologous sequences. HMMs are powerful tools for detecting distant relatives, and for prediction of protein folding patterns. They are the only method based entirely on sequences—that is, without explicitly using structural information—competitive with PSI-BLAST for identifying distant homologues. They also perform well at fold recognition, as assessed in CASP programmes.

Within an HMM is a multiple sequence alignment. However, HMMs are usually presented as *procedures for generating sequences*. A conventional multiple sequence alignment table could also be used to generate sequences, by selecting amino acids at successive positions, each amino acid chosen from a position-specific probability distribution derived from the profile. But HMMs are more general than profiles:

1. They include the possibility of introducing gaps into the generated sequence, with position-dependent gap penalties.
2. Application of profiles requires that the multiple sequence alignment be specified up front; the pattern statistics are then derived from the alignment. HMMs carry out the alignment and the assignment of probabilities together.

The internal structure of an HMM shows the mechanism for generating sequences (Fig. 4.6). Begin at Start, and follow some chain of arrows until arriving at End. Each arrow takes you to a state of the system. At each state (1) you take some action—emit a residue, perhaps—and (2) choose an arrow to take you to the next state. The action and the choice of successor state are governed by sets of probabilities. Associated with each state that emits a residue are: one probability distribution for the twenty amino acids, and a second probability distribution for the choice of successor state. Both of these probability distributions are calibrated to encode information about a particular sequence family. In this way, the same general structure can be specialized to many different sequence families.

The dynamics of the system are such that only the current state influences the choice of its successor—the system has no 'memory' of its history. This is characteristic of First-order Markov processes, studied by the nineteenth century Russian mathematician A. A. Markov. Distinguish the succession of states from the succession of amino acids emitted to form the output sequence. Several paths through the system can generate the same sequence. Only the succession of characters emitted is visible; the state sequence that generated the characters remains internal to the system, i.e. hidden. By the probability distributions associated with the individual states, the system captures—or models—the patterns inherent in a family of sequences. Hence the name, Hidden Markov Model.

Software for applying HMMs to biological sequence analysis can achieve:

1. **Training** Given a set of unaligned homologous sequences, it can align them and adjust the transition and residue output probabilities to define an HMM capturing the patterns inherent in the sequences submitted.

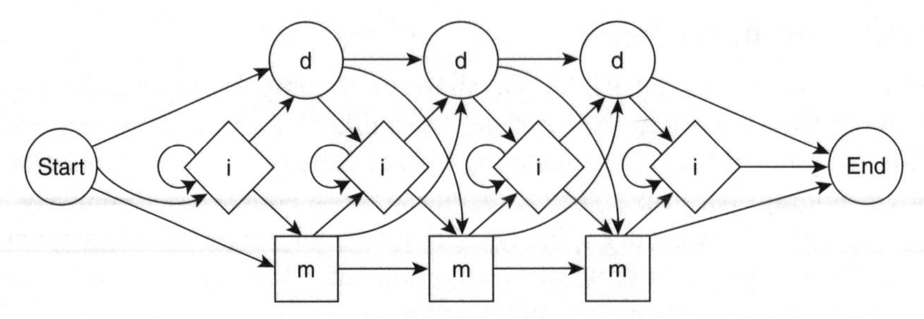

Fig. 4.6 The structure of a Hidden Markov Model (HMM). Corresponding to each residue position in a multiple sequence alignment, the HMM contains a match state (m), and a delete state (d). Insert states (i) appear between residue positions, and at the beginning and end.

- Match states emit a residue. Here the term match means only that there is *some* amino acid both in the model underlying the HMM and in the sequence emitted, not that these are necessarily the *same* amino acid. The probability of emitting each of the twenty amino acids in each of the match states is a property of the model. As with profiles, the probabilities are position dependent.
- Delete states skip a column in the multiple sequence alignment. Arriving at a delete state from a match or insert state corresponds to gap opening, and the probabilities of these transitions reflect a position-specific gap-opening penalty. Arriving at a delete state from a previous delete state corresponds to gap extension.
- Insert states appear between two successive positions in the alignment. If the system enters an insert state, a new residue that does not correspond to a position in the alignment table appears in the emitted sequence. An insert state can be followed by itself, to insert more than one residue. The succession of residues emitted from match and insert states generates the output sequence.

After taking the action appropriate to the type of state (m, d, or i), another probability distribution governs the choice of the next state. In every possible succession of states, every column of the embedded alignment must be visited, and either matched or deleted—there is no way to traverse the network without passing through either an m state or a d state at each position.

2. **Detection of distant homologues** Given an HMM and a test sequence, calculate the probability that the HMM would generate the test sequence. If an HMM trained on a known family of sequences would generate the test sequence with relatively high probability, the test sequence is likely to belong to the family.
3. **Align additional sequences** The probability of any sequence of states can be computed from the individual state-to-state transition probabilities. Finding the most likely sequence of states that the HMM would use to generate one or more test sequences reveals their optimal alignment to the family.

Case Study 4.2: Prediction of transmembrane proteins and signal peptides by Hidden Markov Models

Membranes are the wrapping of cells, and of subcellular compartments and organelles. Membranes contain phospholipid bilayers. The tails of the phospholipids point towards the interior of the membrane, creating an organic medium ~3 nm thick. Polar head groups point outwards. Between the organic region and the external aqueous environment are transitional layers enriched in aromatic sidechains, approximately 1.5 nm thick, on both surfaces of the membrane.

Many proteins are designed to sit within membranes. Among their adaptations are regions containing mostly nonpolar residues, that interact with the organic layer of the membrane. Membrane proteins mediate the exchange of matter, energy, and information between cell interiors and surroundings. Examples of membrane protein functions include energy transduction, via the generation or release of concentration gradients across cell or organelle membranes; and signal reception and transmission.

It is estimated that in the human genome, approximately 30% of genes encode membrane proteins. Given that membrane proteins are so common, it is important to have reliable tools for their annotation. Relatively few membrane protein structures have been experimentally determined. This places a greater burden on computational tools for sequence analysis, to identify and characterize them.

Approximately 70% of known targets of drugs are membrane proteins.

Many membrane proteins contain a set of seven consecutive α-helices that traverse the membrane, oriented approximately perpendicular to the plane of the membrane (see Fig. 4.7). These helices are connected by loops that protrude into the aqueous surroundings. (A second class of membrane protein structures contains a β-barrel.) Transmembrane helices are typically 15–30 residues long. Although enriched in hydrophobic residues—appropriate to their nonpolar environment—they contain some polar sidechains, usually in interfaces between helices packed together in the structure.

A useful clue to the orientation of the helices across the membrane is the '+-inside rule'. The loops between helices lie either entirely inside or entirely outside the cell. Those inside contain a preponderance of positively-charged residues.

A simple approach to prediction of membrane proteins involves looking for amino acid segments 15–30 residues in length, that are rich in hydrophobic residues. However, signal peptides also contain hydrophobic helices: the signal sequence typically comprises a positively-charged n-region, followed by a helical hydrophobic h-region, followed by a polar c-region. Methods for recognizing transmembrane helices in amino acid sequences tend to pick up the h-regions of signal peptides as false positives. Methods for recognizing signal peptides in amino acid sequences tend to pick up transmembrane helices as false positives.

→

Case Study 4.2 *(continued)*

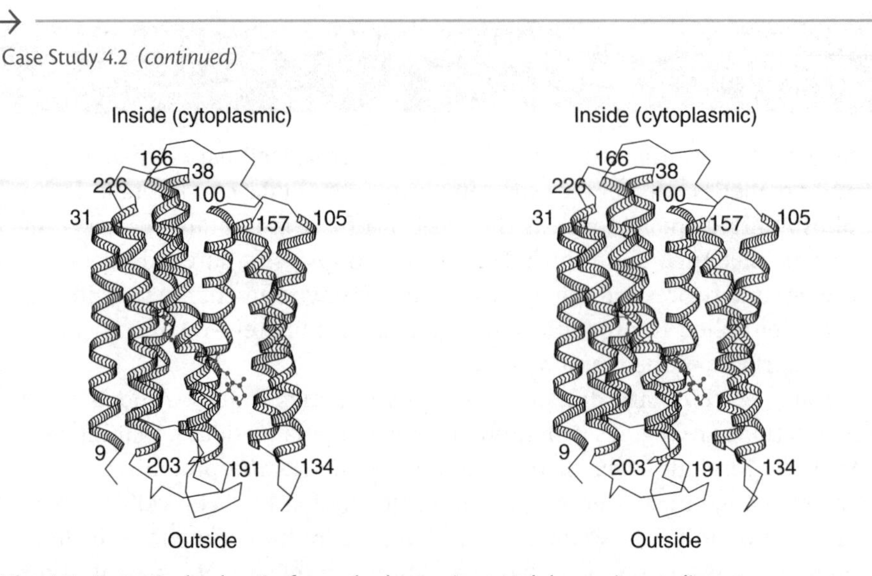

Fig. 4.7 Bacteriorhodopsin from the bacterium *Halobacterium salinarum*, (formerly *Halobacterium halobium*), [2brd] viewed in the plane of the membrane. The ligand shown in red is the chromophore, retinal.

Käll, Krogh and Sonnhammer trained Hidden Markov Models to test simultaneously for transmembrane helices and signal peptides.* The goals are to find both at the same time, to discriminate between them in the results, and to predict not only the positions of the transmembrane helices but the locations—cytoplasmic or interior—of the loops. The method, called **Phobius**, is available at http://phobius.cgb.ki.se/.

Developed from previous work, the HMM for the combined prediction contains separate HMMs for transmembrane helix prediction, and signal peptide prediction, hooked up in parallel. A connection links the models: a transition is possible from the last state of the signal peptide model to the cytoplasmic-side loop state of the helical transmembrane protein model. This makes it possible to treat sequences of proteins that contain a signal peptide sequence *followed* by a transmembrane helical structure. Because a signal peptide must *precede* a transmembrane helical structure in the sequence, no link is necessary *from* the transmembrane helical structure model *to* the signal peptide model.

Phobius was trained on a chosen set of several hundred sequences, retaining only one exemplar of sets of very close relatives. The set included membrane proteins, and control sequences containing neither signal peptides nor transmembrane helices. Annotation of the sequences of the membrane proteins in the training set classified each residue into:

- cytoplasmic loop
- transmembrane helix

* Käll, L., Krogh, A. & Sonnhammer, E. L. L. (2004), A combined transmembrane topology and signal peptide prediction method, *J. Mol. Biol.*, **338**, 1027–1036.

→

- non-cytoplasmic loop
- non-cytoplasmic long loop
- signal peptide cleavage site
- n-region of signal peptide
- h-region of signal peptide
- c-region of signal peptide

The available sequences were split into ten groups. In ten separate calculations, the system was trained on the union of nine of the ten groups with the tenth reserved for testing. After comparison and tuning, a final stage of refinement of the model used all the sequences together.

Application of Phobius to the sequence of bacteriorhodopsin, the protein shown in Fig. 4.7, gives the following results:

```
ID   2brd
FT   DOMAIN       7   11       NON CYTOPLASMIC.
FT   TRANSMEM    12   29
FT   DOMAIN      30   40       CYTOPLASMIC.
FT   TRANSMEM    41   63
FT   DOMAIN      64   82       NON CYTOPLASMIC.
FT   TRANSMEM    83  101
FT   DOMAIN     102  107       CYTOPLASMIC.
FT   TRANSMEM   108  129
FT   DOMAIN     130  134       NON CYTOPLASMIC.
FT   TRANSMEM   135  156
FT   DOMAIN     157  176       CYTOPLASMIC.
FT   TRANSMEM   177  199
FT   DOMAIN     200  204       NON CYTOPLASMIC.
FT   TRANSMEM   205  224
FT   DOMAIN     225  227       CYTOPLASMIC.
```

Readers can compare this prediction with the experimental structure (see Problem 4.9).

Phobius is the most successful algorithm currently available for recognizing signal peptides and helical transmembrane proteins, and for predicting the orientation of the transmembrane segments. Phobius is capable of distinguishing h-domains of signal peptides from transmembrane helices. The number of false classifications of signal peptides was 3.9%, and the number of false classifications of transmembrane helices was 7.7%. These results represent a great improvement over previous methods. It is interesting that addressing the two problems at once proved to be more successful than treating them separately.

Web resources: Hidden Markov Models

Two research groups specializing in biological applications of Hidden Markov Models run web servers and distribute their programs:

R. Hughey, K. Karplus and D. Haussler (University of California at Santa Cruz.):

http://cse.ucsc.edu/research/compbic/sam.html

http://cse.ucsc.edu/research/compbio/HMM-apps/HMM-applications.html

S. Eddy (Washington University, St. Louis, MO, USA)

http://hmmer.wustl.edu/

Results of analysis of known sequences and structures are also available on the Web:

Pfam is a database of multiple sequence alignments and HMMs for many protein domains, developed by A. Bateman, E. Birney, R. Durbin, S. R. Eddy, K. L. Howe and E. L. Sonnhammer:

http://www.sanger.ac.uk/Software/Pfam

J. Gough, K. Karplus, R. Hughey, and C. Chothia have generated HMMs for all PDB superfamilies:

http://stash. mrc-lmb.cam.ac.uk/SUPERFAMILY/

Phylogeny

We have now seen several examples of evolution, in proteins and in genomes. These represent the extension to the molecular level of concerns that have occupied biologists since Darwin and even before. The basic principle is that *the origin of similarity is common ancestry.* Although there are many exceptions, arising from convergent evolution or horizontal gene transfer, the importance of this principle both for rationalizing contemporary observations and giving a window into the history of life cannot be underestimated.

The field of phylogeny has the goals of working out the relationships among species, populations, individuals, or genes. (The general term is 'taxa'.) The *observable* taxa—for instance the extant species for which we wish to work out the pattern of ancestry, are called the 'operational taxonomic units', abbreviated to OTUs.) Relationship is taken in the literal sense of kinship or genealogy, that is, assignment of a scheme of descendants of a common ancestor (see Box). Evolutionary relationships give us a glimpse at the historical development of life (see Box: Time scale of Earth history). Although molecules themselves cannot be dated, the evolutionary events as observed on the molecular level can be calibrated with the fossil record.

Concepts related to biological classification and phylogeny (see page 29)

Homology means, specifically, descent from a common ancestor.
Similarity is the measurement of resemblance or difference, independent of the source of the resemblance. Similarity is observable in data collectable *now*, and involves no historical hypotheses. In contrast, assertions of homology require inferences about historical events which are almost always unobservable.

→

→

Clustering is bringing together similar items, distinguishing classes of objects that are more similar to one another than they are to other objects outside the classes. Most people would agree about degrees of similarity, but clustering is more subjective. When classifying objects, some people prefer larger classes, tolerating wider variation; others prefer smaller, tighter, classes. They are called *groupers* or *splitters*.
Hierarchial clustering is the formation of clusters of clusters of...
Phylogeny is the description of biological relationships, usually expressed as a tree. A statement of phylogeny among objects *assumes* homology and *depends* on classification. Phylogeny states a topology of the relationships based on classification according to similarity of one or more sets of characters, or on a model of evolutionary processes. In many cases, phylogenetic relationships based on different characters are consistent, and support one another. If different characters induce inconsistent phylogenetic relationships, then all the putative relationships are dubious. Conversely, note that the same similarity data may be consistent with different possible topologies or trees.

Time scale of Earth history

Geological ages (e.g. Cenozoic), epochs (e.g. Quaternary), and cataclysmic events (e.g. Asteroid impact: mass extinction) in black. First appearance of, or prevalence of, different life forms in red.

mya = millions of years ago.

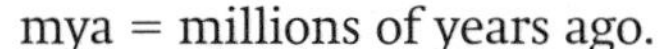

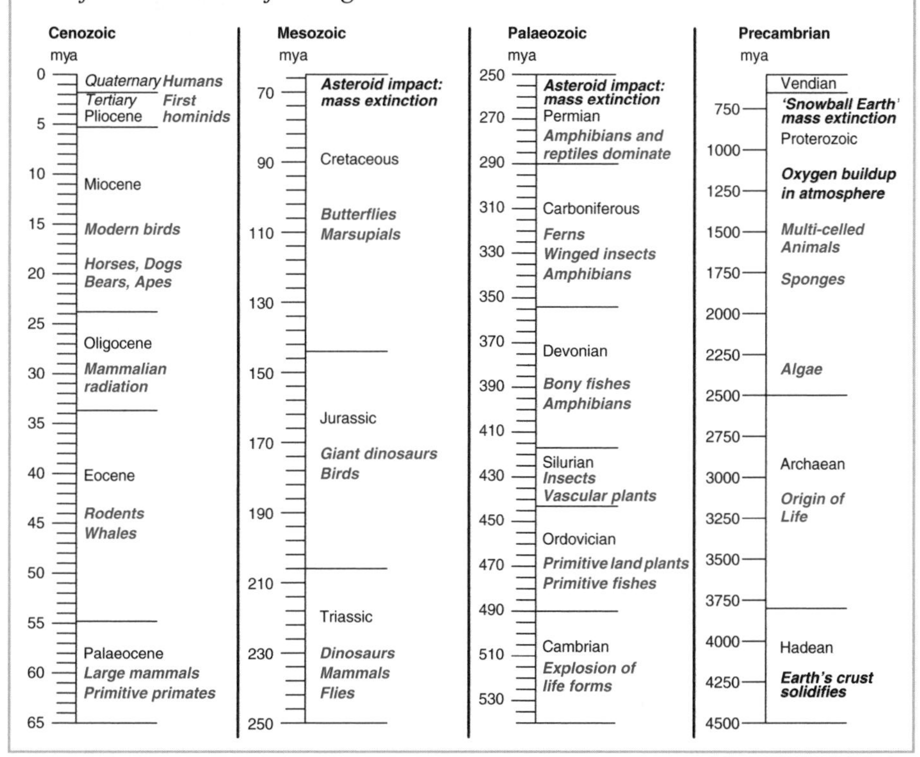

The results of phylogenetic analyses are usually presented in the form of an evolutionary tree. The taxonomy of the ratites—large flightless birds—is a typical example (Fig. 4.8a). The ancestor of the ratites is believed to be a bird that could fly, probably related to the extant tinamous.

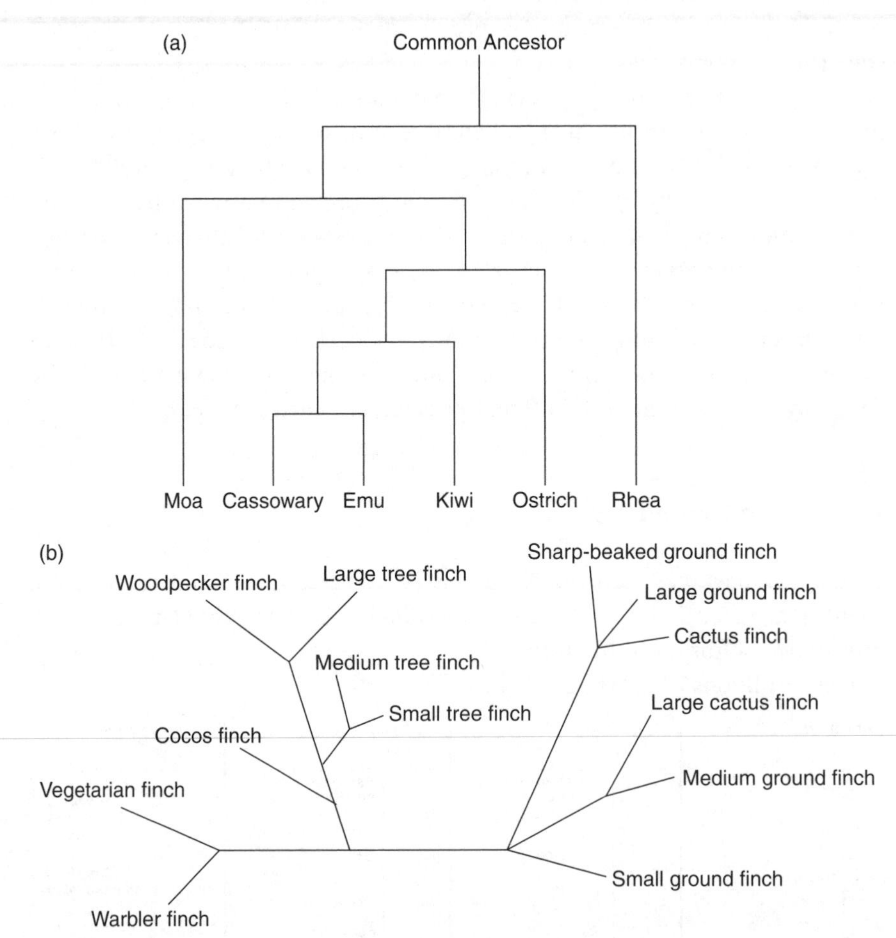

Fig. 4.8 (a) Phylogenetic tree of ratites (large flightless birds) based on mitochondrial DNA sequences. The common ancestor is at the *root* of this tree. A surprising implication of these DNA sequences is that the moa and kiwi are not closest relatives, and therefore that New Zealand must have been colonized twice by ratites or their ancestors. (b) *Unrooted* tree of relationships among finches from the Galapagos and Cocos Islands. Darwin studied the Galapagos finches in 1835, noting the differences in the shapes of their beaks and the correlation of beak shape with diet. Finches that ate fruits had beaks like those of parrots, and finches that ate insects had narrow, prying beaks. These observations were seminal to the development of Darwin's ideas. As early as 1839 he wrote, in *The Voyage of the Beagle*, 'Seeing this gradation and diversity of structure in one small, intimately related group of birds, one might really fancy that from an original paucity of birds in this archipelago, one species had been taken and modified for different ends.'

Such a tree, showing all descendants of a single original ancestral species, is said to be **rooted**. (The root of the tree typically appears at the top or the side; botanists will have to get used to this.) Alternatively, we may be able to specify relationships but not order them according to a history. The relationships among the finches of the Galapagos Islands, studied by Darwin, plus a related species from the nearby Cocos Island, are shown in an **unrooted tree** (Fig. 4.8b). Addition of data from a species on the South American mainland ancestral to the island finches would allow us to **root the tree**.

Statement of a tree of relationships may reveal only the connectivity or topology of the tree, in which case the lengths of the branches contain no information. A more ambitious goal is to show the distances between taxa quantitatively, for instance to label the branches with the time since divergence from common ancestor.

Given a set of data that characterizes different groups of organisms—for example, DNA or protein sequences, or protein structures, or shapes of teeth from different species of animals—how can we derive information about the relationships among the organisms in which they were observed? It is rare for species relationships and ancestry to be directly observable. Evolutionary trees determined from genetic data are often based on inferences from the patterns of similarity, which are all that is observable among species living now. We generally assume that the more similar the characters the more closely related the species, although this is a dangerous assumption. Nevertheless, from the relationships among the characters we wish to infer patterns of ancestry: the *topology* of the phylogenetic relationships (informally, the 'family tree').

To what extent do the topologies of the relationships depend on the choice of character? In particular, are there *systematic* discrepancies between the implications of molecular and palaeontological analysis?

Molecular approaches to phylogeny developed against a background of traditional taxonomy, based on a variety of morphological characters, embryology, and, for fossils, information about the geological context (stratigraphy). The classical methods have some advantages. Traditional taxonomists have much less restricted access to extinct organisms, via the fossil record. They can *date* appearances and extinctions of species by geological methods. Molecular biologists, in contrast, have very limited access to extinct species. Some subfossil remains of species which became extinct as recently as the last century or two have legible DNA, including specimens of the quagga (a relative of the zebra) and the thylacine (Tasmanian 'wolf', a marsupial), and some New Zealand birds (including moas). We have seen, in Chapter 1, an example of a sequence from the mammoth. Some DNA sequences from Neanderthal man have been recovered from an individual who died approximately 30 000 years ago. But *Jurassic Park* remains fiction!

A crucial event in the acceptance of molecular methods occurred in 1967 when V. M. Sarich and A. C. Wilson dated the time of divergence of humans from chimpanzees at 5 million years ago, based on immunological data. At that time palaeontologists dated this split at 15 million years ago, and were reluctant to accept the molecular approach. Reinterpretation of the fossil record led to acceptance of a more recent split, and broke the barrier to general acceptance of molecular methods. (It is now generally accepted that human and chimpanzee lineages diverged between ~6–8 million years ago.)

Indeed, many molecular properties have been used for phylogenetic studies, some surprisingly long ago. Serological cross-reactivity was used from the beginning of the last century until superseded by direct use of sequences. In one of the most premature scientific studies I know of, E. T. Reichert and A. P. Brown published, almost a century ago (in 1909), a phylogenetic analysis of fishes based on haemoglobin crystals. Their work was based on Stenö's law (1669), that although different crystals of the same substance have different dimensions—some are big, some small—they have the same interfacial angles, reflecting the similarity in microscopic arrangement and packing of the atomic or molecular units within the crystals. Reichert and Brown showed that the interfacial angles of crystals of haemoglobins isolated from different species showed patterns of similarity and divergence parallel to the species' taxonomic relationships.

Reichert and Brown's results are replete with significant implications. They show that proteins have definite, fixed shapes, an idea by no means recognized at the time. They imply that as species progressively diverge, the structures of their haemoglobins progressively diverge also. In 1909, no one had a clue about nucleic acid or protein sequences. In principle, therefore, the recognition of evolution of protein structures preceded, by several decades, the idea of evolution of sequences.

Today, DNA sequences provide the best measures of similarities among species for phylogenetic analysis. The data are digital. It is even possible to distinguish selective from non-selective genetic change, using the third position in codons, or untranslated regions such as pseudogenes, or the ratio of synonymous to nonsynonymous codon substitutions. Many genes are available for comparison. This is fortunate, because given a set of species to be studied, it is necessary to find genes that vary at an appropriate rate. Genes that remain almost constant among the species of interest provide no discrimination of degrees of similarity. Genes that vary too much cannot be aligned. There is an analogous situation in radioactive dating requiring choice of an isotope with a half-life of the same general magnitude as the time interval to be determined.

Fortunately genes vary widely in their rates of change. The mammalian mitochondrial genome, a circular double-stranded DNA molecule approximately 16 000 bp long, provides a useful fast-changing set of sequences for the study of evolution among closely-related species. In contrast, ribosomal RNA sequences were used by C. Woese to identify the three major kingdoms: Archaea, Bacteria and Eukarya (see Fig. 1.2).

Conversely, different rates of change of sequences of different genes can lead to different and even contradictory results in phylogenetic studies. This is especially true if what we want is not just the topology of the relationships but the branch lengths. In addition, horizontal gene transfer, and convergent evolution, are competing phenomena—that is, competing with descent—that interfere with the deduction of phylogenetic relationships. In attempts to push the determination of evolutionary relationships farther and farther back in time, it has been found that horizontal gene transfer appears to have been more important early in life history.

Phylogenetic trees

We describe phylogenetic relationships as trees. In computer science, a tree is a particular kind of graph. A graph is a structure containing nodes (abstract points) connected by edges (represented as lines between the points). (See Box; we shall develop the ideas of graphs in connection with the discussion of networks in Chapter 6.) A **path** from one node to another is a consecutive set of edges beginning at one point and ending at the other, like our trip from Malmö to Tromsø. A **connected graph** is a graph containing at least one path between any two nodes. From these we can define a **tree**: a connected graph in which there is *exactly* one path between every two points. A particular node may be selected as a **root**; but this is not necessary—abstract trees may be rooted or unrooted (see Fig. 4.8). Unrooted trees show the topology of relationship but not the pattern of descent. A rooted tree in which every node has two descendants is called a **binary tree** (see PERL program, page 204).

Another special kind of graph is a **directed graph** in which each edge is a one-way street (see page 315). Examples include the Hidden Markov Model diagram shown in Fig. 4.6, and the neural networks illustrated in Chapter 5. Rooted phylogenetic trees are, implicitly, directed graphs, the ancestor-descendent relationship implying the direction of each edge.

It may be possible to assign numbers to the edges of a graph to signify, in some sense, a 'distance' between the nodes connected by the edges. The graph may then be drawn to scale, with the sizes of the edges proportional to the assigned lengths. The length of a path through the graph is the sum of the edge lengths.

In phylogenetic trees, edge lengths signify either some measure of the dissimilarity between two species, or the length of time since their separation. The assumption that differences between properties of living species reflects their divergence times will be true only if the rates of divergence are the same in all branches of the tree. Many exceptions are known. For instance, among mammals many proteins from rodents show relatively fast evolutionary rates (see Weblem 4.8).

PERL Example 4.2 A program to draw binary trees

```
#!/usr/bin/perl
#drawtree.prl -- draws binary trees (root at top)
#usage:  echo '(A((BC)D)(EF))' | drawtree.prl > output.ps

print <<EOF;
%!PS-Adobe-\n%%BoundingBox: atend
/n /newpath load def /m /moveto load def /l /lineto load def
/rm /rmoveto load def /rl /rlineto load def /s /stroke load def
1.0 setlinewidth 50 100 translate 2 2 scale
/Helvetica findfont 10 scalefont setfont
EOF

$tree = <>; chop($tree); $_ = reverse($tree); s/[()]//g;

$x = 0; $y = 0;
while ($nd = chop()) {
   print "$x $y m ($nd) stringwidth pop -0.5 mul 0 rm ($nd) show\n";
   $xx{$nd} = $x; $x+=20; $yy{$nd} = 10;
}

while ($tree =~ s/\(?([A-Z])([A-Z])\)?/\1/) {
    print "n $xx{$1} $yy{$1} m\n";
    ($yy{$1} > $yy{$2}) || {$yy{$1} = $yy{$2}}; $yy{$1} += 20;
    print "$xx{$1} $yy{$1} l $xx{$2} $yy{$1} l $xx{$2} $yy{$2} l s\n";

    $xx{$1} = 0.5*($xx{$1} + $xx{$2});
}
print "n $xx{$tree} $yy{$tree} m 0 20 rl s showpage\n";

$rx = 2*$x + 30; $yt = 2*$yy{$tree} + 146;
print "%%BoundingBox: 40 95 $rx $yt\n";
```

The input: (A((BC)D)(EF)) produces the following output, as a PostScript file, which can be printed on most printers and displayed on most terminals.

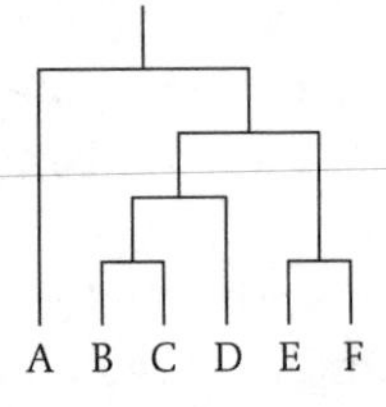

Glossary of terms related to graphs

Graph an abstract structure containing **nodes** (points) and **edges** (lines connecting points).
Path a consecutive set of edges.
Connected graph a graph in which there is at least one path between every two nodes.
Tree a connected graph with exactly one path between every two nodes.
Edge length a number assigned to each edge signifying in some sense the distance between the nodes connected by the edge.
Path length the sum of the lengths of the edges that comprise the path.

Broadly, there are two approaches to deriving phylogenetic trees. One approach makes no reference to any historical model of the relationships. Proceed by measuring a set of distances between species, and generate the tree by a hierarchical clustering procedure. This is called the **phenetic** approach. The alternative, the cladistic approach, is to consider possible pathways of evolution, infer the features of the ancestor at each node, and choose an optimal tree according to some model of evolutionary change. Phenetics is based on similarity; cladistics is based on genealogy.

Clustering methods

Phenetic, or clustering, approaches to determination of phylogenetic relationships are explicitly non-historical. Indeed, hierarchical clustering is perfectly capable of producing a tree even in the absence of evolutionary relationships. A departmental store has goods clustered into sections according to the type of product—for instance, clothing or furniture—and subclustered into more closely-related subdepartments, such as men's and women's shoes. Men's and women's shoes have a common ancestor, but there is no implication that shoes and furniture do.

A simple clustering procedure works as follows: Given a set of species, determine for all pairs a measure of the similarity or difference between them. This could depend on a physical body trait such as the difference between the average adult height of members of two species. Or one could use the number of different bases in alignments of mitochondrial DNA. To create a tree from the set of dissimilarities, first choose the two most closely-related species and insert a node to represent their common ancestor. Then replace the two selected species by a set containing both, and replace the distances from the pair to the others by the average of the distances of the two selected species to the others. Now we have a set of pairwise dissimilarities, not between individual species, but between sets of species. (Regard each remaining individual species as a set containing only one element.) Then repeat the process, as in the following example.

Example 4.7

Consider four species characterized by homologous sequences ATCC, ATGC, TTCG and TCGG. Taking the number of differences as the measure of dissimilarity between each pair of species, use a simple clustering procedure to derive a phylogenetic tree.

The distance matrix is:

	ATCC	ATGC	TTCG	TCGG
ATCC	0	**1**	2	4
ATGC		0	3	3
TTCG			0	2
TCGG				0

Because the matrix is symmetric, we need fill in only the upper half. The smallest nonzero distance is **1** (in boldface), between ATCC and ATGC. Therefore our first cluster is {ATCC, ATGC}. The tree will contain the fragment:

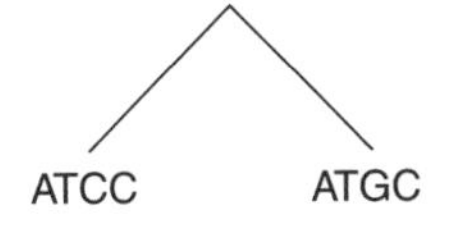

→

→

Example 4.7 *(continued)*

The reduced distance matrix is:

	{ATCC,ATGC}	TTCG	TCGG
{ATCC,ATGC}	0	$\frac{1}{2}(2+3) = 2.5$	$\frac{1}{2}(4+3) = 3.5$
TTCG		0	**2**
TCGG			0

The next cluster is {TTCG, TCGG}, distance **2**. Finally, linking the clusters {ATCC, ATGC} and {TTCG, TCGG} gives the tree:

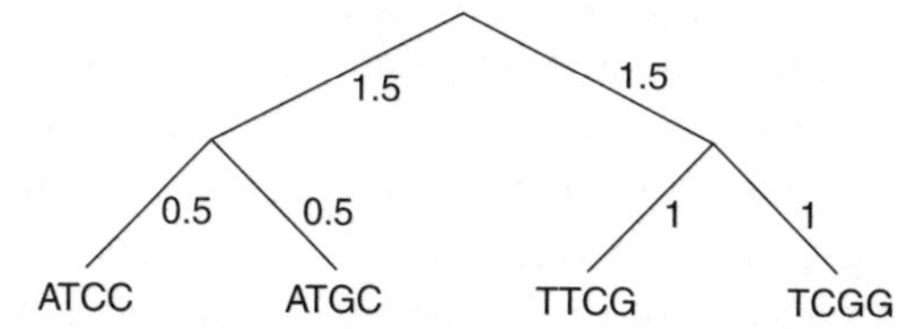

Branch lengths have been assigned according to the rule:

branch length of edge between nodes X and Y = $\frac{1}{2}$ distance between X and Y

Whether the branch lengths are truly proportional to the divergence times of the taxa represented by the nodes must be determined from external evidence.

This process of tree building is called the UPGMA method (Unweighted Pair Group Method with Arithmetic mean). A modification of the UPGMA method by N. Saitou and M. Nei, called Neighbour Joining, is designed to correct for unequal rates of evolution in different branches of the tree.

Cladistic methods

Cladistic methods deal explicitly with the patterns of ancestry implied by the possible trees relating a set of taxa. Their aim is to select the correct tree by utilizing an explicit model of the evolutionary process. The most popular cladistic methods in molecular phylogeny are the **maximum parsimony** and **maximum likelihood** approaches. They are specialized to sequence data, starting from a multiple sequence alignment. Neither maximum parsimony nor maximum likelihood could be applied to anatomic characters such as average adult height.

The *maximum parsimony* method of W. Fitch defines an optimal tree as the one that postulates the fewest mutations. For instance, given species characterized by homologous sequences ATCG, ATGG, TCCA and TTCA, the tree:

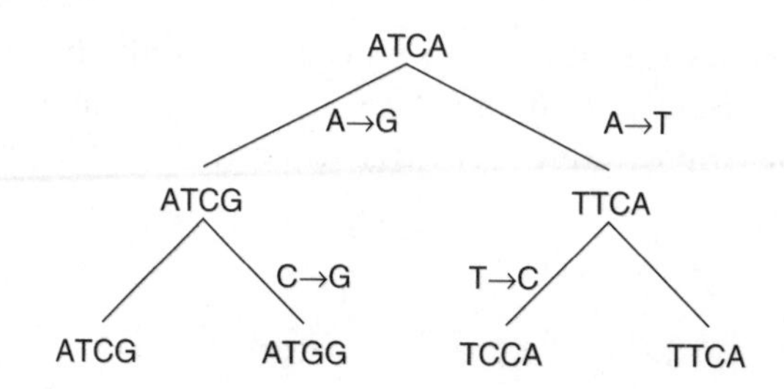

postulates four mutations. An alternative tree:

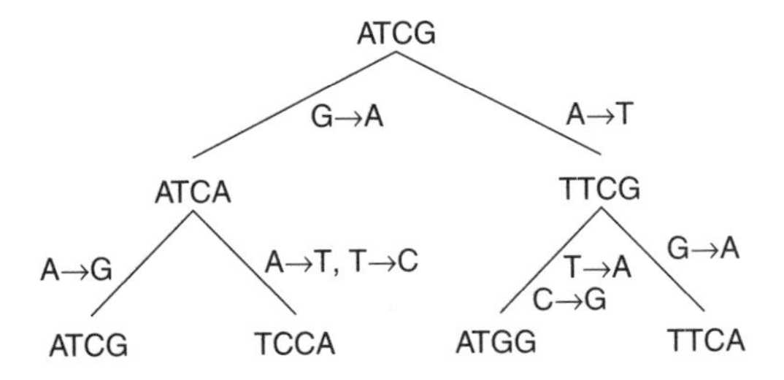

postulates seven mutations. Note that the second tree implies that the G → A mutation in the fourth position occurred twice independently. The former tree is optimal according to the maximum parsimony method, because no other tree involves fewer mutations. In many cases, several trees may postulate the same number of mutations, fewer than any other tree. For such cases the maximum parsimony approach does not give a unique answer.

The *maximum likelihood* method assigns quantitative probabilities to mutational events, rather than merely counting them. Like maximum parsimony, maximum likelihood reconstructs ancestors at all nodes of each tree considered; but it also assigns branch lengths based on the probabilities of the mutational events postulated. For each possible tree topology, the assumed substitution rates are varied to find the parameters that give the highest likelihood of producing the observed sequences. The optimal tree is the one with the highest likelihood of generating the observed data.

Both maximum parsimony and maximum likelihood methods are superior to clustering techniques. This has been demonstrated with cases where independent evidence—for instance, from palaeontology—provides a correct answer, and also with simulated data—computed generation of evolving sequences.

The problem of varying rates of evolution

Suppose that the four species A, B, C and D have the phylogenetic tree:

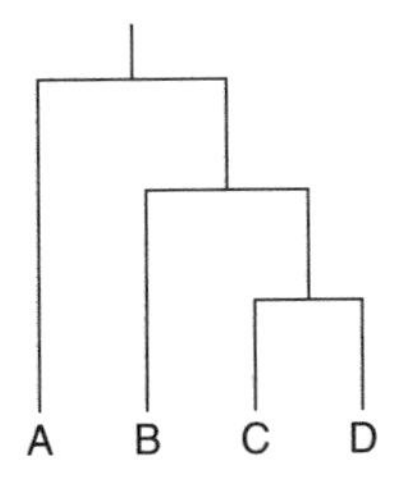

This tree is consistent with the dissimilarity matrix:

	A	B	C	D
A	0	3	3	3
B		0	2	2
C			0	1
D				0

Suppose, however, that taxon D is changing very fast, although the phylogeny is unaltered. The dissimilarity matrix might then be observed to be:

	A	B	C	D
A	0	3	3	20
B		0	2	20
C			0	20
D				0

from which we would derive the incorrect phylogenetic tree:

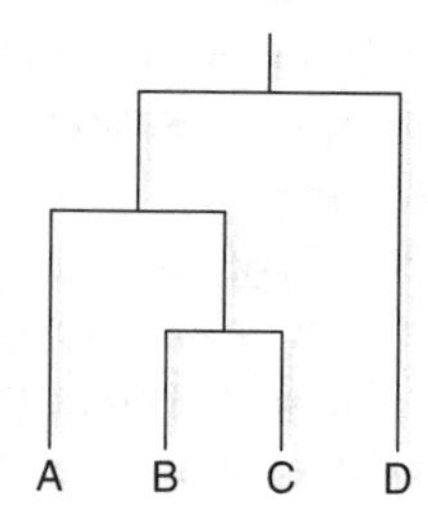

All the methods discussed here are subject to errors of this kind if the rates of evolutionary change vary along different branches of the tree. To test for varying rates, compare the species under consideration with an **outgroup**—a species more distantly related to all the species in question than any pair of them is to each other. For instance, if we are studying species of primates, a non-primate mammal such as the cow would be a suitable outgroup. If the rates of evolution among the primate species were constant, we should expect to observe approximately equal dissimilarity measures between all primate species and the cow. If this is not observed, the suggestion is that evolutionary rates have varied among the primates, and the character being used may well not provide the correct phylogenetic tree.

Computational considerations

Cladistic methods—maximum parsimony and maximum likelihood—are more accurate than simpler clustering methods such as UPGMA, but require large amounts of computer time if the number of species is appreciable. The total number of possible trees, which cladistic methods are committed to considering if they could, increases very rapidly with the number of species. As a result, in many cases of interest these methods can give only approximate answers, even with respect to their intrinsic assumptions.

Because calculated phylogenies are often approximations, it is important to try to test them. Methods include:

1. Comparison of phylogenies obtained from different characters describing the same set of taxa—are they consistent? If trees produced from different characters share a subtree, perhaps that portion of the phylogeny has been determined reliably and other portions have not.

2. Analysis of subsets of taxa should give the same answer—with respect to the subset—as appears within the full tree.
3. Formal statistical tests, involving rerunning the calculation on subsets of the original data, are known as **jackknifing** and **bootstrapping**:
   - **Jackknifing** is calculation with data sets sampled randomly from the original data. For phylogeny calculations from multiple sequence alignments, select different subsets of the positions in the alignment, and rerun the calculation. Finding that each subset gives the same phylogenetic tree lends it credibility. If each subset gives a different tree, none of them is trustworthy.
   - **Bootstrapping** is similar to jackknifing except that the positions chosen at random may include multiple copies of the same position, to form data sets of the same size as the original, to preserve statistical properties of the sampling.
4. If there are very long edges, consider seriously the possibility of unequal variation in evolutionary rate that may have perturbed the calculation. Introduce outgroup taxa to check.

Web resources: Phylogenetic trees

The taxonomic community has expended great effort to produce mature software. The PHYLIP package (PHYLogeny Inference Package) of J. Felsenstein is an integrated collection of many different techniques. The programs work on many different types of computers, and are freely distributed and easily obtained.

Summaries of tools for phylogenetics; includes useful list of web sites, and general listing of phylogeny software:

http://evolution.genetics.washington.edu/phylip/software.html

and Whelan, S., Liò, P. & Goldman, N. (2001), Molecular phylogenetics: State-of-the-art methods for looking into the past, *Trends in Genetics*, 17, 262–272.

Some multiple sequence alignment packages, such as CLUSTAL-W, provide facilities to launch a phylogenetic tree calculation from the alignments they produce.

Recommended reading

Altschul, S. F. & Koonin, E. V. (1998), Iterated profile searches with PSI-BLAST—a tool for discovery in protein databases, *Trends Biochem. Sciences*, **23**, 444–447. [Description of one of the most important tools for database searching for sequence similarity.]

Altschul, S. F., Boguski, M. S., Gish, W. & Wootton, J. C. (1994), Issues in searching molecular sequence databases, *Nature Genetics*, **6**, 119–129. [General background to challenges in designing information retrieval methods and interpreting the results.]

Eddy, S. (1996), Hidden Markov Models, *Current Opinion in Struct. Biol.*, **6**, 361–365. [Readable introduction to an important mathematical technique providing powerful tools for detection of distantly-related sequences, and protein fold recognition.]

Efron, B. & Gong, G. (1983), A leisurely look at the bootstrap, the jackknife, and cross-validation, *The American Statistician*, **37**, 36–48. [Classic paper on statistical methods for calibrating pattern recognition procedures.]

Gura, T. (2000), Bones, molecules . . . or both? *Nature*, **406**, 230–233. [Discussion on the congruences and conflicts between evolutionary trees based on molecules and morphology.]

Penny, D., Hendy, M. D., Zimmer, E. A & Hamby, R. K. (1990), Trees from sequences: Panacea or Pandora's box? *Aust. Syst. Bot.*, **3**, 21–38. [Cautionary notes about determination of phylogenetic trees.]

Whelan, S., Liò, P. & Goldman, N. (2001), Molecular phylogenetics: State-of-the-art methods for looking into the past, *Trends in Genetics*, **17**, 262–272. [Review, with links to software.]

Exercises, Problems, and Weblems

Exercises

4.1 What is the Hamming distance between the words DECLENSION and RECREATION?

4.2 What is the Levenshtein distance between the words BIOINFORMATICS and CONFORMATION?

4.3 The Levenshtein distance between the strings `agtcc` and `cgctca` is 3, consistent with the following alignment:

```
ag-tcc
cgctca
```

Provide a sequence of three edit operations that convert `agtcc` to `cgctca`.

4.4 'I wasted time and now doth time waste me.' (a) First sketch the expected appearance of a dotplot of this character string against itself. (b) Then calculate the dotplot exactly, recording only character identities as dots in the matrix, and compare with (a).

4.5 What values of window and threshold (see program, page 164) would you use to eliminate the singletons in the DOROTHYHODGKIN dotplot, but retain the other matches shown?

4.6 For each of the matrices (a) PAM250 and (b) BLOSUM62, which substitution is more probable, W↔F or H↔R?

4.7 To what alignment does the path through the following dotplot correspond?

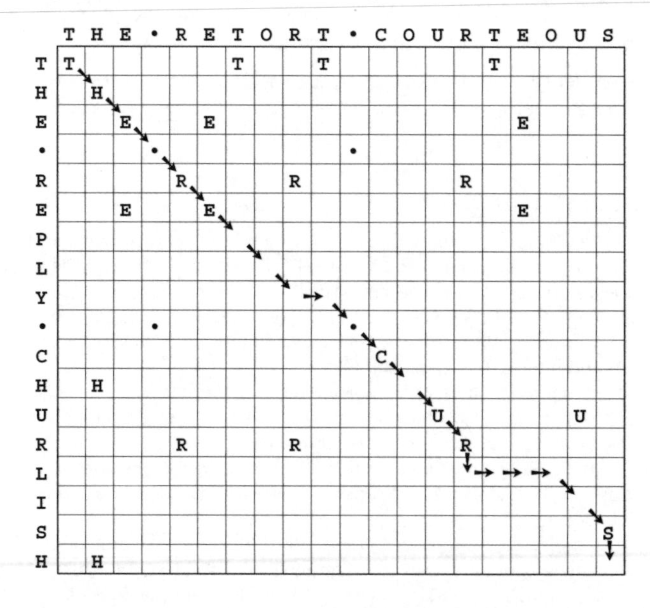

4.8 In planning your trip from Malmö to Tromsø (see page 177), suppose that for personal reasons you wanted to include a visit to Uppsala. How could you adjust the costs of the segments to ensure that the minimal-cost route passes through Uppsala?

4.9 How would you use a dotplot to pick up palindromic DNA sequences of the type that appear partly on each strand, as in the specificity sites of restriction endonucleases?

4.10 Modify the PERL program on page 164 that draws dotplots to accept sequences in FASTA format.

4.11 To what value of *P* would a *Z*-score of 1 correspond in a normal distribution?

4.12 For each of the alignments in Fig. 4.2, state whether it is in the twilight zone, more similar than the twilight zone or less similar than the twilight zone.

4.13 Figure 4.2a shows the sequence alignment of papaya papain and kiwi fruit actinidin, and the corresponding dotplot. The sequence alignment shows two places at which one or more residues are deleted from the papain sequence, and one place at which a residue is deleted from the actinidin sequence. On a photocopy of Fig. 4.2a, indicate in the dotplot the positions of these insertions and deletions.

4.14 Suppose it were argued that randomizing a sequence is not an appropriate way to generate a control population for analysis of the statistical significance of pairwise sequence alignments, because natural sequences have nonrandom dipeptide or tripeptide frequencies. What improved way to generate a control population would you suggest?

4.15 Comparisons of DNA sequences of homologous chromosomes in different people show that, on average, 1 of every 700 bp of noncoding DNA is different. 95% of the human genome is noncoding. Estimate the number of polymorphisms in the human genome, to give some idea of the number of potential DNA markers.

4.16 Show the calculations that lead to the entry with value 65 in Example 4.6. What is the significance of the observation that there two arrows coming from it?

4.17 The α-helix formed by residues 32–49 in *E. coli* thioredoxin is interrupted. On a photocopy of Fig. 4.5 indicate where this interruption appears. At what residue is this distortion likely to occur?

4.18 At what positions in the *E. coli* thioredoxin sequence do turns occur in the structure that are *not* associated with insertions or deletions in the sequence alignment?

4.19 (a) Using simple 'inventory' scoring, what hexapeptide gives the greatest possible value for a match to positions 25–30 in thioredoxin scoring table (page 189). (b) Using a scoring scheme distributed among all 20 amino acids according to the BLOSUM62 matrix, compare the score of this hexapeptide with the score of the hexapeptide VDFSAE.

4.20 (a) Make an inventory of the region from residues 90–95, similar to the table on page 189. What contribution would the following sequences aligned to these residues make to a simple profile score using inventories as weights? (b) ISSAVK (c) FVGAKE.

4.21 (a) Is the following pair of trees identical in topology?

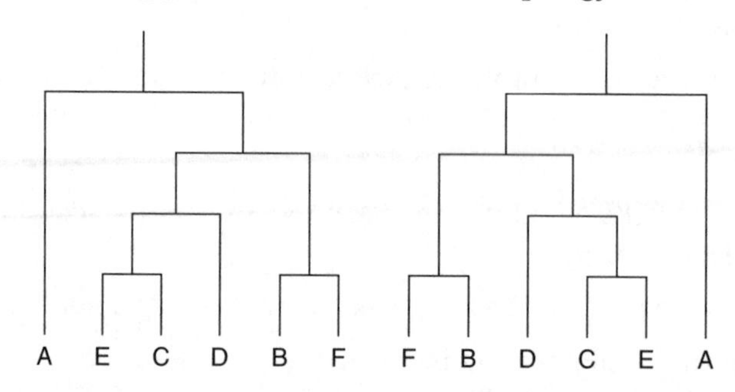

(b) Is the following pair of trees identical in topology?

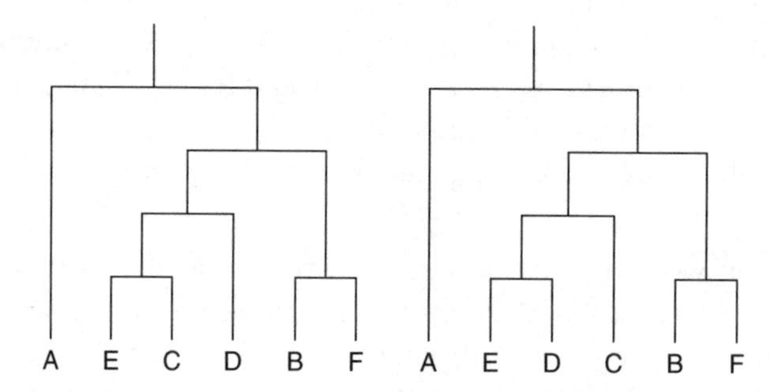

4.22 Draw all possible rooted trees relating three taxa. How many are there?

4.23 For the final graph in Example 4.7, how was the branch length 1.5 of the nodes joining the clusters {ATCC, ATGC} and {TTCG, TCGG} arrived at?

4.24 Using the original matrix of distances, and the final tree, in Example 4.7, for each pair of species compare the original distance between them with the sum of the lengths of the paths joining them in the tree.

4.25 Draw an example of a completely connected graph that is not a tree.

4.26 Mitochondrial DNA sequences of European, African and Asian cattle suggest that European and African breeds are more closely related to each other than to Indian breeds. To exclude the appearance of this result as the spurious result of differential rates of evolution in the two lineages, suggest a reasonable choice of a species as an outgroup.

4.27 Draw the final tree in Example 4.7 to scale, with sizes of the edges proportional to the assigned branch lengths.

4.28 For the dynamic programming method for alignment of two sequences of length n, we noted that the execution *time* requirements scale as n^2. In a naive implementation of the algorithm, how would the storage *space* requirements scale with n : (a) If we want to determine an optimal alignment, so that traceback information must be stored? (b) If we want only the score and not an alignment, so that traceback need not be stored? (Note: subtle ways of implementing the algorithm substantially reduce the space requirements over the naive implementation.)

Problems

4.1 Draw a dotplot of the following sequence from the Wheat dwarf virus genome: `ttttcgtgagtgcgcggaggctttt` against itself. In what respects is it not a perfect palindrome?

4.2 (a) How would you change the algorithm in the section on dynamic programming (starting on page 176) to find the optimal matches of a relatively short pattern $A = a_1a_2 \ldots a_n$ in a long sequence $B = b_1b_2 \ldots b_m$ with $n \ll m$. (No gap penalty in the regions of B that precede and follow the region matched by A.) This corresponds to motif matching as described in Chapter 1. (b) Redo the calculations of Example 4.6 for aligning the strings `ggaatgg` and B = `atg` as a motif matching problem, using the same scoring scheme: match = 0, mismatch = 20, *internal* gap initiation 25, gap extension 22. (c) How do the results that you get differ from those derived in the example?

4.3 How could you modify the profile method to retain its ability to pick up non-mammalian thioredoxins if a large number of additional mammalian, closely-related, sequences were added to the table? Consider (a) methods that attempt to remove redundancy by ignoring certain sequences, and (b) methods that retain all the sequences but include a weighting scheme to balance the representation of the closely-related ones.

4.4 Write a PERL program to make profile inventories from a multiple sequence alignment, and to score the matching of query sequences using BLOSUM62. Assume that the query sequence has already been aligned before it is presented to the program.

4.5 (a) Write a PERL program to read in two character strings and report all matches of contiguous five-character regions. Test this on the strings:

`My.care.is.loss.of.care,.by.old.care.done,` and

`Your.care.is.gain.of.care,.by.new.care.won`

(b) Develop this program to extend and combine the matches found to the longest regions that contain perfectly-matching 5–mers, without gaps, with no more than 25% mismatches overall.

4.6 Extend the previous problem by writing a PERL program to illustrate, in a form based on dotplots, the progress of a BLAST-type algorithm at the stages when it (a) detects all matching substrings of length 5, (b) extends them to maximal contiguous matches, (c) combines them to form a match with no more than k mismatches. You may make use of the PERL program for dotplots in the text.

4.7 Write a program to *animate* the progress of a BLAST-type algorithm as described in the previous problem. As background, look on the Web for examples of animation of string search algorithms. (This problem requires a relatively high level of experience with computing.)

4.8 Single-stranded RNAs, such as tRNA, adopt conformations containing *stem-loop* regions, in which a region of the chain loops back on itself to form a double-stranded helix from complementary base pairs, with antiparallel strands. How would a program that would detect palindromes be useful in analysing RNA sequences to detect regions capable of forming perfect (i.e. no mismatched bases) stem-loop structures?

4.9 Compare the predictions of the residue limits of the transmembrane helices of bacteriorhodopsin by the Phobius method with those observed in the experimental structure (Fig. 4.7).

4.10 Suppose that you have a pair of dice, one red and one green. Define a *state* of the pair of dice as the following pair of numbers: the number appearing on the upper face of the red one followed by the number appearing on the upper face of the green one. Instead of rolling the dice, pass from state to state by tipping each of the dice by 90° in any direction with equal probability. Then a state in which 6 is up on one of the dice can be followed, with equal probability, by 2, 3, 4, or 5 up. (Dice are constructed so that the sum of the numbers on each of the three sets of opposite faces is 7. Therefore the probability that 1 follows 6 is zero, because this would require a 180° rotation.) The probability of the sequence 6, 2, 6, 4 is $(1/4)^4 = 1/256$. The probability of generating the sequence 6, 2, 5, 4 is zero, because the transition from $2 \to 5$ is not allowed, and the probability of the sequence 6, 6, 2, 3, 4 is zero because the system must change its state, so 6 cannot be followed by 6.

This procedure defines a first-order Markov process.

Write a program to answer the following questions: Suppose the initial state has a four at the top of the red die and a three at the top of the green die. (a) What is the probability of another state in which the numbers add up to 7 appearing within 5 moves? (b) If the initial state is an 8, what is the probability that another 8 appears before a 7?

4.11 Show that any (undirected) graph that has either of the following properties must also have the other: (1) There is a unique path between any two nodes. (2) The graph contains no cycles.

4.12 How many paths are there altogether from Start to Finish in Fig. 4.9? Count them in each of the following ways:

(a) Brute force—write down all the possibilities. This is actually less of a mindless exercise than it appears. It demonstrates that it is really not so difficult to do as it first seems. Second, it shows how you will sense patterns as you do it.

(b) In Fig. 4.9, count the number of paths from Start to A and from A to Finish. Multiply these numbers together to get the total number of paths from Start to Finish that pass through A. Then count the number of paths from Start to B and from B to Finish. What is the relationship between these numbers? Multiply them together to get the total number of paths from Start to Finish that pass through B. Compute the total number of paths from Start to Finish as the sum of the number of paths from Start to Finish that pass through A, B, C, and D.

(c) Recognize that to go from Start to Finish requires six steps, including exactly three left turns and three right turns (else you won't end up at the right place). Different choices of the order of right and left turns correspond to different

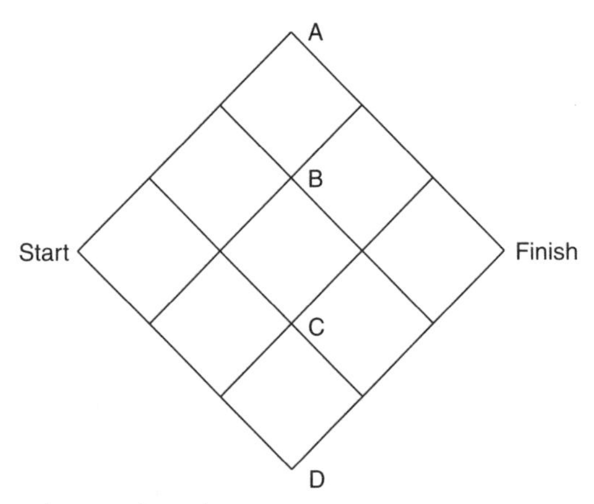

Fig. 4.9 Counting paths on a finite lattice.

paths. To count the number of paths, recognize that you need decide only how many ways there are to choose the three steps at which you turn left (for then you must turn right at the other three steps). To assign three left turns to six steps, first we can choose one of six steps for one left turn, then one of the five remaining steps for the next left turn, and then one of the four remaining steps for the final left turn. However, the product of these numbers overcounts the number of possibilities, because it includes the same sets of steps assigned in different order. Each triple can arise in six different ways, and it is necessary to correct for this. The result is equal to the binomial coefficient $\binom{6}{3} = 6!/(3!3!)$.

4.13 For the final tree in Example 4.7, derive possible ancestors at internal nodes chosen from a maximum parsimony criterion. Are there any ambiguities?

4.14 A convenient notation for trees uses nested parentheses to indicate the clusters. (a) Expand the following into a rooted tree: (A(BC)D). (b) Write the parenthesis notation for the trees shown in Exercise 4.21.

4.15 Add an adequate amount of comments to the PERL program for drawing trees.

4.16 Write a PERL program for the UPGMA method of deriving a phylogenetic tree from a matrix of distances. You may use the program for tree drawing to produce graphical output.

Weblems

4.1 Retrieve the gene sequence of mitochondrial ATPase subunit 6 from Atlantic hagfish (*Myxine glutinosa*). Draw a dotplot against the homologous gene from sea lamprey (*Petromyzon marinus*). Comment on the similarity observed and compare with the similarity between the lamprey and dogfish sequences shown on page 162.

4.2 Submit the amino acid sequence of papaya papain to a BLAST search and to a PSI-BLAST search. Which of the homologues appearing in Fig. 4.2 are successfully detected by BLAST? Which by PSI-BLAST?

4.3 Submit the amino acid sequence of papaya papain to a PSI-BLAST search (see previous weblem). In the results for the match to human procathepsin L, indicate on a photocopy of the dotplot (Fig. 4.2b) the regions of local matches reported.

4.4 Find structures of thioredoxins appearing in the alignment table in Plate III, from organisms other than *E. coli*. On a photocopy of the alignment table, indicate the regions of helix and strands of sheet as assigned in the Protein Data Bank entries, and compare with the helices and strands of *E. coli* thioredoxin.

4.5 Align the amino acid sequence of papaya papain and the homologues shown in Fig. 4.2 using CLUSTAL-W or T-Coffee. Compare the results with the alignment table in Pfam based on Hidden Markov Models, and with the structural alignments in Fig. 4.2.

4.6 Can PSI-BLAST identify the homology between immunoglobulin domains and the domains of *Cellulomonas fimi* endogluconase C and *Streptococcus agalactiae* IgA receptor?

4.7 (a) Can PSI-BLAST identify the relationship between *Klebsiella aerogenes* urease, *Pseudomonas diminuta* phosphotriesterase and mouse adenosine deaminase? (b) Compare the alignments of these three sequences produced by DALI and by CLUSTAL-W or T-Coffee.

4.8 The growth hormones in most mammals have very similar amino acid sequences. (The growth hormones of the Alpaca, Dog, Cat, Horse, Rabbit, and Elephant each differ from that of the Pig at no more than 3 positions out of 191.) Human growth hormone is very different, differing at 62 positions. The evolution of growth hormone accelerated sharply in the line leading to humans. By retrieving and aligning growth hormone sequences from species closely related to humans and our ancestors, determine *where* in the evolutionary tree leading to humans the accelerated evolution of growth hormone took place.

The next series of weblems is designed to place the human species in its biological context by analysis of sequences from near and distant relatives, and to illustrate some of the variety of genetic information that has been used to investigate phylogenetic relationships.

4.9 The living species most closely related to humans are apes and monkeys. Alu elements are a type of SINE (Short INterspersed Element) useful as species markers. Although part of the repetitive noncoding portion of the genome, some Alu elements function in gene regulation. On the basis of Alu elements that regulate the genes for parathyroid hormone, the haematopoietic cell-specific FcϵRI-γ receptor, the central nervous system-specific nicotinic acetylcholine receptor α3, and the T-cell-specific CD8α, derive a phylogenetic tree for human, chimpanzee, gorilla, orangutan, baboon, rhesus monkey and macaque monkey.

4.10 Humans are primates, an order that we, apes, and monkeys share with lemurs and tarsiers. On the basis of the β-globin gene cluster of human, a chimpanzee, an old-world monkey, a new-world monkey, a lemur, and a tarsier, derive a phylogenetic tree of these groups.

4.11 Primates are mammals, a class we share with marsupials and monotremes. Extant marsupials live primarily in Australia and neighbouring islands, except for the oppossum, found in North and South America. Extant monotremes are limited to three species: the platypus and two related species of echidna. Collect the nuclear genes for mannose 6-phosphate/insulin-like growth factor II receptor from mammalian species including placentals, marsupials and monotremes. From them draw an evolutionary tree, indicating branch lengths. Are monotremes more closely related to placental mammals or to marsupials?

4.12 Mammals are vertebrates, a subphylum that we share with fishes, sharks, birds and reptiles, amphibia, and primitive jawless fishes (example: lampreys). For the coelacanth (*Latimeria chalumnae*), the great white shark (*Carcharodon carcharias*), skipjack tuna (*Katsuwonus pelamis*), sea lamprey (*Petromyzon marinus*), frog (*Rana pipens*), and Nile crocodile (*Crocodylus niloticus*), using sequences of cytochromes *c* and pancreatic ribonucleases, derive evolutionary trees of these species.

4.13 Tetrapods are gnathostomes, a superclass that we share with fishes. The traditional view of fish → tetrapod evolution is that jawed vertebrates split into one group containing cartilaginous fishes, *Chondrichthyes*, including sharks and rays, and another containing both ray-finned fishes, *Actinopterygii*, including modern bony fishes such as cod and salmon, and lobe-finned fishes, *Sarcopterygii*, including coelacanths, lungfishes and tetrapods (see Fig. 4.10a). Test this hypothesis using at least 12 mitochondrial protein-coding genes from at least 30 species including sharks, lungfish, bony fishes, amphybia, reptiles, birds, and mammals, using a lamprey as an outgroup. Of the three groups—cartilaginous fishes, bony fishes and tetrapods—which pair appears to be most closely related: bony fishes and tetrapods as in the traditional view, or a different pair? Consider specifically a phylogeny according to which the tetrapods split off first, and cartilaginous fishes, bony fishes, and even lungfishes are sister taxa (see Fig. 4.10b).

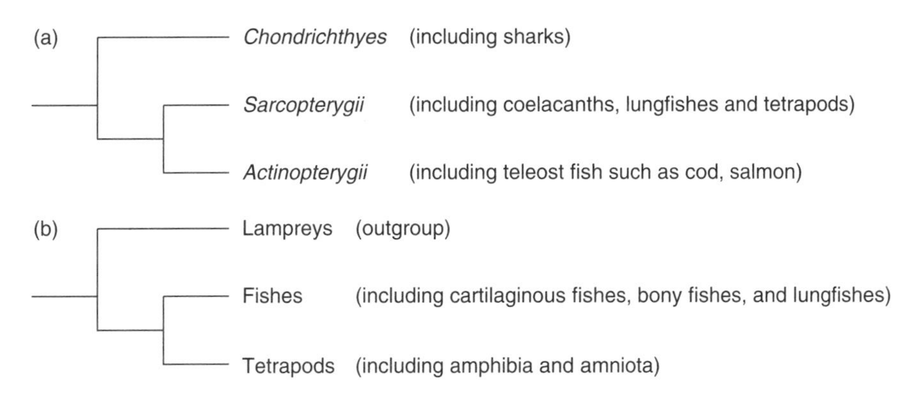

Fig. 4.10 (a) Traditional view of gnathostome evolution: The first split was between cartilaginous fishes such as sharks and others; then tetrapods emerged from a group containing coelecanths, lungfishes and tetrapods split off. An alternative to be considered is that the split leading to the tetrapods was the earliest. (b) Alternative tree, rooted using lamprey as an outgroup, in which the lineage leading to tetrapods split off first.

4.14 Vertebrates are chordates, a phylum that also includes lancelets (small fish-like marine animals; example: amphioxus), and jawless vertebrates (lacking a true vertebral column (example: lamprey). As in other organisms with bilateral symmetry (including insects), vertebrate HOX genes encode a family of DNA-binding proteins. The expression of these genes varies along the head-to-tail body axis, and controls the setting out of the body plan. Indeed there is an amazing mapping between the order of the genes on the chromosome, the order of their action along the body, and the relative times during development of the onset of their activity.

During the course of vertebrate evolution there have been large scale genomic duplications, associated presumably with the development of greater complexity of body architecture, as presciently suggested by S. Ohno in 1970. The genomes of insects and amphioxus have a single HOX cluster. Zebrafish have seven HOX clusters, interpretable in terms of a series of duplications: $1 \rightarrow 2 \rightarrow 4 \rightarrow 8$ followed by loss of one to reduce $8 \rightarrow 7$.

Find the number of HOX clusters in the human and the lamprey, perform a multiple sequence alignment to assign correspondences among the individual genes, and derive from the results a phylogenetic tree for amphioxus, lamprey, fishes, and mammals.

4.15 Chordates are deuterostomes (see Fig. 1.3), a grouping we share with urochordates (example: sea squirts), hemicordates (example: amphioxus), and echinoderms (example: starfish). There are systematic differences between these four phyla in their mitochondrial genetic codes. Determine, for examples of organisms in each phylum, the amino acids that correspond to the codons ATA and AGA. Derive from these results a phylogenetic tree of the four deuterostome phyla.

CHAPTER 5

Protein structure and drug discovery

Chapter contents

Introduction *220*

Protein stability and folding *223*

The Sasisekharan-Ramakrishnan-Ramachandran plot describes allowed mainchain conformations *223*

The sidechains *225*

Protein stability and denaturation *225*

Protein folding *228*

Applications of hydrophobicity *229*

Superposition of structures, and structural alignments *233*

DALI (Distance-matrix ALIgnment) *235*

Evolution of protein structures *236*

Classifications of protein structures *238*

SCOP *239*

Protein structure prediction and modelling *240*

Critical Assessment of Structure Prediction (CASP) *242*

Secondary structure prediction *244*

Homology modelling *250*

Fold recognition *252*

Conformational energy calculations and molecular dynamics *255*

ROSETTA *259*

LINUS *259*

Assignment of protein structures to genomes *263*

Prediction of protein function *265*

Divergence of function: orthologues and paralogues *266*

Drug discovery and development *269*

The lead compound *271*

Bioinformatics in drug discovery and development *273*

Recommended reading 284

Exercises, Problems, and Weblems 285

Learning goals

1. To understand the concept of protein folding: the process by which the one-dimensional amino acid sequence encoded by a gene takes up a definite and biologically active three-dimensional conformation.
2. To recognize that steric considerations severely limit the conformations of the polypeptide chain, with the Sasisekharan-Ramakrishnan-Ramachandran plot showing the allowed states of the mainchain.
3. To get to know the 20 sidechains—the actors that play all the roles in all the proteins.
4. To understand the hydrophobic effect and its implications for the structures and energetics of folded proteins.
5. To generalize the ideas of sequence alignment to the alignment of protein sequences by structural superposition.
6. To know the relationship between divergence of sequence and divergence of structure in protein evolution.
7. To become familiar with classification of protein folding patterns, as presented for example, by the Structural Classification of Proteins (SCOP) database and web site.
8. To know some basic approaches to the prediction of protein structure from amino acid sequence, and the state of the art as revealed in the Critical Assessment of Structure Prediction (CASP) programmes.
9. To understand the basis of Hidden Markov Models, the most powerful methods now available for deducing affinities between proteins from their sequences.
10. To know the basic requirements for a successful drug, and understand some approaches to drug discovery and design.

Introduction

The great variety of three-dimensional structures and functions of proteins arise in molecules that share underlying common features. Chemically, proteins are like strings of Christmas tree lights: Each protein consists of a linear (that is, unbranched) polymer mainchain with different amino acid sidechains attached at regular intervals (Fig. 1.6). The wire linking the string of lights corresponds to the repetitive mainchain or backbone, and the variable sequence of colours of the lights corresponds to the individuality of the sequence of sidechains.

The amino acid sequence of a protein is specified by the nucleotide sequence of a gene. The three-dimensional structures of protein molecules are determined, without further participation of nucleic acids, by the one-dimensional sequences of their amino acids. Proteins fold spontaneously to their native conformations.

How does the amino acid sequence encode the three-dimensional structure? Any possible folding of the mainchain places different residues into contact. The interactions of the sidechains and mainchain, with one another and with the solvent, and the restrictions placed on sidechain mobility, determine the relative stabilities of different conformations. This is a consequence of the second law of thermodynamics, which states that systems at constant temperature and pressure find an equilibrium state that is a compromise between comfort (low enthalpy, H) and freedom (high entropy, S), to give a minimum Gibbs free energy $G = H - TS$, in which T is the absolute temperature. (In human relationships, marriage is just such a compromise.)

Proteins have evolved so that one folding pattern of the mainchain is thermodynamically significantly better than other conformations. This is the native state. If we could calculate sufficiently accurately the energies and entropies of different conformations, and if we could computationally examine a large enough set of possible conformations to be sure of including the correct one, it would be possible consistently to predict protein structures from amino acid sequences on the basis of a priori physicochemical principles. There has been progress towards this goal but it has not yet been achieved.

The mainchain of each protein in its native state describes a curve in space. We now know the structures of 30 000 proteins (including many replicates or single-site mutants), and see in them a great variety of spatial patterns. The first problem in analysing these structures is one of presentation. Figure 5.1 illustrates, for the small protein acylphosphatase, the difficulty in interpreting a fully-detailed, literal representation, and the kind of simplified pictures that computer programs produce to give us visual access to the material. An active cottage industry has produced many different simplified representations. A skilled molecular illustrator will combine them to show different parts of a structure in finely-tuned degrees of detail.

The central frame of Fig. 5.1 shows the course through space of the mainchain of acylphosphatase. Two regions at the front of the picture have the form of helices—like classic barber's poles—with their axes almost vertical in the orientation shown. Acylphosphatase also contains four strands of sheet. These too are approximately vertical in orientation. The four strands interact laterally to stabilize their assembly into a β-sheet. In the bottom frame, helices and strands are represented as 'icons': helices as cylinders and strands of sheet as large arrows. The top frame of Fig. 5.1, showing the most detailed representation of the structure, including mainchain and sidechains, indicates the importance of simplification in producing an intelligible picture of even a small protein.

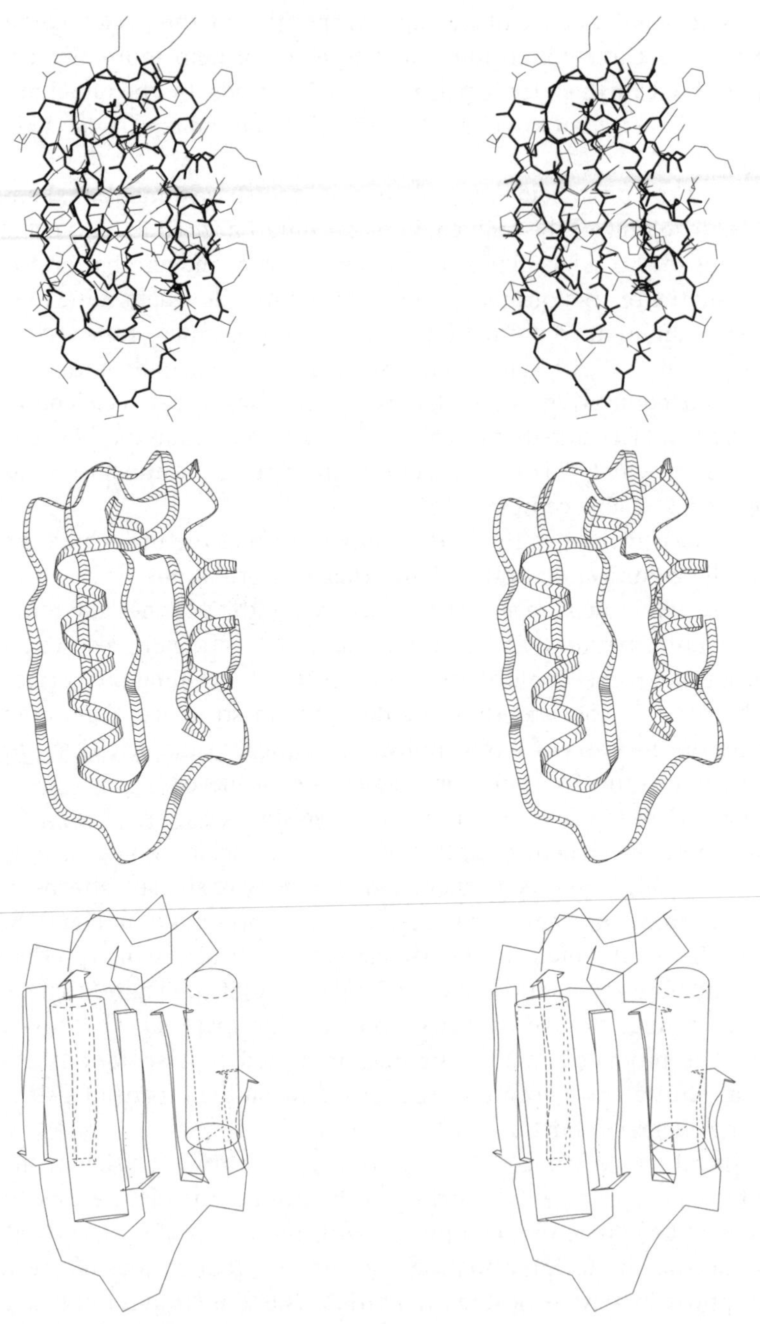

Fig. 5.1 Proteins are sufficiently complex structures that it has been necessary to develop specialized tools to present them. This figure shows a relatively small protein, acylphosphatase, at three different degrees of simplification. Top: complete skeletal model; mainchain bolder than sidechains. Centre: the course of the chain is represented by a smooth interpolated curve, the chevrons indicating the direction of the chain. Bottom: schematic diagram, in which cylinders represent helices and arrows represent strands of sheet. The solid objects in the picture are represented as 'translucent' by altering lines that pass behind them to broken lines. It is possible to superpose different representations visually by rotating the page 90° and viewing in stereo (but not for too long!).

Protein stability and folding

Although it is not yet possible to predict the structures of proteins from basic physical principles alone, we do understand the general nature of the interactions that determine protein structures.

To form the native structure, the protein must optimize the interactions within and between residues, subject to constraints on the space curve traced out by the mainchain. Preferred conformations of the mainchain bias the folding pattern towards recurrent structural patterns: helices, extended regions that interact to form sheets, and several standard types of turns.

The Sasisekharan-Ramakrishnan-Ramachandran plot describes allowed mainchain conformations

To a good approximation, the mainchain conformation of each non-glycine residue is restricted to two discrete conformational states.

A fragment of the linear polypeptide chain common to all protein structures is shown in Fig. 5.2. Rotation is permitted around the N-Cα and Cα-C single bonds of all residues (with one exception: proline). The angles ϕ and ψ around these bonds, and the angle of rotation around the peptide bond, ω, define the conformation of a residue. The peptide bond itself tends to be planar, with two allowed states: *trans*, $\omega \approx 180°$ (usually) and *cis*, $\omega \approx 0°$ (rarely, and in most cases at a proline residue). The sequence of ψ, ϕ and ω angles of all residues in a protein defines the backbone conformation.

The principle that two atoms cannot occupy the same space limits the values of conformational angles. The allowed ranges of ϕ and ψ, for $\omega = 180°$, fall into

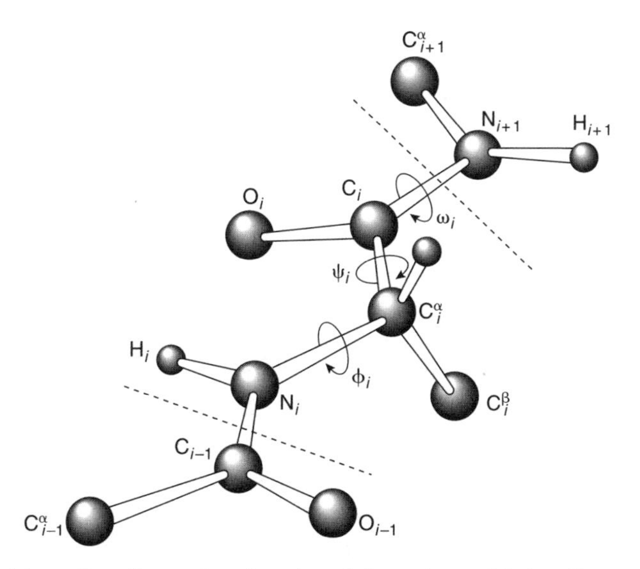

Fig. 5.2 Definition of conformational angles of the polypeptide backbone.

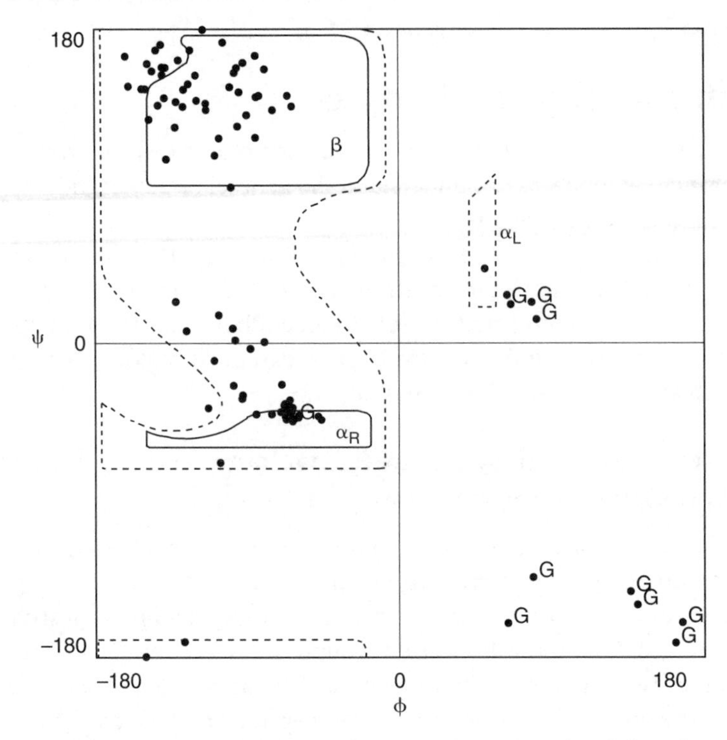

Fig. 5.3 A Sasisekharan-Ramakrishnan-Ramachandran plot of acylphosphatase (PDB code 2ACY). Note the clustering of residues in the α and β regions, and that most of the exceptions occur in Glycine residues (labelled G).

defined regions in a graph called a Sasisekharan-Ramakrishnan-Ramachandran plot—usually shortened to 'Ramachandran plot' (see Fig. 5.3). Solid lines in the figure delimit energetically-preferred regions of φ and ψ; broken lines in the figure delimit sterically-disallowed regions. The conformations of most amino acids fall into either the α_R or β regions. Glycine has access to additional conformations. In particular it can form a left-handed helix: α_L. Figure 5.3 shows the typical distribution of residue conformations in a well-determined protein structure. Most residues fall in or near the allowed regions, although a few are forced by the folding into energetically less-favourable states.

The allowed regions generate standard conformations. A stretch of consecutive residues in the α conformation (typically 6–20 in native states of globular proteins) generates an α-helix. Repeating the β conformation generates an extended β-strand. Two or more β-strands can interact laterally to form β-sheets, as in acylphosphatase (Fig. 5.1). Helices and sheets are 'standard' or 'prefabricated' structural pieces that form components of the conformations of most proteins. They are stabilized by relatively weak interactions, **hydrogen bonds**, between mainchain atoms. In some fibrous proteins virtually all of the residues belong to one of these types of structure: wool contains α-helices; silk β-sheets. Amyloid fibrils, formed in disease states by many proteins, also contain extensive β-sheets. Steric interactions permit a stretch of consecutive residues to be all in the α_R conformation or all in the β conformation, but disallow a helix followed by a strand and vice versa.

Typical globular proteins contain several helix and/or sheet regions, connected by *turns*. Usually the ends of helix or strand regions appear on the surface of a domain of a protein structure. They are connected by turns, or loops—regions in which the chain alters direction to point back into the structure. Many but not all turns are short, surface-exposed regions that tend to contain charged or polar residues.

How does the mainchain choose among the possible allowed conformations? What is unique about each protein is the sequence of its sidechains. Therefore interactions involving sidechains must determine the mainchain conformation.

The sidechains

Sidechains offer the physicochemical versatility required to generate all the different folding patterns. The sidechains of the twenty amino acids vary in:

- **Size** The smallest, glycine, consists of only a hydrogen atom; one of the largest, phenylalanine, contains a benzene ring.
- **Electric charge** Some sidechains bear a net positive or negative charge at normal pH. Asp and Glu are negatively charged. Lys and Arg are positively charged. (Charged residues of opposite sign can form attractive pairwise interactions called **salt bridges**.)
- **Polarity** Some sidechains are polar; they can form hydrogen bonds to other polar sidechains, or to the mainchain, or to water. Other sidechains are electrically neutral. Some of these contain chemical groups related to ordinary hydrocarbons such as methane or benzene. Because of the thermodynamically unfavourable interaction of hydrocarbons with water, these are called 'hydrophobic' residues. Congregation of hydrophobic residues in protein interiors, predicted by W. J. Kauzmann before the first protein structures were determined, is an important contribution to protein stability. This effect is analogous to the formation of droplets of oil in salad dressing (see Box: The hydrophobic effect).
- **Shape and rigidity** The overall shape of a sidechain depends on its chemical structure and on its degrees of internal conformational freedom.

Protein stability and denaturation

What are the chemical forces that stabilize native protein structures? What is the process by which a protein folds from an ensemble of denatured conformations to a unique native state?

To address these questions, biochemists have studied the denaturation of proteins in response to heat, or to increasing concentrations of urea or guanidinium hydrochloride (commonly-used denaturants). Some measurements are *static*—determination of the amount of native and denatured states at equilibrium under different conditions, or the heat released at points along the transition. Others are *kinetic*—measurement of rates of folding or unfolding, or identification of structures that appear transiently during the process.

The hydrophobic effect

The difference among the different amino acid sidechains in their preferences for aqueous or oil-like environments is one of the governing principles of protein structure.

What is the hydrophobic effect? It is the sparing solubility of non-polar solutes in water, arising from the microscopic structure of liquid water around such solutes. Phase separation in oil-water mixtures—for instance, salad dressing—is one common example. Another is that gases (unlike most solids) are less soluble in water as the temperature increases. Readers with whistling tea kettles will have heard low levels of sound prior to proper boiling, as the dissolved air comes out of solution as the water is heated.

What is the origin of the hydrophobic effect? Cold water is a highly structured liquid. It contains many hydrogen bonds, which account for its high heat of vaporization and low density. But water is even more highly ordered around solutes than in the pure liquid. Methane dissolved in water—it is only slightly soluble, but soluble enough to study—is surrounded by a cage of water molecules called a clathrate complex. As a result, dissolving methane in water makes the solvent even *more* ordered, lowering the entropy. The natural tendency toward states of higher entropy inhibits the dissolving of methane in water. This is why methane and other hydrocarbons are only very slightly water-soluble. The solubilities of nonpolar gases decrease upon heating—from an already small value in cold water—because as the temperature increases entropy plays an even more important role in determining the equilibrium state.

The hydrophobic effect in aqueous solutions of simple nonpolar solutes was well known to physical chemists when W. J. Kauzmann, in 1959, recognized its importance for protein structure.

The nonpolar sidechains of proteins are similar to oil-like solutes. Their interaction with water is unfavourable. Kauzmann predicted that they would be sequestered in protein interiors, away from the solvent. This *oil-drop model of protein interiors* was confirmed by the X-ray crystal structures of globular proteins. We now recognize also the importance of high packing densities in protein interiors, and that it is better to regard the interior of a folded protein as more like a crystal than like an organic liquid. But the hydrophobic effect has lost none of its significance.

The backbone must traverse the interiors of the protein, and carries with it the polar N and O atoms of the peptide groups, which can interact with other polar mainchain atoms and with polar sidechains such as threonine or asparagine. Thus the interior is not completely oil-like. However, charged residues are almost completely excluded from protein interiors; in rare cases they form internal salt bridges. Conversely, the surface of a protein is not exclusively charged or polar. About half the residues on the surface of a protein are nonpolar.

One important message is that proteins are only marginally stable. The native state of globular proteins is typically only 20–60 kJ mol^{-1} (5–15 kcal mol^{-1}) more stable than the denatured state. This is the equivalent of about one or two water–water hydrogen bonds.

Precisely why proteins have marginal stability is unclear. Some people believe that it facilitates protein turnover. Others suggest that proteins are as stable as they need to be so 'why bother' (less informally: there is no selective advantage in) further optimizing the stabilizing interactions. We do know that the interactions that stabilize native proteins are capable of producing protein structures with much higher stabilities.

Suppose you are a globular protein in aqueous solution, and you want to achieve a stable native state. Your major problem is the great loss of conformational freedom, relative to the ensemble of denatured states, that is exacted from you in adopting a unique conformation. This entails a large reduction in entropy, which is thermodynamically unfavourable. One way in which you can compensate is to form a compact globular state, burying many residues in the interior away from contact with water. The release of water from interaction with the nonpolar atoms of the protein produces a compensating *increase* in entropy arising from the hydrophobic effect (see Box).

That's fine, but now you discover that to form the compact state you have buried many polar atoms, including but not limited to mainchain nitrogen and carbonyl oxygens. In the denatured state, these atoms make hydrogen bonds to water. When buried in the interior, their hydrogen-bonding potential must somehow be satisfied. (Don't forget: one or two uncompensated hydrogen bonds and you've blown it; your native state would be unstable.) A fairly general-purpose solution that satisfies mainchain hydrogen-bonding potential is to form helices or sheets.

There is a bonus: Formation of secondary structure also ensures that the mainchain is in a stereochemically acceptable conformation, as limited by the Sasisekharan-Ramakrishnan-Ramachandran plot. Residues in α-helices are all in the α conformation; residues in strands of β-sheet are all in the β conformation.

How do you decide which regions should form helices or strands? Enthalpically, helix and sheet are reasonably similar for most residues. However, entropically, some sidechains are more hindered in helices than in strands; these prefer strands. These effects bias the formation of secondary structures. Specific sequences providing sidechain-mainchain hydrogen bonds form **helix caps**, governing where α-helices begin and end.

How compact is the globular state required to be? You could achieve exclusion of water from your interior by fairly loose packing—as long as no channel is larger than 1.4 Å in radius (the size of a water molecule). But the closer together you can squeeze your atoms, the better advantage you can take of Van der Waals forces, general forces of attraction between atoms that give matter its general cohesion. Protein interiors are densely packed: the fitting together of the sidechains is like a solved jigsaw puzzle. However, the puzzle pieces (the residues) are deformable, so the folding process is more complicated than the rigid matching of pieces in ordinary jigsaw puzzles.

In summary, you have to find a conformation of the chain that simultaneously solves all the following problems:

1. All residues must have stereochemically allowed conformations. This applies to both the mainchain and the sidechains. Steric collisions would raise the energy of the conformation and render it unstable.
2. Buried polar atoms must be hydrogen-bonded to other buried polar atoms. If you miss out a few hydrogen bonds, the protein will prefer to form the denatured state in order to allow these polar atoms to hydrogen-bond to solvent.
3. Enough hydrophobic surface must be buried, and the interior must be sufficiently densely packed, to provide thermodynamic stability.

For most proteins, there is a unique solution of all these problems, and this defines the native state. Some proteins change conformation when they bind ligands, or pass through metastable states, as part of their mechanisms of function.

The fact that one conformation of a protein—the native state—has substantially greater stability than other conformations is complex but not mysterious. It is a question of optimizing the available interactions, and selecting sequences for which this optimum is unique and substantially lower than others. For most regions the local structure is determined by local interactions. Therefore if the native state were not unique there would have to be more than one way to fit a given set of pieces together. Given the chain constraints it is easy for evolution to avoid this.

Protein folding

Suppose again that you are a protein, and that you are denatured. Now that you understand how your native state is stabilized, how would you go about finding it? Clearly you can't try all conformations—many years ago C. Levinthal calculated that a simple conformational search, using reasonable numbers for speeds of internal rotations, would require much too much time. Two circumstances conspire to make the *process* by which proteins fold to their native states mysterious as well as complex.

First is the fact that proteins are only marginally stable. This implies that any quasi-stable intermediate in protein folding must be even less stable, else the folding process would get trapped in the intermediates. Indeed, for many proteins, measurements of fractions of molecules in native and denatured states as a function of temperature or denaturant concentration imply simple, two-state, Native ↔ Denatured equilibria in which undetectably few molecules are anything but native or denatured. This confirms that any putative intermediates can have no more than marginal stability. But this makes it difficult to follow the folding transition structurally.

The second circumstance that makes protein folding mysterious is that the denatured state is so heterogeneous that in the absence of stable intermediates there is no convenient way to visualize the complete pathway.

Contrast protein folding with two other types of structure formation:

1. In assembling do-it-yourself furniture, one passes through a succession of well-defined intermediate states. First one screws A to B in the native-like conformation. The structure of the A–B fragment is determined and stabilized purely by the interactions between A and B. Were it not for gravity, a stable A–B intermediate would be formed. But proteins don't have the luxury of forming stable intermediates.
2. In assembling an arch from its voussoirs, the structure as a whole has no stability until the keystone is inserted. Only the completed arch has independent stability, there are no stable intermediates, and the only way to assemble the structure is by using scaffolding which is subsequently removed. But proteins don't have the luxury of using external scaffolding.

What proteins have to do is to work with unstable intermediates—like do-it-yourself furniture in the *presence* of gravity—and to get the job finished before the intermediates fall apart, or else to keep reforming them and trying again.

Identification of transient structure during protein folding can be achieved experimentally by isotope exchange measurements. Prepare a sample of denatured protein in which all hydrogen atoms are replaced by deuterium. (It is possible to separate signals from H and D in NMR experiments.) At various times during refolding, in separate experiments, expose the sample to a pulse of protons. After the native state is formed, detect where in the structure D $\leftrightarrow$ H exchange occurred and when. Such studies justify the model that many proteins fold by initial formation of a 'molten globule' containing some native secondary structure, but without the tertiary structural interactions that lock the molecule into its final conformation. This is followed by a hierarchical condensation to form supersecondary structure, etc., leading eventually to accretion of the native state. For most proteins, there is no evidence for non-native structures as intermediates along productive folding pathways, although non-native structures—such as incorrect proline isomers—can divert and thereby slow down the folding process.

The conclusion is that structures of local regions are determined primarily by local interactions, and, although these interactions may be inadequate to stabilize local regions to the point where they can be isolated, they are good enough to provide a low-energy pathway for structure assembly.

Applications of hydrophobicity

Using a **hydrophobicity scale** that assigns a value to each amino acid, we can plot the variation of hydrophobicity along the sequence of a protein. This is called a **hydrophobicity profile**. Analysis of hydrophobicity profiles has been used to predict the positions of turns between elements of secondary structure, exposed and buried residues, membrane-spanning segments, and antigenic sites.

Example 5.1 Use of hydrophobicity profiles to predict the positions of turns between helices and strands of sheet

Figure 5.4a shows the hydrophobicity profile of hen egg white lysozyme. It has pronounced minima at the following residues: 17, 44, 70, 100, and 117. Figure 5.4b shows the structure of hen egg white lysozyme, from which it is possible to check the correlation between turns in the structure and the positions of the minima in the hydrophobicity profile.

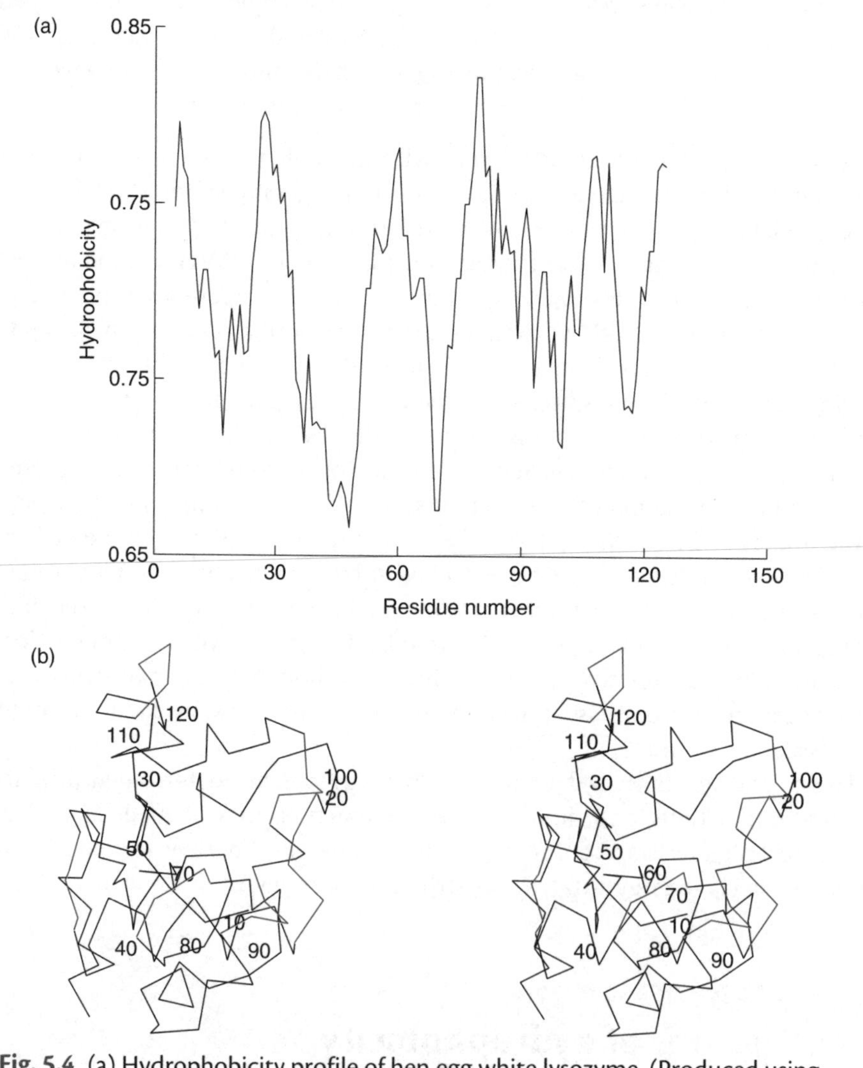

Fig. 5.4 (a) Hydrophobicity profile of hen egg white lysozyme. (Produced using the Primary Structure Analysis tools available through http://www.expasy.org.) (b) Structure of hen egg white lysozyme. Regions corresponding to minima in the hydrophobicity plot are shown in red.

→

→

Four of the major minima in the hydrophobicity profile appear at or near the positions of turns. Another minimum occurs in a surface-exposed region, but in the structure this corresponds to a strand of a β-sheet rather than to a turn. One of the minima is within a helix. Conversely, many of the turns do not correspond to pronounced minima in the hydrophobicity plot. Hydrophobicity profiles provide useful information, but do not unambiguously predict all turns in a protein structure.

Example 5.2 The helical wheel

O. B. Ptitsyn observed that α-helices in globular proteins often have a 'hydrophobic face' turned inwards towards the protein interior, and a 'hydrophilic face' turned outwards towards the solvent. Each residue in an α-helix appears at a position 100° around the circumference from its predecessor. Therefore, to achieve Ptitsyn's effect, the sequence of residues should alternate between hydrophobic and hydrophilic with a periodicity of approximately four.

To check this relationship, the residues can be projected onto a plane perpendicular to a helix axis, a diagram called a **helical wheel.** This example shows the sequence of an α-helix of sperm whale myoglobin. Charged and polar residues appear in boldface type; others in ordinary type.

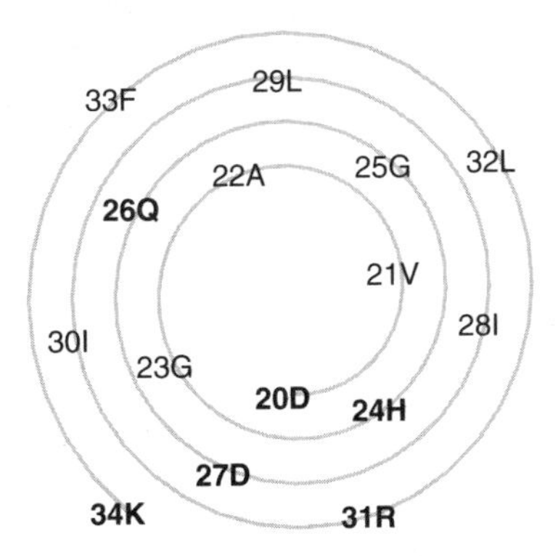

The helix has a hydrophobic face—which points to the inside of the structure, and a hydrophilic face—which points outside. From such a pattern of hydrophobicity we can predict whether a region of an amino acid sequence is likely to form an α-helix in the native protein structure.

The next box shows a PERL program to draw helical wheels.

PERL Example 5.1 A program to draw helical wheels

```
#!/usr/bin/perl
#helwheel.pl -- draw helical wheel
#usage: echo DVAGHGQDILIRLFKSH | helwheel.prl > output.ps
# or    echo 20DVAGHGQDILIRLFKSH | helwheel.prl > output.ps
#       the numerical prefix sets the first residue number

# The output of this program is in PostScript (TM),
#      a general-purpose graphical language

# The next section prints a header for the PostScript file

print <<EOF;
%!PS-Adobe-
%%BoundingBox: (atend)
%1 0 0 setrgbcolor
%newpath
%37.5 161 moveto 557.5 161 lineto 557.5 681 lineto 37.5 681 lineto
%closepath stroke
297.5 421. translate 2 setlinewidth 1 setlinecap
/Helvetica findfont 20 scalefont setfont 0 0 moveto
EOF

#  Define fonts to associate with each amino acid

$font{"G"} = "Helvetica";      $font{"A"} = "Helvetica";      $font{"S"} = "Helvetica";
$font{"T"} = "Helvetica";      $font{"C"} = "Helvetica";      $font{"V"} = "Helvetica";
$font{"I"} = "Helvetica";      $font{"L"} = "Helvetica";      $font{"F"} = "Helvetica";
$font{"Y"} = "Helvetica";      $font{"P"} = "Helvetica";      $font{"M"} = "Helvetica";
$font{"W"} = "Helvetica";      $font{"H"} = "Helvetica-Bold"; $font{"N"} = "Helvetica-Bold";
$font{"Q"} = "Helvetica-Bold"; $font{"D"} = "Helvetica-Bold"; $font{"E"} = "Helvetica-Bold";
$font{"K"} = "Helvetica-Bold"; $font{"R"} = "Helvetica-Bold";

$_= <>;                                    #  read line of input
chop();$_ =~ s/\s//g;                      #  remove terminal carriage return and blanks

if ($_ =~ s/^(\d+)//)                      #  if input begins with integer
     {$resno = $1;}                        #  extract it as initial residue number
else {$resno = 1}                          #  if not, set initial residue number = 1

$radius = 50;                              #  initialize values for radius,
$x = 0; $y = -50; $theta = -90;            #  x, y and angle theta

#  print light gray spiral arc as succession of line segments, 10 per residue

$npoints = 10*(length($_) - 1);

print "0.8 0.8 0.8 setrgbcolor\n";         #  set colour to light gray
print "newpath\n";                         #  draw spiral arc
printf("%8.3f %8.3f moveto\n",$x,$y);
foreach $d (1 .. $npoints) {               #  10 points per residue
    $theta += 10; $radius += 0.6;          #  increase radius and theta
    $x = $radius*cos($theta*0.01747737);   #  calculate new value of x
    $y = $radius*sin($theta*0.01747737);   #    and y

    printf("%8.3f %8.3f lineto\n",$x,$y);
}
print "stroke\n";

#  print residues and residue numbers

$radius = 50;                              #  reinitialize values for radius,
$x = 0; $y = -50; $theta = -90;            #  x, y and angle theta
print "0 setgray\n";                       #  set colour to black
```

→

→

```
foreach (split ("",$_)) {                    # loop over characters from input line
    print "/$font{$_} findfont ";            # set font appropriate
    print "20 scalefont setfont\n";          # for this amino acid
    printf("%8.3f %8.3f moveto\n",$x,$y);    # move to current point
    print " ($resno$_) stringwidth";         # adjust position to center residue
    print " pop -0.5 mul -7 rmoveto\n";      #    identification on point on spiral
    print " ($resno$_) show\n";              # print residue number and id
    print "% $theta $resno$_\n";
    $theta += 100; $radius += 6;             # set new values of angle, radius
    $x = $radius*cos($theta*0.01747737);     # compute new values of x
    $y = $radius*sin($theta*0.01747737);     #    and y
    $resno++;                                # increase residue number
}

print "showpage\n";                          # postscript signals to
print "%%BoundingBox:";                      # print
$xl = 297.5 - 1.05*$radius;                  #   x
$xr = 297.5 + 1.05*$radius;                  #    and
$yb = 421.  - 1.05*$radius;                  #     y
$yt = 421.  + 1.05*$radius;                  #      limits
printf("%8.3f %8.3f %8.3f %8.3f\n",$xl,$xr,$yb,$yt);


print "showpage\n";
print "%%EOF\n";                             # and wind up
```

Superposition of structures, and structural alignments

Some aspects of sequence analysis carry over fairly directly into structural analysis, some must be generalized, and others have no analogues at all.

As in the case of sequences, a fundamental question in analysing structures is to devise and compute a measure of similarity. If two molecules have identical or very similar structures, we can imagine superposing them so that corresponding points are as close together as possible. Then the average distance between corresponding points is a measure of the structural similarity. In practice it is conventional to report the root-mean-square deviation of the corresponding atoms:

$$\text{r.m.s. deviation} = \sqrt{\sum d_i^2/n}$$

where d_i is the distance between the ith pair of atoms (one atom from each structure) after optimal fitting, and n is the number of points.

This assumes that we have prespecified the correspondence between the points; that is, the alignment.

If the correspondence is not known, we must first determine it and only then calculate the r.m.s. deviation of the alignable substructures. If each point corresponds to an atom representing the successive residues of a protein or nucleic acid structure (the Cα atoms of proteins or the phosphorus atoms of nucleic acids), the problem is literally a question of alignment (= assignment of residue-residue correspondences) (see Box, page 235). Indeed, determination of residue-residue correspondences via structural superposition of two or more proteins is a powerful method of sequence alignment. Because structure tends to diverge more conservatively than sequence during evolution, structure alignment is a more powerful method than pairwise sequence alignment for detecting homology, and aligning the sequences and measuring the structural similarity of distantly-related proteins. (See Box, page 235.)

Example 5.3 Structural alignment of γ-chymotrypsin and *Staphylococcus aureus* epidermolytic toxin A

Chymotrypsin and *S. aureus* epidermolytic toxin A are both members of the chymotrypsin family of proteinases. Figure 5.5 shows a structural superposition of PDB entries 8GCH (γ-chymotrypsin) (black) and 1AGJ (*S. aureus* epidermolytic toxin A) (red). The molecules share the common chymotrypsin-family serine proteinase folding pattern, and the Ser-His-Asp catalytic triad (thicker lines).

A sequence alignment derived from the superposition follows:

```
8gch CGVPAIQPVLIVNG------------------------------EEAVP--GS----WPWQVSLQ-DKTG
1agj --------------EVSAEEIKKHEEKWNKYYGVNAFNLPKELFSKVDEKDR-QKYPYNTIGNVFVK-G-

8gch FH--FCGGSLINE-NWVVTAAHC-GV-T---T-SDVVVAGEFDQG---SSSEKI--QKLKIAKVFK-NS-
1agj --QTSATGVLIG-KNTVLTNRHIAK-FANGDPSKVSFRPSI-NTDDNGNT-E-TPYGEYEVKEILQEP-F

8gch KYNSLTINNDITLLKLST-----AAS--FSQTVSAVCLPSASD--DFAAGTTCVTTGWG-LTRYNTPD-R
1agj GAG-----VDLALIRLKPDQNGVSL-GDK---ISPAKIGT---SNDLKDGDKLELIGYPFDH----KVNQ

9gch LQQASLPLL-SNTNCKKYWGTKIKDAM--ICAGASGV-SSCMGDSGGPLVCKKNGAWTLVGIVSWGSSTC
1agj MHRSEIELTTLS---------------RGLRYY----GFTVPGNSGSGIFNSN---GELVGIHSSK----

8gch STST---------PGVYARVTA-LVNWVQQTLAAN-
1agj ----VSHLDREHQINYGVGIGNYVKRIINEKN---E
```

The resemblance between these two sequences is well within the 'twilight zone.' It could not be derived correctly from standard pairwise alignment of the two sequences alone.

Fig. 5.5 Structural superposition of γ-chymotrypsin [8GCH] (black) and *S. aureus* epidermolytic toxin A [1AGJ] (red). The sidechains of the catalytic triads are shown. Observe that the region around the active site is the best-conserved part of the protein.

Determination of similarity and alignment in computational chemistry

1. Similarity of two sets of atoms with known correspondences:

$$p_i \longleftrightarrow q_i,\ i = 1, \ldots N.$$

The analogue, for sequences, is the Hamming distance: mismatches only.

2. Similarity of two sets of atoms with unknown correspondences, but for which the molecular structure—specifically the linear order of the residues—restricts the possibilities. In the case of proteins or nucleic acids we are limited to correspondences in which we retain the order along the chain:

$$p_{i(k)} \longleftrightarrow q_{j(k)},\ k = 1, \ldots K \leq N, M$$

with the constraint that: $k_1 > k_2 \Rightarrow i(k_1) > i(k_2)$ and $j(k_1) > j(k_2)$. This can be thought of as analogous to the Levenshtein distance, or to sequence alignment with gaps. The result of such a calculation is an alignment of parts or all of the sequences.

3. Similarities between two sets of atoms with unknown correspondence, with no restrictions on the correspondence:

$$p_{i(k)} \longleftrightarrow q_{j(k)}$$

This problem arises in the following important case: Suppose two (or more) molecules have similar biological effects, such as a common pharmacological activity. It is often the case that the structures share a common constellation of a relatively small subset of their atoms that is responsible for the biological activity. These atoms are called a **pharmacophore**. The problem is to identify them: to do so it is useful to be able to find, within two or more molecules, the maximal subsets of atoms that have a similar structure. (See Case Study 5.1.)

DALI (Distance-matrix ALIgnment)

As proteins evolve, their structures change. Among the subtle details that evolution has strongly tended to conserve are the patterns of contacts between residues. That is, if two residues are in contact in one protein, the residues aligned with these two in a related protein are also likely to be in contact. This is true even in very distant homologues, and even if the residues involved change in size. Mutations that change the sizes of packed buried residues cause adjustments in the packing of the helices and sheets against one another.

L. Holm and C. Sander applied these observations to the problem of structural alignment of proteins. If the interresidue contact pattern is preserved in distantly-related proteins, then it should be possible to *identify* distantly-related proteins by detecting conserved contact patterns.

Computationally, one makes matrices of residue-residue contact patterns in two proteins (this is very easy), and then seeks the maximal matching submatrices

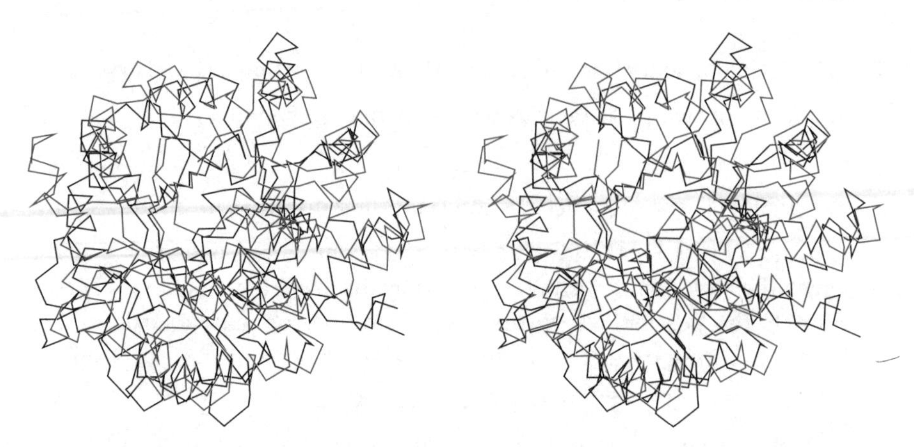

Fig. 5.6 The regions of common fold, as determined by the program DALI by L. Holm and C. Sander, in the TIM-barrel proteins mouse adenosine deaminase [1FKX] (black) and *Pseudomonas diminuta* phosphotriesterase [1PTA] (red). In the alignment shown in this figure, the sequences have only 13% identical residues—closer to midnight than to the twilight zone.

(this is hard). Using carefully chosen approximations, Holm and Sander wrote an efficient program called DALI (for Distance-matrix ALIgnment) that is now in common use for identifying proteins with folding patterns similar to that of a query structure. The program runs fast enough to carry out routine screens of the entire Protein Data Bank for structures similar to a newly-determined structure, and even to perform a classification of protein domain structures from an all-against-all comparison. Holm and Sander have found several unexpected similarities not detectable at the level of pairwise sequence alignment.

An example of DALI's 'reach' into recognition of very distant structural similarities is its identification of the relation between mouse adenosine deaminase, *Klebsiella aerogenes* urease, and *Pseudomonas diminuta* phosphotriesterase (see Fig. 5.6).

DALI is available over the Web. You can submit coordinates to the site `http://www2.ebi.ac.uk/dali/`, and receive the set of similar structures and their alignments with the query.

Evolution of protein structures

Included in the 30 000 protein structures now known are several families in which the molecules maintain the same basic folding pattern over ranges of sequence similarity from near-identity down to well below 20% conservation. The serine proteinases (γ-chymotrypsin and *S. aureus* epidermolytic toxin A, Fig. 5.5) and the adenosine deaminase-phosphotriesterase family, (Fig. 5.6) are examples.

The general response to mutation is structural change. It is characteristic of biological systems that the objects we observe to have a certain form arose by evolution from related objects with similar but not identical form. They must, therefore, be robust, in having the freedom to tolerate some variation. We can take advantage of this robustness in our analysis: By identifying and comparing related objects, we can determine conserved and variable features, and thereby distinguish that which is crucial to structure and function (and therefore conserved) from that which can survive change (and therefore available to vary).

Natural variations in families of homologous proteins that retain a common function reveal how structures accommodate changes in amino acid sequence. Surface residues not involved in function are usually free to mutate. Loops on the surface can often accommodate changes by local refolding. Mutations that change the volumes of buried residues generally do not change the conformations of individual helices or sheets, but produce distortions of their spatial assembly. The nature of the forces that stabilize protein structures sets general limitations on these conformational changes; particular constraints derived from function vary from case to case.

Families of related proteins tend to retain common folding patterns. However, although the general folding pattern is preserved, there are distortions which increase as the amino acid sequences progressively diverge. These distortions are not uniformly distributed throughout the structure. Usually, a large central *core* of the structure retains the same qualitative fold, and other parts of the structure change conformation more radically. Consider the letters B and R. As structures, they have a common core which corresponds to the letter P. Outside the common core they differ: at the bottom right B has a loop and R has a diagonal stroke.

Systematic studies of the structural differences between pairs of related proteins have defined a quantitative relationship between the divergence of the amino acid sequences of the core of a family of structures and the divergence of structure. As the sequence diverges, there are progressively increasing distortions in the mainchain conformation, and the fraction of the residues in the core usually decreases. Until the fraction of identical residues in the sequence drops below about 40–50%, these effects are relatively modest. Almost all the structure remains in the core, and the deformation of the mainchain atoms is on average no more than 1.0 Å. With increasing sequence divergence, some regions refold entirely, reducing the size of the core, and the distortions of the residues remaining within the core increase in magnitude.

A correlation between the divergence of sequence and structure applies to all families of proteins. Figure 5.7a shows the changes in structure of the core, expressed as the root-mean-square deviation of the mainchain atoms after optimal superposition, plotted against the sequence divergence: the percentage of conserved amino acids of the core after optimal alignment. The points correspond to pairs of homologous proteins from many related families. (Those at 100% residue identity are proteins for which the structure was determined in two or more crystal environments, and the deviations show that crystal packing forces—and, to a lesser extent, solvent and temperature—can modify slightly the conformation of the proteins.) Figure 5.7b shows the changes in the fraction of residues in the core as a function of sequence divergence. The fraction of residues in the cores of distantly-related proteins can vary widely: in some cases the fraction of residues in the core remains high, in others it can drop to below 50% of the structure.

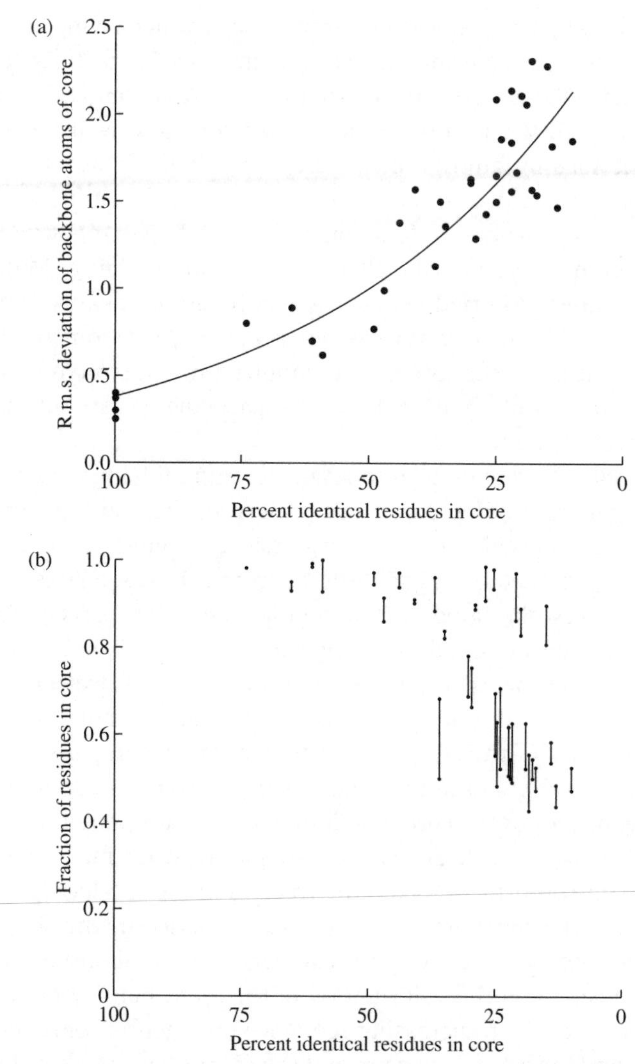

Fig. 5.7 Relationships between divergence of amino acid sequence and three-dimensional structure of the core, in evolving proteins. (a) Variation of r.m.s. deviation of the core with the per cent identical residues in the core. (b) Variation of size of the core with the per cent identical residues in the core. This figure shows results calculated for 32 pairs of homologous proteins of a variety of structural types. (Adapted from Chothia, C. & Lesk, A. M. (1986), Relationship between the divergence of sequence and structure in proteins, *The EMBO Journal*, **5**, 823–826.)

Classifications of protein structures

Being able to measure protein structural differences quantitatively allows us to cluster and classify protein folding patterns. Organization of protein structures according to folding pattern imposes a very useful logical structure on the entries in the

Protein Data Bank. It affords a basis for structure-oriented information retrieval. Several databases derived from the PDB are built around classifications of protein structures. They offer useful features for exploring the protein structure world, including: search for keyword or sequence, navigation among similar structures at various levels of the classification hierarchy, presentation of structure pictures, probing the databank for structures similar to a new structure, and links to other sites. These databases include SCOP (Structural Classification of Proteins), CATH (Class, Architecture, Topology, Homologous superfamily), FSSP/DDD (Fold classification based on Structure-Structure alignment of Proteins/Dali Domain Dictionary), and CE (The Combinatorial Extension Method, by I. N. Shindyalov and P. Bourne).

SCOP

SCOP, by A. G. Murzin, L. Lo Conte, B. G. Ailey, S. E. Brenner, T. J. P. Hubbard, and C. Chothia, organizes protein structures in a hierarchy according to evolutionary origin and structural similarity. At the lowest level of the SCOP hierarchy are individual **domains** (see page 42), extracted from the Protein Data Bank entries. Sets of domains are grouped into **families** of homologues, for which the similarities in structure, sequence, and sometimes function imply a common evolutionary origin. Families containing proteins of similar structure and function, but for which the evidence for evolutionary relationship is suggestive but not compelling, form **superfamilies**. Superfamilies that share a common folding topology, for at least a large central portion of the structure, are grouped as **folds**. Finally, each fold group falls into one of the general **classes**. The major classes in SCOP are α, β, α + β, α/β, and miscellaneous 'small proteins,' which often have little secondary structure and are held together by disulphide bridges or ligands.

The box shows the SCOP classification of flavodoxin from *Clostridium beijerinckii* (Plate V). For illustrations of the degree of similarities of proteins grouped together at different levels of the hierarchy, and discussion of other classification schemes, see *Introduction to Protein Architecture: The Structural Biology of Proteins*, Chapter 4.

SCOP classification of Flavodoxin from *Clostridium beijerinckii*

1. **Root** SCOP
2. **Class** α and β proteins (α/β)
   Mainly parallel β-sheets (β-α-β units)
3. **Fold** Flavodoxin-like
   3 layers, α/β/α; parallel β-sheet of 5 strands, order 21345
4. **Superfamily** Flavoproteins
5. **Family** Flavodoxin-related binds FMN
6. **Protein** Flavodoxin
7. **Species** *Clostridium beijerinckii*

The SCOP release of January 2004 contained 13 220 PDB entries, split into 31 474 Domains. The distribution of entries at different levels of the hierarchy is:

Class	Number of		
	families	superfamilies	folds
All-α proteins	337	224	138
All-β proteins	276	171	93
α/β proteins	374	167	97
α + β proteins	391	263	184
Multi-domain proteins	35	28	28
Membrane and cell surface proteins	28	17	11
Small proteins	116	77	54
Total	1557	947	605

Numerous other web sites offering classifications of protein structures are indexed at: http://www.bioscience.org/urllists/protdb.htm .

Protein structure prediction and modelling

The observation that each protein folds spontaneously into a unique three-dimensional native conformation implies that nature has an algorithm for predicting protein structure from amino acid sequence. Some attempts to understand this algorithm are based solely on general physical principles; others appeal to known amino acid sequences and protein structures. A proof of our understanding would be the ability to reproduce the algorithm in a computer program that could predict protein structure from amino acid sequence.

Most attempts to predict protein structure from basic physical principles alone try to reproduce the interatomic interactions in proteins, to define a computable energy associated with any conformation. Computationally, the problem of protein structure prediction then becomes a task of finding the global minimum of this conformational energy function. So far this approach has not succeeded, partly because of the inadequacy of the energy function and partly because the minimization algorithms tend to get trapped in local minima.

Other a priori approaches to structure prediction are based on attempts to simplify the problem, to capture somehow the essentials.

The alternative to a priori methods are approaches based on assembling clues to the structure of a target sequence by finding similarities to known structures. These empirical or 'knowledge-based' methods are becoming very powerful.

We are coming closer and closer to saturating the set of possible folds with known structures. This is the stated goal of **structural genomics** projects (see Box). Once we have a complete set of folds and sequences, and powerful methods for relating them, empirical methods will provide pragmatic solutions of many

Structural genomics

In analogy with full-genome sequencing projects, structural genomics has the commitment to deliver the structures of the complete protein repertoire. X-ray crystallographic and NMR experiments will solve a 'dense set' of proteins, such that all proteins are within modelling range of one or more known experimental structures. More so than genomic sequencing projects, structural genomics projects combine results from different organisms. The human proteome is of course of special interest, as are proteins unique to infectious micro-organisms.

The goals of structural genomics have become feasible partly by advances in experimental techniques, which make high-throughput structure determination possible; and partly by advances in our understanding of protein structures, which define reasonable general goals for the experimental work, and suggest specific targets.

The theory and practice of homology modelling suggests that at least 30% sequence identity between target and some experimental structure is necessary. This means that experimental structure determinations will be required for an exemplar of every sequence family, including many that share the same basic folding pattern. Experiment will have to deliver the structures of something like 10 000 domains. In the year 2004, approximately 5000 structures were deposited in the PDB, so the throughput rate is not far from what is required.

Methods of bioinformatics can help select targets for experimental structure determination that offer the highest pay-off in terms of useful information. Goals of target selection include:

- elimination of redundant targets—proteins too similar to known structures.
- identification of sequences with undetectable similarity to proteins of known structure.
- identification of sequences with similarity only to proteins of unknown function, or
- proteins of unknown structure with 'interesting' functions; for example, human proteins implicated in disease, or bacterial proteins implicated in antibiotic resistance.
- proteins with properties favourable for structure dermination—likely to be soluble, contain methonine (which facilitates solving the phase problem of X-ray crystallography).

The machinery for carrying out the modelling is already up and running. MODBASE (http://alto.compbio.ucsf.edu/modbase.cgi/index.cgi) and 3DCrunch (http://www.expasy.org/swissmod/SWISS-MODEL.html) collect homology models of proteins of known sequence.

Structural genomics projects are supported by large-scale initiatives from the US National Institutes of Health and private industry.

problems. What will be the effect of this on attempts to predict protein structure a priori? The intellectual appeal of the problem will still be there. After all, nature folds proteins without searching databases. But it is unlikely that the problem will continue to command interest of the same intensity, and support of the same largesse, once a pragmatic solution has been found.

However, there is a paradox: The methods being developed for identifying folding patterns in sequences are more than exercises in tuning parameters in scoring functions. They are experiments that explore and expose the essential features of amino acid sequences that determine protein structures. When they succeed, we will have a far sounder basis for understanding sequence-structure relationships than we do now. It may be that a posteriori understanding will provide the clues that will make a priori prediction possible.

Methods for prediction of protein structure from amino acid sequence include:

- **Attempts to predict secondary structure** without attempting to assemble these regions in three-dimensions. The results are lists of regions of the sequence predicted to form α-helices and regions predicted to form strands of β-sheet.
- **Homology modelling**: prediction of the three-dimensional structure of a protein from the known structures of one or more related proteins. The results are a complete coordinate set for mainchain and sidechains, intended to be a high-quality model of the structure, comparable to at least a low-resolution experimental structure.
- **Fold recognition**: given a library of known structures, determine which of them shares a folding pattern with a query protein of known sequence but unknown structure. If the folding pattern of the target protein does not occur in the library, such a method should recognize this. The results are a nomination of a known structure that has the same fold as the query protein, or a statement that no protein in the library has the same fold as the query protein.
- **Prediction of novel folds**, either by a priori or knowledge-based methods. The results are a complete coordinate set for at least the mainchain and sometimes the sidechain also. The model is intended to have the correct folding pattern, but would not be expected to be comparable in quality to an experimental structure. D. Jones has likened the distinction between fold recognition and a priori modelling to the difference between a multiple-choice question on an exam and an essay question.

Critical Assessment of Structure Prediction (CASP)

The CASP programmes were introduced briefly in Chapter 1. CASP organizes blind tests of protein structure predictions, in which participating crystallographers and NMR spectroscopists make public the amino acid sequences of the proteins they are investigating, and agree to keep the experimental structures secret until predictors have had a chance to submit their models. CASP runs on a two-year cycle. At the end of the year a gala meeting brings the predictors together to discuss the current results and to gauge progress.

Predictions in CASP have traditionally fallen into three main categories: (1) comparative modelling—in effect homology modelling, (2) fold recognition, and (3) modelling of novel folds:

CASP Category	Nature of target
Comparative modelling	Close homologues of known structure are available; homology modelling methods are applicable.
Fold recognition	Structures with similar folds are available, but no sufficiently close relative for homology modelling; the challenge is to identify structures with similar topology.
New Fold	No structure with same folding pattern known; requires either a genuine a priori method or a knowledge-based method that can combine features of several known structures.

Assessors, one for each category, compare the predicted and experimental structures, and judge the predictions. Speakers at the end-of-year meeting include the organizers, the assessors, and selected predictors, including those who have been particularly successful, or who have an interesting novel method to present.

The latest CASP programme took place in 2004. Departures from past practice include: (1) secondary structure prediction is no longer separarately assessed, and (2) a new category, prediction of function, was introduced. There were 87 targets. In all categories, 201 groups of predictors submitted a total of 28 965 models. This was approximately equal to the number of entries in the PDB at the time!

Many predictions are prepared by groups of researchers who inspect the results generated by their computer programs, and select and edit them before submission. In addition, the target sequences are sent to web servers that return predictions without human intervention. The CAFASP: Critical Assessment of Fully Automated Structure Prediction programme monitors the quality of these predictions. It is thereby possible to determine to what extent successful procedures could be made fully automatic. CASP thus comprises three challenges:

Human against protein	CASP
Computer against protein	CAFASP
Human against computer	CASP *v.* CAFASP

A separate programme of blind tests of prediction evaluates methods for predicting protein-protein interactions, or 'docking'. This is CAPRI—Critical Assessment of PRedicted Interactions. Both CASP and CAPRI held assessment meetings in December 2004.

Structure predictions of the sixth CASP programme showed continued improvements. For the most part progress has been incremental rather than spectacular, with one notable exception: David Baker's group predicted and redifined the structure of a small (70-residue) protien from *Thermus thermophilus*, producing a model that deviated by 1.59 Å from the X-ray structure! Indeed, improvements in knowledge-based methods originally developed for novel folds threatens to supersede traditional methods for fold recognition, such as threading, that make explicit reference to libraries of complete structures.

Results at CAPRI show that complexes between partners that do not undergo major conformational changes can now be predicted accurately from the structures of the components. Large conformational changes upon complex formation still present difficulties. However, progress could be seen in at least one case, the trimeric TBE envelope protien.

For both CASP and CAPRI, the best results ae very impressive. One observer commented that the current state of protien structure prediction is that 'failure can no longer be guaranteed.' Consistency is the challenge.

Secondary structure prediction

It seems obvious that (1) it should be easier to predict secondary structure than tertiary structure, and (2) to predict tertiary structure, a sensible way to proceed would be first to predict the helices and strands of sheet and then to assemble them. Whether or not these propositions are correct, many people have believed and acted upon them. Given the amino acid sequence of a protein of unknown structure, they produce **secondary structure predictions**, the assignment of regions in the sequence as helices or strands of sheet.

To assess the quality of a secondary structure prediction, classify the residues in the experimental three-dimensional structure into three categories (helix = H, strand = E (extended), and other = -). The per cent of residues predicted correctly is denoted Q3. At the 2000 CASP programme, the PROF server by B. Rost achieved a good prediction of a domain from the *Thermus aquaticus* mismatch repair protein MutS. The value of Q3 for Rost's prediction is 81%:

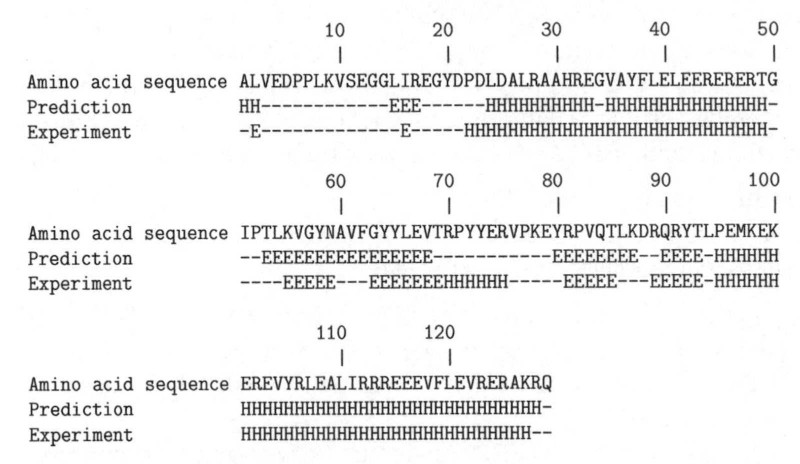

```
                            10        20        30        40        50
                             |         |         |         |         |
Amino acid sequence ALVEDPPLKVSEGGLIREGYDPDLDALRAAHREGVAYFLELEERERERTG
Prediction          HH------------EEE------HHHHHHHHHH-HHHHHHHHHHHHHHH-
Experiment          -E-------------E-----HHHHHHHHHHHHHHHHHHHHHHHHHHHH-

                            60        70        80        90       100
                             |         |         |         |         |
Amino acid sequence IPTLKVGYNAVFGYYLEVTRPYYERVPKEYRPVQTLKDRQRYTLPEMKEK
Prediction          --EEEEEEEEEEEEEEEE-----------EEEEEEEE--EEEE-HHHHHH
Experiment          ----EEEEE---EEEEEEEHHHHHH-----EEEEE---EEEEE-HHHHHH

                           110       120
                             |         |
Amino acid sequence EREVYRLEALIRRREEEVFLEVRERAKRQ
Prediction          HHHHHHHHHHHHHHHHHHHHHHHHHHHH-
Experiment          HHHHHHHHHHHHHHHHHHHHHHHHHHH--
```

Figure 5.8 shows the experimental structure, with the *predicted* secondary structures distinguished. Except for a short 3_{10} helix, the secondary structural elements are predicted correctly except for some minor discrepancies in the positions at

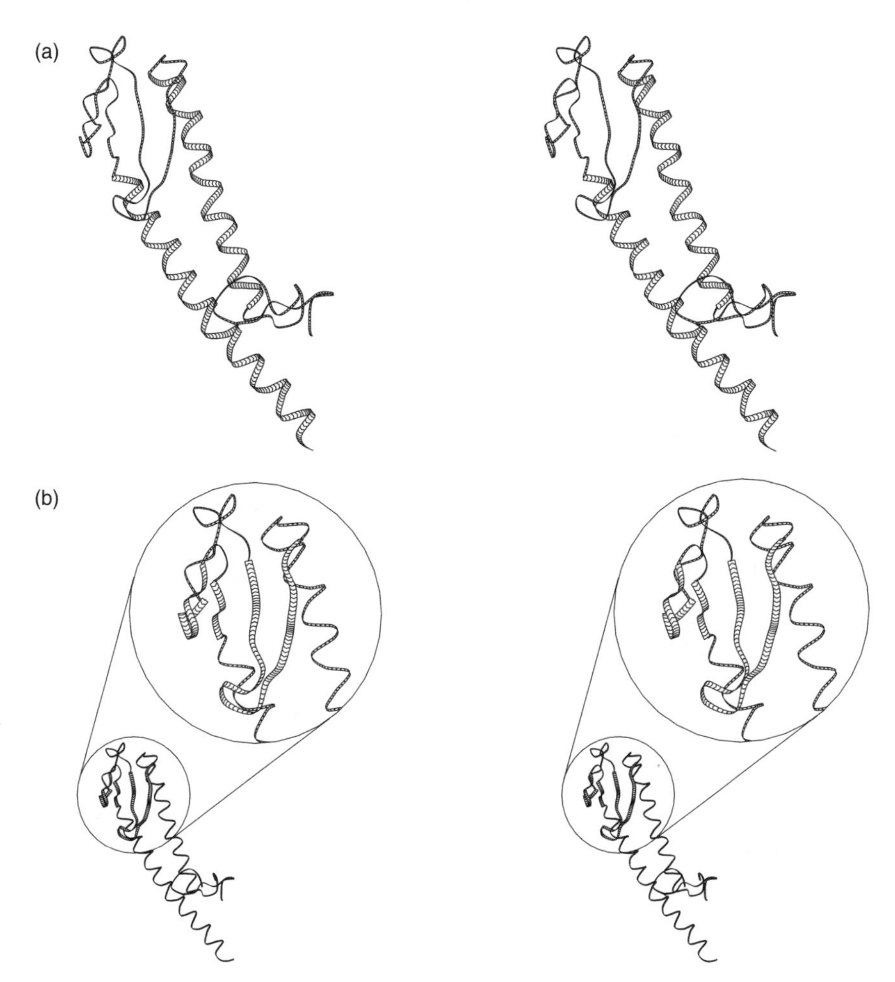

Fig. 5.8 The structure from the *Thermus aquaticus* mismatch repair protein MutS [1EWQ]. (a) The regions predicted by the PROF server of Rost to be helical are shown as wider ribbons. The prediction missed only a short 3_{10} helix, at the top left of the picture. (b) The regions predicted to be in strands are shown as wider ribbons.

which they start and end. (Other scoring schemes that check for segment overlap are less sensitive to end effects.) The quality of this result is very high but not exceptionally rare. This target was classified as being of *medium* difficulty by the CASP4 assessors. At present, PROF is running at an average accuracy of Q3 ~ 77%. Other secondary structure prediction methods are also doing comparably well.

The most powerful methods of secondary structure prediction are based on **neural networks**.

Neural networks

Neural networks are a class of general computational structures based loosely on the anatomy and physiology of biological nervous systems. They have been applied successfully to a wide variety of pattern recognition, classification, and decision problems.

A single neuron, in the computational scheme, is a node in a directed graph, with one or more entering connections designated as input, and a single leaving connection called the output:

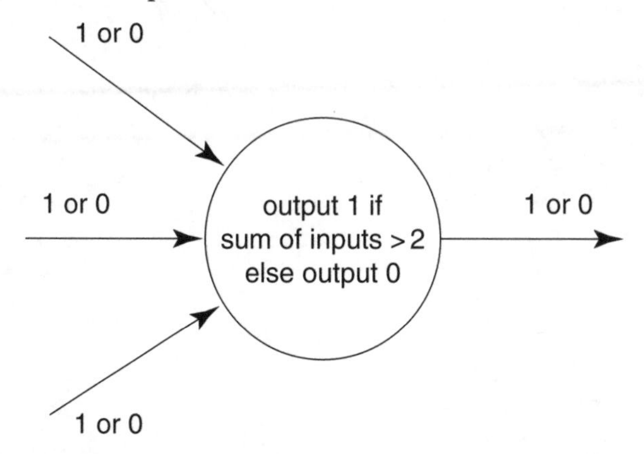

In the physiological metaphor, one says that the neuron 'fired' if the output is 1, and that the neuron 'didn't fire' if the output is 0. Simulated neurons can differ in the number of input and output connections, and in the formula for deciding whether to fire (see Box).

To form a network, assemble several neurons and connect the outputs of some to the inputs of others. Some nodes contain connections that provide input to the entire network; some deliver output information from the network to the outside world; and others, that do not interact directly with the outside, are called **hidden layers**.

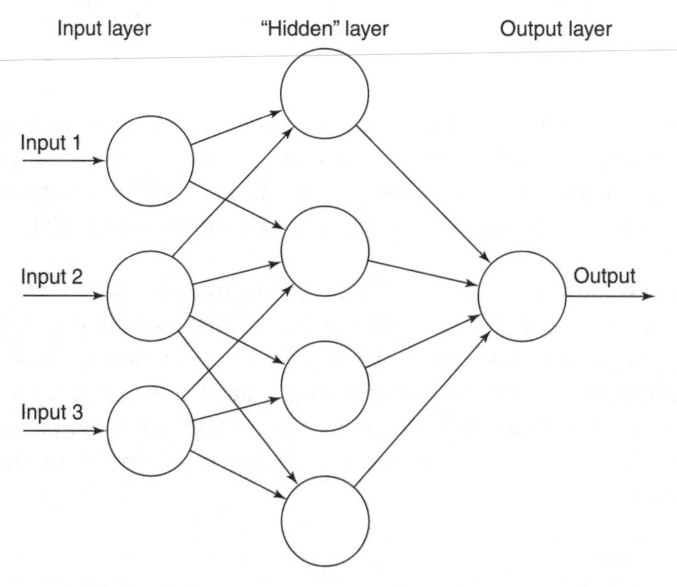

An unlimited degree of complexity is available by assembling and connecting neurons, and by varying the strengths of the connections. That is, instead of taking a simple sum of inputs $i_1 + i_2 + i_3$, take a weighted sum—for instance, $10i_1 + 5i_2 + i_3$, which would make the neuron most sensitive to input 1 and least sensitive to input 3. Biologically, this corresponds to changing the strengths of synapses.

Logic of neural networks

For a single neuron, a linear decision process governing the output has a geometric interpretation in terms of lines and planes. The neuron in the following figure has two inputs. If we interpret the inputs as the coordinates of a point (x, y) in the plane, the neuron 'decides' on which side of a line the input point lies. The output will be 1 if and only if $x + y \leq 2$;that is, if the point is below and to the left of the line $x + y = 2$.

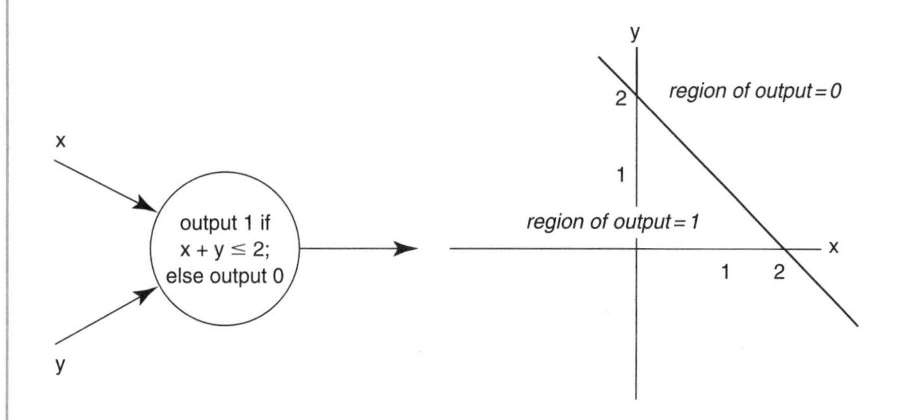

A *neural network* is specified by the topology of its connections, and the weights and decision formulas of its nodes. A network can make more complex decisions than a single neuron. Thus, if one neuron with two inputs can decide on which side of a line a point lies, three neurons can select points that lie within a triangle:

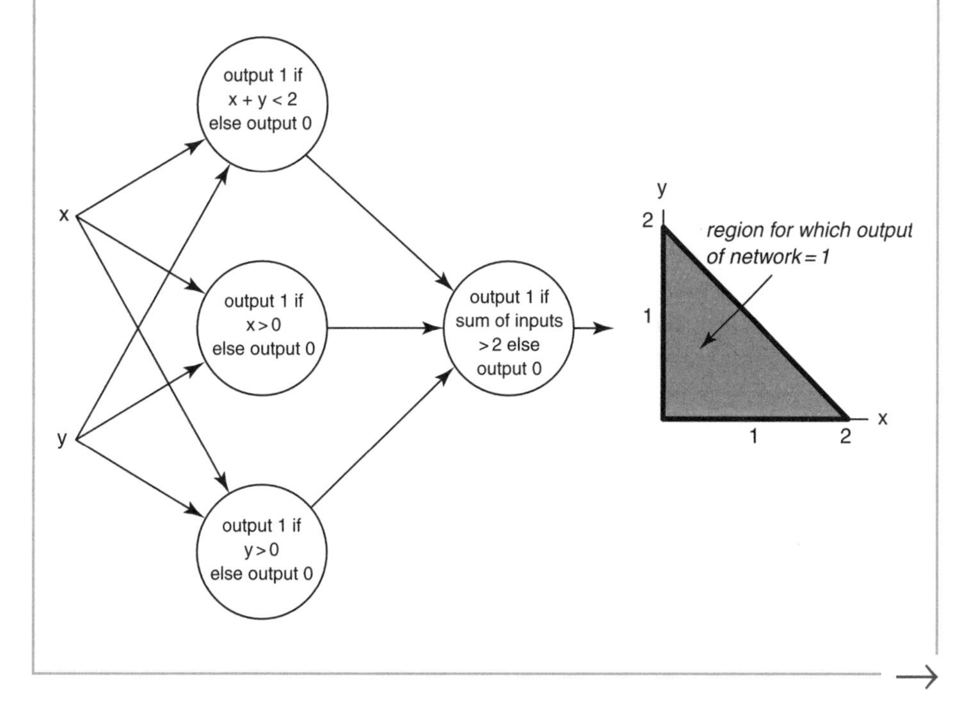

→

Logic of neural networks (*continued*)

Neural networks are more powerful and robust if the output is a smoothly-varying function of the inputs. Such networks can perform more general kinds of computations and are better at pattern recognition. Also, for training the network it is useful if the output is a differentiable function of the parameters. To this end, a sharp threshold function for the output of a neuron is replaced by a smoothed-out step, or sigmoidal, function:

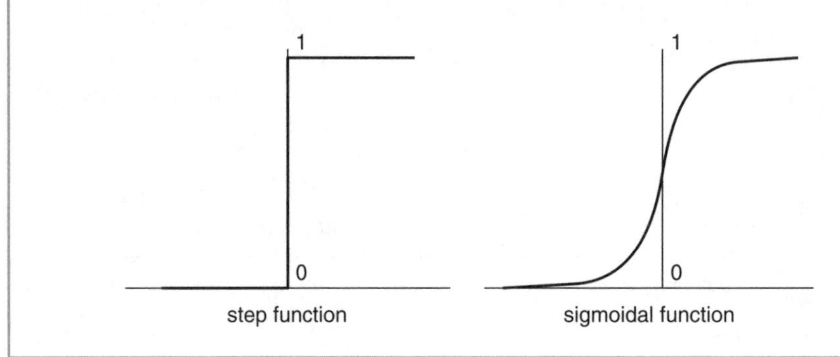

A property of a neural network that gives it great power is that the weights may be regarded as *variables*, and a calculation or learning process may determine the weights appropriate for a particular decision or pattern identifier. To train a network, feed the system sets of sample input for which the desired output is known, and compare the output with the correct answer. If the observed output differs from the desired one, adjust the parameters. The topology of the network remains invariant during the training process, although of course setting a weight to 0 has the effect of detaching an input.

The type of neural network that has been applied to secondary structure prediction is shown in Fig. 5.9.

A major advance in secondary structure prediction occurred with the application of evolutionary information, the recognition that multiple sequence alignment tables contain much more information than individual sequences. The conservation of secondary structure among related proteins means that the sequence-structure correlations are much more robust when a family as a whole is taken into account. Most neural network-based methods for secondary structure prediction now feed the input layer not simply the identities of the amino acid at successive positions, but a profile derived from a multiple sequence alignment.

It has also proved useful to run two neural networks in tandem, to make use of observed correlations among conformations of residues at neighbouring positions. Predictions of the states of several successive residues, by a network similar to the one shown in Fig. 5.9, are combined by a second network into a final prediction.

A test of the maturity of a prediction method is whether it can be made fully automatic. Some computational methods produce only rough drafts of a protein

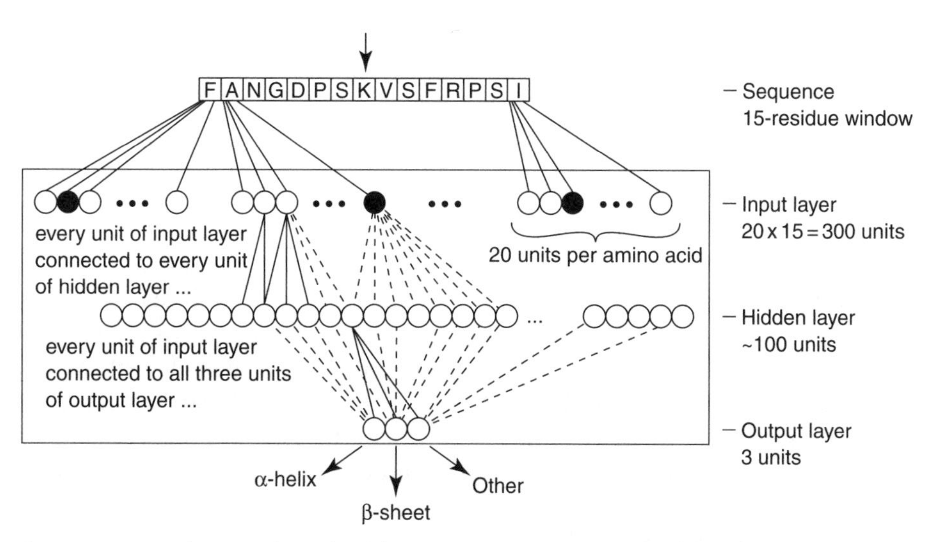

Fig. 5.9 A neural network applicable to secondary structure prediction contains three layers:

- The input layer sees a sliding 15-residue window in the sequence. That is, it treats a 15-residue region, predicts the secondary structure of the central residue (marked by an arrow, at the top) and then moves the window one residue along the amino acid sequence and repeats the process. To each of the 15 residues in the current window there correspond 20 nodes in the input layer of the network, one of which will be triggered according to the amino acid in that position.
- A hidden layer of ~100 units connects the input with the output. Each node of the hidden layer is connected to *all* input and output units; not all the connections are shown.
- The output layer consists of only three nodes, that signify prediction that the central residue in the window be in a helix, strand, or other conformation.

structure prediction, requiring human intervention to bring them to final form. Others are automatic, and there are many web servers that will accept sequences and return the predictions. PROF, the system that predicted the secondary structure of MutS, is one of these.

A continuous, fully-automatic, analysis of protein structure prediction web servers, including but not limited to secondary structure predictions, is called EVA. It is a collaboration among groups in New York, San Francisco, and Madrid. The Protein Data Bank supplies sequences shortly in advance of release of the corresponding structures, and the software implementing EVA submits them to prediction servers and analyses the results. It can be thought of as a continuous CASP programme, restricted to methods that can be both applied *and* judged automatically. See: http://cubic.bioc.columbia.edu/eva .

The goals of EVA are to monitor progress in the field, and to indicate to users the best protein structure prediction servers in different categories. EVA has access to much more data than has been available in the CASP programmes themselves. Therefore its conclusions are in principle less vulnerable to statistical fluctuations in the nature and difficulty of the targets than the tests reported in CASP programmes.

Homology modelling

Model-building by homology is a useful technique when one wants to predict the structure of a target protein of known sequence, when the target protein is related to at least one other protein of known sequence *and* structure. If the proteins are closely related, the known protein structures—called the parents—can serve as the basis for a model of the target. Although the quality of the model will depend on the degree of similarity of the sequences, it is possible to specify this quality before experimental testing (see Fig. 5.7). In consequence, knowing the quality of the model required for the intended application permits intelligent prediction of the probable success of the exercise.

Steps in homology modelling are:

1. Align the amino acid sequences of the target and the protein or proteins of known structure. It will generally be observed that insertions and deletions lie in the loop regions between helices and sheets.
2. Determine mainchain segments to represent the regions containing insertions or deletions. Stitching these regions into the mainchain of the known protein creates a model for the complete mainchain of the target protein.
3. Replace the sidechains of residues that have been mutated. For residues that have not mutated, retain the sidechain conformation. Residues that have mutated tend to keep the same sidechain conformational angles, and could be modelled on this basis. However, computational methods are now available to search over possible combinations of sidechain conformations.
4. Examine the model—both by eye and by programs—to detect any serious collisions between atoms. Relieve these collisions, as far as possible, by manual manipulations.
5. Refine the model by limited energy minimization. The role of this step is to fix up the exact geometrical relationships at places where regions of mainchain have been joined together, and to allow the sidechains to wriggle around a bit to place themselves in comfortable positions. The effect is really only cosmetic—energy refinement will not fix serious errors in such a model.

To a great extent, this procedure produces 'what you get for free' in that it defines the model of the protein of unknown structure by making minimal changes to its known relative. Unfortunately it is not easy to make substantial improvements. A rule of thumb (referring again to Fig. 5.8) is that if the two sequences have at least 40–50% identical amino acids in an optimal alignment of their sequences, the procedure described will produce a model of sufficient accuracy to be useful for many applications. For very distantly-related proteins, neither the procedure described nor any other currently available method will produce a model, correct in detail, of the target protein from the structure of its relative.

In most families of proteins the structures contain relatively constant regions and more variable ones. The core of the structure of the family retains the folding topology, although it may be distorted, but the periphery can entirely refold. A single parent structure will permit reasonable modelling of the conserved portion of the target protein, but will fail to produce a satisfactory model of the variable portion. Moreover, it will not be easy to predict which are the variable and constant regions. A more favourable situation occurs when several related proteins of known structure can serve as parents for modelling a target protein. These reveal the regions of constant and variable structure in the family. The observed distribution of structural variability among the parents dictates an appropriate distribution of constraints to be applied to the model.

Mature software for homology modelling is available. SWISS-MODEL is a web site that will accept the amino acid sequence of a target protein, determine whether a suitable parent or parents for homology modelling exist, and, if so, deliver a set of coordinates for the target. SWISS-MODEL was developed by T. Schwede, M. C. Peitsch and N. Guex, now at The Geneva Biomedical Research Institute. Another program in widespread use, MODELLER, was originally developed by A. Šali.

Web resources: Homology modelling

SWISS-MODEL (homology modelling server):
http://www.expasy.ch/swissmod/SWISS-MODEL.html

Results of application of SWISS-MODEL to proteins of known sequence are available through 3DCrunch:
http://www.expasy.org/swissmod/SWISS-MODEL.html

MODELLER (homology modelling software):
http://salilab.org/modeller/modeller.html

Results of application of MODELLER to proteins of known sequence are available through MODBASE:
http://alto.compbio.ucsf.edu/modbase.cgi/index.cgi

For a description of web sites in structural genomics: Wixon, J. (2001), Structural genomics on the web, *Comp. Funct. Genomics*, 2, 103–113.

An example of the automatic prediction by SWISS-MODEL is the prediction of the structure of a neurotoxin from red scorpion (*Buthus tamulus*) from the known structure of the neurotoxin from the related North African yellow scorpion (*Androctonus australis hector*). These two proteins have 52% identical residues in their sequence alignment. With such a close degree of similarity it is not surprising that the model fits the experimental result very closely, even with respect to the sidechain conformation (Fig. 5.10).

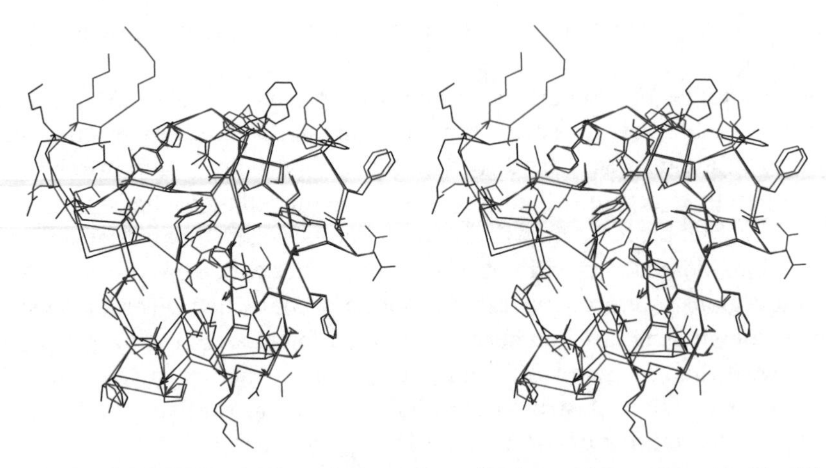

Fig. 5.10 SWISS-MODEL predicts the structure of red scorpion neurotoxin [1DQ7] (red) from a closely-related protein [1PTX] (black). The prediction was done *automatically*. Observe that most of the buried sidechains have not mutated, and have very similar conformations. Some sidechains on the surface have different conformations, and the mainchain of the C-terminus is in a different position (upper left). Not shown is a network of disulphide bridges, which constrain the structure. However, a model of this high quality would be expected for two such closely-related proteins, even without the extra constraints.

Fold recognition

Searching a sequence database for a probe sequence and searching a structure database with a probe structure are problems with known solutions. The mixed problems—probing a sequence database with a structure, or a structure database with a sequence, are less straightforward. They require a method for evaluating the compatibility of a given sequence with a given folding pattern.

The goal is to abstract the essence of a set of sequences or structures. Other proteins that share the pattern are expected to adopt similar structures.

3D profiles

We have discussed patterns and profiles derived from multiple sequence alignments and their application to detection of distant homologues. One way to take advantage of available structural information to improve the power of these methods is a type of profile derived from the available sequences *and* structures of a family of proteins.

J. U. Bowie, R. Lüthy and D. Eisenberg analysed the *environments* of each position in known protein structures and related them to a set of preferences of the 20 amino acids for these structural contexts.

Given a protein structure, they classified the environment of each amino acid in three separate categories:

1. Its mainchain hydrogen-bonding interactions, that is, its secondary structure.

2. The extent to which it is buried within or on the surface of the protein structure.
3. The polar/nonpolar nature of its environment.

The secondary structure may be one of three possibilities: *helix*, *sheet* and *other*. A sidechain is considered buried if the accessible surface area is less than 40 Å^2, partially buried if the accessible surface area is between 40 and 114 Å^2, and exposed if the accessible surface area is greater than 114 Å^2. The fraction of sidechain area covered by polar atoms is also measured. The authors define six classes on the basis of accessibility and polarity of the surroundings. Sidechains in each of these six classes may be in any of three classes according to the secondary structures. This gives a total of 18 classes.

Assigning each sidechain to one of 18 class means that it is possible to write a coded description of a protein structure as a message in an alphabet of 18 letters, called a **3D structure profile**. Algorithms developed for sequence searches can thereby be applied to 'sequences' of encoded structures. For example, one could try to align two distantly-related sequences by aligning their 3D structure profiles rather than their amino acid sequences. The 3D profile method translates protein structures into one-dimensional probe (or probe-able) objects that do not explicitly retain either the sequence or structure of the molecules from which they were derived.

Next, how can one relate the 3D structure profile to the corpus of known sequences and structures? It is clear that some amino acids will be unhappy in certain kinds of sites; for example, a charged sidechain would not be buried in an entirely nonpolar environment. Other preferences are not so clear-cut, and it is necessary to derive a preference table from a statistical survey of a library of well-refined protein structures.

Suppose now that we are given a sequence and want to evaluate the likelihood that it takes up, say, the globin fold. From the 3D structure profile of the known sperm whale myoglobin structure we know the environment class of each position of the sequence. Consider a particular alignment of the unknown sequence with sperm whale myoglobin, and suppose that the residue in the unknown sequence that corresponds to the first residue of myoglobin is phenylalanine. The environment class in the 3D structure profile of the first residue of sperm whale myoglobin is: exposed, no secondary structure. One can score the probability of finding phenylalanine in this structural environment class from the table of preferences of particular amino acids for this 3D structure profile class. (The fact that the first residue of the sperm whale myoglobin sequence is actually valine is not used, and in fact that information is not directly accessible to the algorithm. Sperm whale myoglobin is represented only by the sequence of environment classes of its residues, and the preference table is averaged over proteins with many different folding patterns.) Extension of this calculation to all positions and to all possible alignments not allowing gaps within regions of secondary structure gives a score that measures how well the given unknown sequence fits the sperm whale myoglobin profile.

A particular advantage of this method is that it can be automated, with a new sequence being scored against every 3D profile in the library of known folds, in essentially the same way as a new sequence is routinely screened against a library of known sequences.

Use of 3D profiles to assess the quality of structures

The 3D profile derived from a structure depends only very indirectly on the amino acid sequence. It is therefore meaningful to ask, not only whether it is possible to identify other amino acid sequences compatible with the given fold, but whether the score of a 3D profile for its own parent sequence is a measure of the compatibility of the sequence with the structure. Naturally, if real sequences did not generally appear to be compatible with their own structures, one would be forced to conclude that a useful method for examining the relationship between sequence and structure had not been achieved. Two interesting results are observed: (1) Protein structures determined correctly do fit their own profiles well, although other, related, proteins may give *higher* scores. The profile is abstracting properties of the family, not of individual sequences. (2) When a sequence does *not* match a profile computed from an experimental structure of that protein, there is likely to have been an error in the structure determination. The positions in the profile that do not match can identify the regions of error.

Threading

Threading is a method for fold recognition. Given a library of known structures, and a sequence of a query protein of unknown structure, does the query protein share a folding pattern with any of the known structures? The fold library could include some or all of the Protein Data Bank, or even hypothetical folds.

The basic idea of threading is to build many rough models of the query protein, based on each of the known structures and using different possible alignments of the sequences of the known and unknown proteins. This systematic exploration of the many possible alignments gives threading its name: Imagine trying out all alignments by pulling the query sequence gently through the three-dimensional framework of any known structure. Gaps must be allowed in the alignments, but if the thread is thought of as being sufficiently elastic the metaphor of threading survives.

Both threading and homology modelling deal with the three-dimensional structure induced by an alignment of the query sequence with known structures of homologues. Homology modelling focuses on one set of alignments and the goal is a very detailed model. Threading explores many alignments and deals with only rough models usually not even constructed explicitly:

Homology modelling	Threading
First, identify homologues	Try all possible parents
Then, determine optimal alignment	Try many possible alignments
Optimize one model	Evaluate many rough models

Successful fold recognition by threading requires:

1. A method to score the models, so that we can select the best one.
2. A method for calibrating the scores, so that we can decide whether the best-scoring model is likely to be correct.

Several approaches to scoring have been tried. One of the most effective is based on empirical patterns of residue neighbours, as derived from known structures. Observe the distribution of interresidue distances in known protein structures, for all 20 × 20 pairs of residue types. For each pair, derive a probability distribution, as a function of the separation in space and in the amino acid sequence. For instance, for the pair Leu–Ile, consider every Leu and Ile residue in known structures, and, for each Leu–Ile pair, record the distance between their Cβ atoms, and the difference in their positions in the sequence. Collecting these statistics permits estimation of how well the distributions observed in a model agree with the distributions in known structures.

The Boltzmann equation relates probabilities and energies. Usual applications of the Boltzmann equation start from an energy function and predict a probability distribution. (A standard example is the prediction of the density of the atmosphere as a function of altitude, from the gravitational potential energy function of the air molecules.) For threading, one turns this on its head, and *derives* an energy function *from* the probability distribution. This energy function is then used to score threading models.

For each structure in the fold library, the procedure finds the assignment of residues that produces the lowest energy score. Although this is an alignment problem, the nonlocal interactions mean that it can't be solved by dynamic programming.

Fold recognition at CASP

The best methods for fold recognition are consistently effective. These include but are not limited to methods based on threading.

Figures 5.11 and 5.12 show a prediction by A. G. Murzin, and another prediction by Bonneau, Tsai, Ruczinski and Baker, of targets from the 2000 CASP programme, both proteins of unknown function from *H. influenzae*.

Conformational energy calculations and molecular dynamics

A protein is a collection of atoms. The interactions between the atoms create a unique state of maximum stability. Find it, that's all!

The computational difficulties in this approach arise because (a) the model of the interatomic interactions is not complete or exact, and (b) even if the model were exact we should face an optimization problem in a large number of variables, involving nonlinearities in the objective function and the constraints, creating a very rough energy surface with many local minima. Like a golf course with many bunkers, such problems are very difficult.

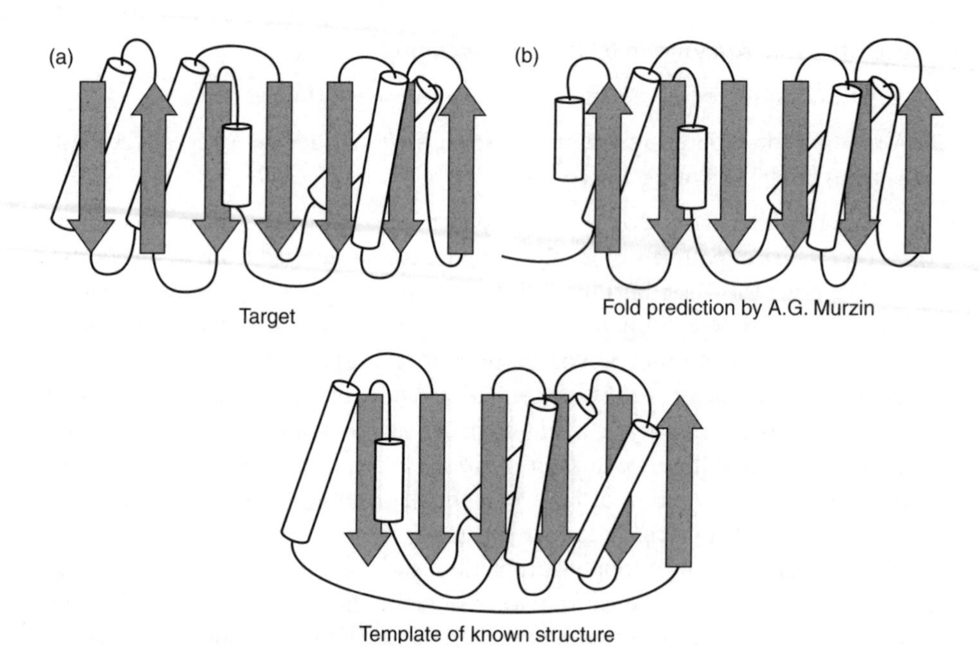

Fig. 5.11 Prediction of structure of *H. influenzae*, hypothetical protein. (a) The folding pattern of the target. (b) Prediction by A. G. Murzin. (c) Folding pattern of the closest homologue of known structure: an N-ethylmaleimide-sensitive fusion protein involved in vesicular transport (PDB entry 1NSF). The topology of Murzin's prediction is closer to the target than that of the closest single parent.

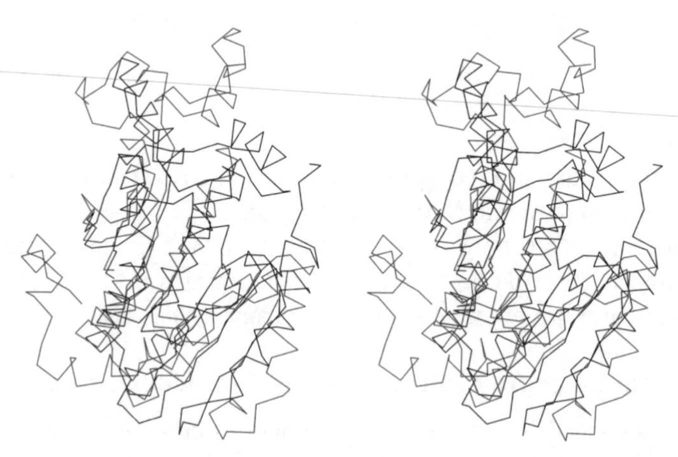

Fig. 5.12 Prediction by Bonneau, Tsai, Ruczinski and Baker of another hypothetical protein from *H. influenzae*, based on glycine N-methyltransferase [1XVA]. Experimental structure, black; prediction, red. Note that much of the prediction superposes well on the experimental structure and that the parts that do not superpose well have similar local structures but improper orientation and packing against the main body of the protein.

The interactions between atoms in a molecule can be divided into:

(a) Primary chemical bonds—strong interactions between atoms that must be close together in space. These are regarded as a fixed set of interactions that are not broken or formed when the conformation of a protein changes, but which, however, are equally consistent with a large number of conformations.

(b) Weaker interactions that depend on the conformation of the chain. These can be significant in some conformations and not in others—they affect sets of atoms that are brought into proximity by different folds of the chain.

The conformation of a protein can be specified by giving the list of atoms in the structure, their coordinates, and the set of primary chemical bonds between them (this can be read off, with only slight ambiguity, from the amino acid sequence). Terms used in the evaluation of the energy of a conformation typically include:

- **Bond stretching** $\Sigma_{bonds}\, K_r(r - r_0)^2$. Here r_0 is the equilibrium interatomic separation and K_r is the force constant for stretching the bond. r_0 and K_r depend on the type of chemical bond.
- **Bond angle bend** $\Sigma_{angles}\, K_\theta(\theta - \theta_0)^2$. For any atom *i* that is chemically bonded to two (or more) other atoms *j* and *k*, the angle *i–j–k* has an equilibrium value θ_0 and a force constant for bending K_θ.
- Other terms to enforce proper stereochemistry penalize deviations from planarity of certain groups, or enforce correct chirality (handedness) at certain centres.
- **Torsion angle** $\Sigma_{dihedrals}\, \frac{1}{2}V_n[1+\cos n\theta]$. For any four connected atoms: *i* bonded to *j* bonded to *k* bonded to *l*, the energy barrier to rotation by an angle θ of atom *l* with respect to atom *i* around the *j–k* bond is given by a periodic potential. V_n is the height of the barrier to internal rotation; *n* barriers are encountered during a full 360° rotation. The mainchain conformational angles φ, ψ and ω are examples of torsional rotations (see Fig. 5.2).
- **Van der Waals interactions** $\Sigma_i\, \Sigma_{j<i}\, (A_{ij}R_{ij}^{-12} - B_{ij}R_{ij}^{-6})$. For each pair of non-bonded atoms *i* and *j*, the first term accounts for a short-range repulsion and the second term for a long-range attraction between them. R_{ij} is the distance between atoms *i* and *j*. The parameters *A* and *B* depend on atom type.
- **Hydrogen bond** $\Sigma_i\, \Sigma_{j<i}\, (C_{ij}R_{ij}^{-12} - D_{ij}R_{ij}^{-10})$. The hydrogen bond is a weak chemical/electrostatic interaction between two polar atoms. Its strength depends on distance and also on the bond angle. This approximate hydrogen bond potential does not explicitly reflect the angular dependence of hydrogen bond strength; other potentials attempt to account for hydrogen bond geometry more accurately.

- **Electrostatics** $\Sigma_i \Sigma_{j<i} Q_iQ_j/(\epsilon R_{ij})$. Q_i and Q_j are the effective charges on atoms i and j, R_{ij} is the distance between them, and ϵ is the dielectric 'constant'. This formula applies only approximately to media that are not infinite and isotropic, including proteins.
- **Solvent** Interactions with the solvent, water, and cosolutes such as salts and sugars, are crucial for the thermodynamics of protein structures. Attempts to model the solvent as a continuous medium, characterized primarily by a dielectric constant, are approximations. With the increase in available computer power, it is now possible to include solvent explicitly, simulating the motion of a protein in a box of water molecules.

There are numerous sets of conformational energy potentials of this or closely-related forms, and a great deal of effort has gone into the tuning of parameter sets. The energy of a conformation is computed by summing these terms over all appropriate sets of interacting atoms.

The potential functions satisfy necessary but not sufficient conditions for successful structure prediction. One test is to take the right answer—an experimentally determined protein structure—as a starting conformation, and minimize the energy starting from there. In general most energy functions produce a minimized conformation that is about 1 Å (root-mean-square deviation) away from the starting model. This can be thought of as a measure of the resolution of the force field. Another test has been to take deliberately misfolded proteins and minimize their conformational energies, to see whether the energy value of the local minimum in the vicinity of the correct fold is significantly lower than that of the local minimum in the vicinity of an incorrect fold. Such tests reveal that multiple local minima cannot be reliably distinguished from the correct one on the basis of calculated conformational energies.

Attempts to predict the conformation of a protein by minimization of the conformational energy have so far not provided a method for predicting protein structure from amino acid sequence. In order to overcome the problems both of getting trapped in local minima, and of the absence of a good model for protein-solvent interactions, molecular dynamics models have been developed. The protein plus explicit solvent molecules are treated—via the force field—by classical Newtonian mechanics. It is true that this permits exploration of a much larger segment of phase space. However, as an a priori method of structure prediction, it has still not succeeded consistently. However, these are calculations that are extremely computationally intensive and here, perhaps more than anywhere else in this field, advances deriving from the increasing 'brute force' power of processors will have an effect.

In the meantime, molecular dynamics, if supplemented by experimental data, regularly makes extremely important contributions to structure determinations by both X-ray crystallography (usually) and nuclear magnetic resonance (always). How is molecular dynamics integrated into the process of structure determination? For any conformation, one can measure the consistency of the model with the experimental data. In crystallography, the experimental data are the absolute values of the Fourier coefficients of the electron density of the molecule. In nuclear magnetic resonance, the experimental data provide constraints on the distances between certain pairs of residues. But in both X-ray crystallography (almost always) and nuclear magnetic resonance, the experimental data underdetermine the protein structure. To solve a structure one must seek a set of coordinates that minimizes a combination of the deviation from the experimental data and the conformational energy. Molecular dynamics is successful at determining such coordinate sets: the dynamics provides adequate coverage of conformation space, and the bias derived from the experimental data channels the calculation towards the correct structure.

ROSETTA

ROSETTA is a program by D. Baker and colleagues that predicts protein structure from amino acid sequence by assimilating information from known structures. In recent CASP programmes, ROSETTA showed consistent success on targets in the Novel Fold categories. At present, it leads the field.

ROSETTA predicts a protein structure by first generating structures of fragments using known structures, and then combining them. First, for each contiguous region of 3 and 9 residues, instances of that sequence and related sequences are identified in proteins of known structure. For fragments this small, there is no assumption of homology to the target protein. The distribution of conformations of the fragments serves as a model for the distribution of possible conformations of the corresponding fragments of the target structure.

ROSETTA explores the possible combinations of fragments using Monte Carlo calculations (see Box). The energy function has terms reflecting compactness, paired β-sheets and burial of hydrophobic residues. The procedure carries out 1000 independent simulations, with starting structures chosen from the fragment conformation distribution pattern generated previously. The structures that result from these simulations are clustered, and the centres of the largest clusters presented as predictions of the target structure. The idea is that a structure that emerges many times from independent simulations it is likely to have favourable features.

Figure 5.13 shows successful predictions by ROSETTA of two targets from the 2000 CASP programme.

ROSETTA is available by License or as a webserver.*

LINUS

LINUS (Local Independently Nucleated Units of Structure) is a program for prediction of protein structure from amino acid sequence, by G. D. Rose and R. Srinivasan. It is a completely a priori procedure, making no explicit reference to any known structures or sequence-structure relationships. LINUS folds the

* http://depts.washington.edu/bakerpg

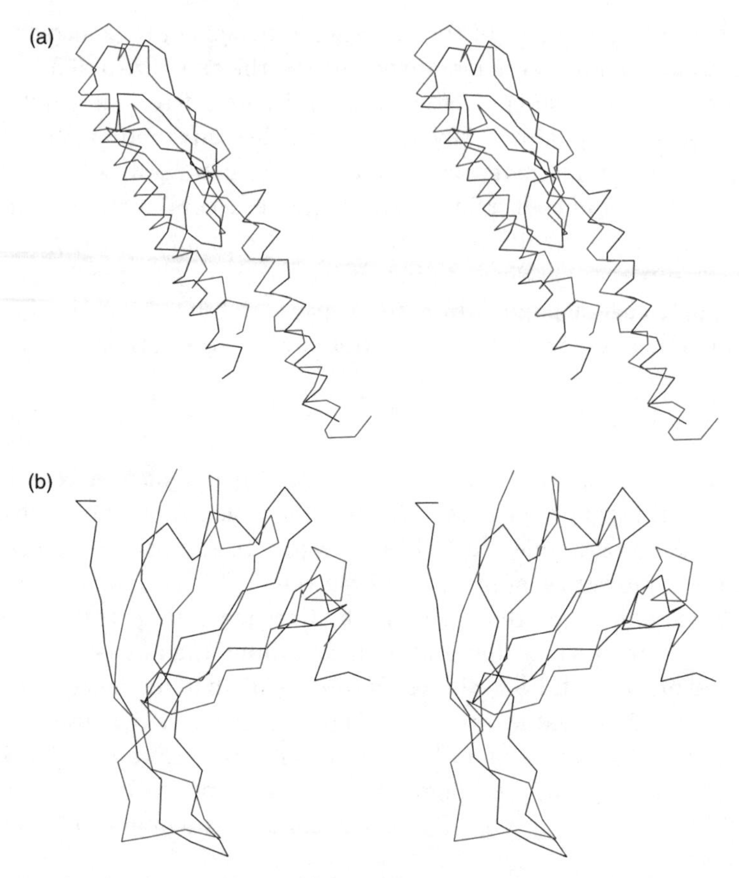

Fig. 5.13 Predictions by ROSETTA of (a) *H. influenzae*, hypothetical protein, (b) The N-terminal half of domain 1 of human DNA repair protein XRCC4. Part b shows a selected substructure containing a 55-residue N-terminal segment (out of a total of 116 residues). Experimental structures, black; predicted structures, red.

polypeptide chain in a *hierarchical* fashion, first producing structures of short segments, and then assembling them into progressively larger fragments.*

An insight underlying LINUS is that the structures of local regions of a protein—short segments of residues consecutive in the sequence—are controlled by local interactions within these segments. During natural protein folding, each segment will preferentially sample its most favourable conformations. However, these preferred conformations of local regions, even the one that will ultimately be adopted in the native state, are below the threshold of stability. Local structure will form transiently and break up many times before a suitable interacting partner stabilizes it. But in the computer one is free to anticipate the results. In a LINUS simulation, favourable structures of local fragments, as determined by their frequent recurrence during the simulation, transmit their

* LINUS is freely available from www.roselab.jhu.edu

Monte Carlo algorithms

Monte Carlo algorithms are used very widely in protein structure calculations, to explore conformations efficiently, and also in many other optimization problems, to search for the minimum of a complicated function. Simple minimization methods based on moving 'downhill' in energy fail because the calculation gets trapped in a local minimum far from the native state.

In general, Monte Carlo methods make use of random numbers to solve problems for which it is difficult to calculate the answer exactly. The name was invented by J. von Neumann, referring to the applications of random number generators in the famous gambling casino.

To apply Monte Carlo techniques to find the minimum of a function of many variables—for instance, the minimum energy of a protein as a function of the variables that define its conformation—suppose that the configuration of the system is specified by the variables x, and that for any values of these variables, we can calculate the energy of the conformation, $\varepsilon(x)$. (x stands for a whole set of variables—perhaps the set of atomic coordinates of a protein, or the mainchain and sidechain torsion angles.)

Then the Metropolis procedure (invented in 1953, allegedly at a dinner party in Los Alamos) prescribes:

1. Generate a random set of values of x, to provide a starting conformation. Calculate the energy of this conformation, $\varepsilon = \varepsilon(x)$.
2. Perturb the variables: $x \to x'$, to generate a neighbouring conformation.
3. Calculate the energy of the new conformation, $\varepsilon(x')$
4. Decide whether to *accept* the step: to move $x \to x'$, or to stay at x and try a different perturbation:
   (a) If the energy has decreased; that is, $\varepsilon = \varepsilon(x) > \varepsilon(x')$—in other words the step went *downhill*—always accept it. The perturbed conformation becomes the new current conformation: set $x' \to x$ and $\varepsilon = \varepsilon(x')$.
   (b) If the energy has increased or stayed the same; that is $\varepsilon(x) \leq \varepsilon(x')$—in other words the step goes *uphill—sometimes* accept the new conformation. If $\Delta = \varepsilon(x') - \varepsilon(x)$, accept the step with a probability $\exp[-\Delta/(kT)]$, where k is Boltzmann's constant, and T is an effective temperature.
5. Return to step 2.

It is step 4b that is the ingenious one. It has the potential to get over barriers, out of traps in local minima. The effective temperature, T, controls the chance that an uphill move will be accepted. T is not the physical temperature at which we wish to predict the protein conformation, but simply a numerical parameter that controls the calculation. For any temperature, the

→

→

higher the uphill energy difference, the less likely that the step will be accepted. For any value of ε, if T is low, then $\varepsilon(x)/(kT)$ will be high, and $\exp[-\varepsilon(x)/(kT)]$ will be relatively low. If T is high, then $\varepsilon(x)/(kT)$ will be low, and $\exp[-\varepsilon(x)/(kT)]$ will be relatively high. The higher the temperature, the more probable the acceptance of an uphill move.

This relatively simple idea has proved extremely effective, with successful applications including but by no means limited to protein structure calculations.

Simulated annealing is a development of Monte Carlo calculations in which T varies—first it is set high to allow efficient exploration of conformations, then it is reduced to drop the system into a low-energy state.

preferred conformations as biases that influence subsequent steps. The procedure applies the principle of a rachet to direct the calculation along productive lines.

LINUS begins by building the polypeptide from the sequence as an extended chain. The simulation proceeds by perturbing the conformations of a succession of randomly-chosen three-residue segments, and evaluating the energies of the results. Structures with steric clashes are rejected out of hand; other energetic contributions are evaluated only in terms of local interactions. A Monte Carlo procedure (see Box) is used to decide whether to accept a perturbed structure or revert to its predecessor. LINUS performs a large number of such steps. It periodically samples the conformations of the residues, to accumulate statistics of structural preferences.

Subsequent stages in the simulation assemble local regions into larger fragments, using the conformational biases of the smaller regions to guide the process. The window within the sequence controlling the range of interactions is progressively opened, from short local regions, to larger ones, and ultimately to the entire protein.

The LINUS representation of the protein folding process is realistic in essential respects, although approximate. All non-hydrogen atoms of a protein are modelled, but the energy function is approximate and the dynamics simplified. The energy function captures the ideas of: (1) steric repulsion preventing overlap of atoms, (2) clustering of buried hydrophobic residues, (3) hydrogen bonding, and (4) salt bridges.

Currently LINUS is generally successful in getting correct structures of small fragments (size between supersecondary structure and domain), and in some cases can assemble them into the right global structure. Figure 5.14 shows the LINUS prediction of the C-terminal domain of rat endoplasmic reticulum protein ERp29.

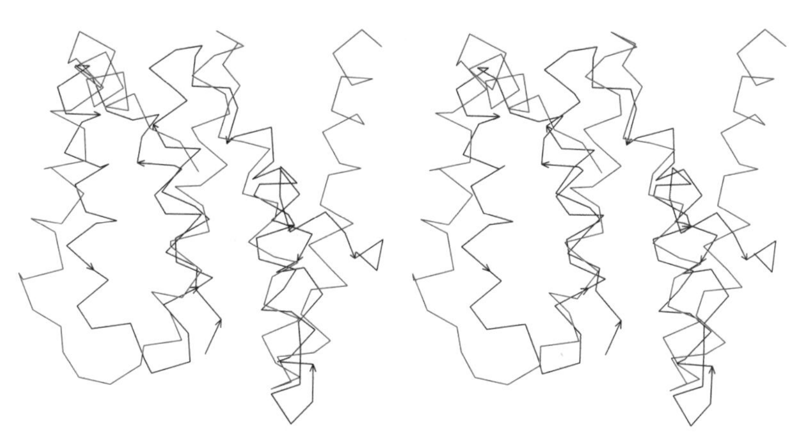

Fig. 5.14 A LINUS prediction of the C-terminal domain of rat endoplasmic reticulum protein ERp29. Experimental structure, black; prediction, red.

Assignment of protein structures to genomes

A genome sequence is the complete statement of a potential life. Assignment of structures to gene products is a first step in understanding how organisms implement their genomic information.

We want to understand the structures of the molecules encoded in a genome, their individual activities and interactions, and the organization of these activities and interactions in space and time during the lifetime of the organism. We want to understand the relationships among the molecules encoded in the genome of one individual, and their relationships to those of other individuals and other species.

For individual proteins, knowing their structure is essential for understanding the mechanism of their function and interactions. For entire organisms, knowing the structures tells us how the repertoire of possible protein folds is used, and how it is distributed among different functional categories in different species. For interspecies comparisons, protein structures can reveal relationships invisible in highly-diverged sequences.

Several methods have been applied to structure assignment:

- **Experimental structure determination.** The best way of all!
- **Detection of homology in sequences.** Sophisticated sequence comparison methods such as PSI-BLAST or Hidden Markov Models can identify relationships between proteins, both within an organism and between species. If the structure of any homologue is known experimentally, at least the general fold of the family can be inferred.
- **Fold-recognition methods** can assign folds to some proteins even in the absence of evidence for homology.
- Specialized techniques detect **membrane proteins**, and **coiled-coils.**

The results of structure assignments provide partial inventories of proteins in the different genomes, and, for the subset of proteins with sufficiently close relatives of known structure, detailed three-dimensional models. The degree of coverage of assignments is changing very fast, primarily because of the rapid growth of sequence and structural data. The table contains a current scorecard.

Species	Number of sequences	Structures assigned	%
E. coli	4289	916	21
M. jannaschi	1773	262	15
S. cerevisiae	6289	1109	18
D. melanogaster	13687	2990	22

(From: GeneQuiz, http://jura.ebi.ac.uk:8765/ext-genequiz/)

What do these results tell us about the usage of the potential protein repertoire? At present, proteins of known structure fall into approximately 750 fold classes, out of an estimated total of 1000. A comparison of folds deduced from the genomes of an archaeon *Methanococcus jannaschii*, a bacterium *Haemophilus influenzae*, and a eukaryote *Saccharomyces cerevisiae*, revealed that out of a total of 148 folds, 45 were common to all three species—and by implication, probably common to most forms of life. The archaeon, *M. jannaschii*, had the fewest unshared folds (see Fig. 5.15).

An inventory of the structures common to all three species showed that the five most common folding patterns of domains are: (1) the P-loop-containing NTP hydrolase fold, (2) the NAD-binding domain, (3) the TIM-barrel fold, (4) the flavodoxin fold,

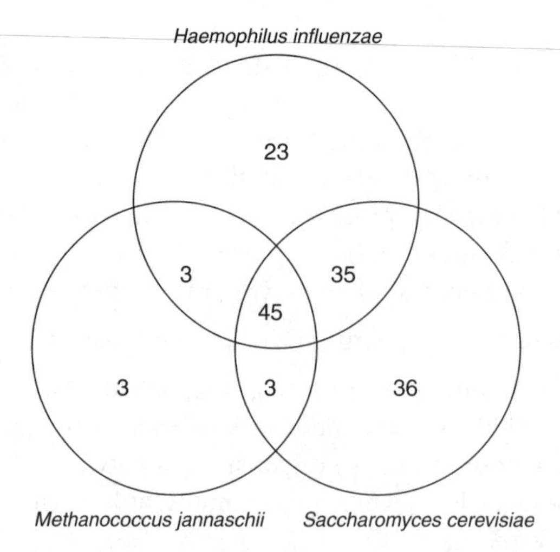

Fig. 5.15 Shared protein folds in an archaeon *Methanococcus jannaschii*, a bacterium *Haemophilus influenzae*, and a eukaryote *Saccharomyces cerevisiae*. (From: Gerstein, M. (1997), A structural census of genomes: comparing bacterial, eukaryotic, and archaeal genomes in terms of protein structure, *J. Mol. Biol.*, **274**, 562–576.)

and (5) the thiamin-binding fold. Plate VI shows the structure and a simplified schematic diagram of the topology of the last of these (see also Weblems 5.3 and 5.4). All are of the α/β type.

Prediction of protein function

The cascade of inference should ideally flow from sequence → structure → function. However, although we can be confident that similar amino acid sequences will produce similar protein structures, the relation between structure and function is more complex. Proteins of similar structure and even of similar sequence can be recruited for very different functions. Conversely, very widely diverged proteins may retain similar functions. Moreover, just as many different sequences are compatible with the same structure, proteins with different folds can carry out the same function.

As proteins evolve they may:

1. retain function and specificity,
2. retain function but alter specificity,
3. change to a related function, or a similar function in a different metabolic context,
4. change to a completely unrelated function.

People often ask: How much must a protein sequence or structure change before the function changes? The answer is: Some proteins have multiple functions, so they need not change at all!

- In the duck, an active lactate dehydrogenase and an enolase serve as crystallins in the eye lens, although they do not encounter the substrates *in situ*. In other cases crystallins are closely related to enzymes, but some divergence has already occurred, with loss of catalytic activity. (This proves that the enzymatic activity is not necessary in the eye lens.)
- A protein from *E. coli*, called Do or DegP or HtrA, acts as a chaperone (catalysing protein folding) at low temperatures, but at 42 °C turns into a proteinase. The rationale seems to be: under normal conditions or moderate heat stress the goal is to salvage proteins that are having difficulty folding; under more severe heat stress, when salvage is impossible, to recycle them.
- We have mentioned already the *E. coli* enzyme lipoate dehydrogenase that is *also* an essential subunit of pyruvate dehydrogenase, 2-oxoglutarate dehydrogenase and the glycine cleavage complex.

These examples of structure-function relationships are at the extreme end of a spectrum that can include a wide range of behaviour.

One problem is that it is not easy to define the idea of difference in function quantitatively. When are two different functions more similar to each other than two other different functions? In some cases altered function may conceal similarity of mechanism. For example, the enolase superfamily contains several homologous enzymes that catalyse different reactions with shared mechanistic features. This group includes enolase itself, mandelate racemase, muconate lactonizing enzyme I, and D-glucarate dehydratase. Each acts by abstracting an α proton from a carboxylic acid to form an enolate intermediate. The subsequent reaction pathway, and the nature of the product, vary from enzyme to enzyme. These proteins have very similar overall structures, a variant of the TIM-barrel fold. Different residues in the active site produce enzymes that catalyse different reactions.

Divergence of function: orthologues and paralogues

The family of chymotrypsin-like serine proteinases includes closely-related enzymes in which function is conserved, and widely diverged homologues that have developed novel functions. Trypsin, a digestive enzyme in mammals, catalyses the hydrolysis of peptide bonds adjacent to a positively-charged residue, Arg or Lys. (A **specificity pocket**, a surface cleft in the active site, is complementary in shape and charge distribution to the sidechain of the residue adjacent to the scissile bond.) Enzymes with similar sequence, structure, function, and specificity exist in many species, including human, cow, Atlantic salmon, and even *Streptomyces griseus* (Fig. 5.16). The similarity of the *S. griseus* enzyme to vertebrate trypsins suggests a lateral gene transfer. For the three vertebrate enzymes, each pair of sequences has ≥64% identical residues in the alignment, and the bacterial homologue has ≥30% identical residues with the others; all have very similar structures. These enzymes are called **orthologues**—homologous proteins in different species. (Other bacterial homologues are very different in sequence.)

Evolution has also created related enzymes in the *same* species with different specificities. Chymotrypsin and pancreatic elastase are other digestive enzymes that, like trypsin, cleave peptide bonds, but next to different residues: Chymotrypsin cleaves adjacent to large flat hydrophobic residues (Phe, Trp) and elastase cleaves adjacent to small residues (Ala). The change in specificity is effected by mutations of residues in the specificity pocket. Another homologue, leukocyte elastase (the object of database searching in Chapter 3) is essential for phagocytosis and defence against infection. Under certain conditions it is responsible for lung damage leading to emphysema.

Homologous proteins in the same species are called **paralogues**. Trypsin, chymotrypsin and pancreatic elastase function in digestion of food. Another set of paralogues mediates the blood coagulation cascade. Although all are proteinases, the requirements for activation and control are very different for digestion and blood coagulation, and the families have diverged and become specialized for these respective roles.

Some homologues of trypsin have developed entirely new functions:

- Haptoglobin has lost its proteolytic activity. It acts as a chaperone, preventing unwanted aggregation of proteins. Haptoglobin forms a tight complex with

```
                           10        20        30        40        50        60        70        80
                            |         |         |         |         |         |         |         |
Human             IVGGYNCEENSVPYQVSLNSGYHFCGGSLINEQWVVSAGHCYKSR---IQVRLGEHNIEVLEGNEQFINAAKIIRHPQYD
Cow               IVGGYTCGANTVPYQVSLNSGYHFCGGSLINSQWVVSAAHCYKSG---IQVRLGEDNINVVEGNEQFISASKSIVHPSYN
Atlantic salmon   IVGGYECKAYSQAHQVSLNSGYHFCGGSLVNENWVVSAAHCYKSR---VEVRLGEHNIKVTEGSEQFISSSRVIRHPNYS
S. griseus        VVGGTRAAQGEFPFMVRLSMG---CGGALYAQDIVLTAAHCVSGSGNNTSITATGGVVDLQSGAAVKVRSTKVLQAPGYN

                  iVGGy c     p qVsLnsGyhfCGGsL n  wVvsA HCyks      vrlge ni v eG eqfi   k i hP Y

                           90       100       110       120       130       140       150       160
                            |         |         |         |         |         |         |         |
Human             RKTLNNDIMLIKLSSRAVINARVSTISLPTAPPATGTKCLISGWGNTASSGADYPDELQCLDAPVLSQAKCEASYP-GKI
Cow               SNTLNNDIMLIKLKSAASLNSRVASISLPTSCASAGTQCLISGWGNTKSSGTSYPDVLKCLKAPILSDSSCKSAYP-GQI
Atlantic salmon   SYNIDNDIMLIKLSKPATLNTYVQPVALPTSCAPAGTMCTVSGWGNTMSSTADS-NKLQCLNIPILSYSDCNNSYP-GMI
S. griseus        --GTGKDWALIKLAQPINQ----PTLKIATTTAYNQGTFTVAGWGANREGGSQQRYLLKAN-VPFVSDAACRSAYGNELV

                       nDimLIKL   a  n  v    lpT     gt c  sGWGnt ssg      L cl  P lS   C   YP g i

                          170       180       190       200       210       220       230
                            |         |         |         |         |         |         |
Human             TSNMFCVGFLE-GGKDSCQGDSGGPVVCNG-----QLQGVVSWGDGCAQKNKPGVYTKVYNYVKWIKNTIAANS
Cow               TSNMFCAGYLE-GGKDSCQGDSGGPVVCSG-----KLQGIVSWGSGCAQKNKPGVYTKVCNYVSWIKQTIASN-
Atlantic salmon   TNAMFCAGYLE-GGKDSCQGDSGGPVVCNG-----ELQGVVSWGYGCAEPGNPGVYAKVCIFNDWLTSTMASY-
S. griseus        ANEEICAGYPDTGGVDTCQGDSGGPMFRKDNADEWIQVGIVSWGYGCARPGYPGVYTEVSTFASAIASAARTL-

                  t  mfC G le GGkDsCQGDSGGPvvc g      lqG VSWG GCA    PGVYtkV     wi  t a
```

Fig. 5.16 Alignment of sequences of trypsins from human, cow, Atlantic salmon and *Streptomyces griseus*. In the lines under the blocks, uppercase letters indicate absolutely conserved residues and lowercase letters indicate residues conserved in three of the four sequences (in most but not all cases the *S. griseus* sequence is the exception).

haemoglobin fragments released from erythrocytes, with several useful effects including preventing the loss of iron.

- The serine proteinase of rhinovirus has developed a separate, independent function, of forming the initiation complex in RNA synthesis, using residues on the opposite side of the molecule from the active site for proteolysis. This is not a modification of an active site—it is the creation of a new one.
- Subunits homologous to serine proteinases appear in plasminogen-related growth factors. The role of these subunits in growth factor activity is not yet known, but it cannot be a proteolytic function because essential catalytic residues have been lost.
- An antifreeze glycoprotein in Antarctic fish is homologous to trypsinogen.
- The insect 'immune' protein scolexin is a distant homologue of serine proteinases that induces coagulation of haemolymph in response to infection.

In the chymotrypsin family we see a retention of structure with similar functions in closely-related proteins, and progressive divergence of function in some but not all distantly-related ones.

The message is that the overall folding pattern of a protein is an unreliable guide to predicting function, especially for very distant homologues. For correct prediction of function in distantly-related proteins it is necessary to focus on the active site. For example:

- J. F. Bazan and R. Fletterick, and, independently, P. Argos, G. Kamer, M. J. Nicklin, and E. Wimmer, recognized that viral 3C proteinases are homologues of chymotrypsin, despite the fact that the serine of the catalytic triad is changed to cysteine.
- W. R. Taylor and L. Pearl recognized the distant homology between retroviral and aspartic proteinases from conserved Asp, Thr, and Gly residues.

Like motif libraries such as PROSITE, such approaches go directly from signature patterns of active-site residues in the sequence to conserved function, even in the absence of an experimental structure.

In focussing on the active site, there is opportunity to use methods similar to those used in drug design in designing ligands, to predict ligands that might bind to the proteins. It is important to make use of other experimental information available, such as tissue distribution patterns of expression, and catalogues of proteins that interact. Attempts to measure function directly, for instance by means of 'gene knockouts', will sometimes provide an answer, but are unproductive if the knockout phenotype is lethal or if there are multiple proteins that share a function.

It seems likely that the contribution of bioinformatics to prediction of protein function from sequence and structure will not be a simple algorithm that provides an unambiguous answer (as there is hope will someday be the case for prediction of structure from sequence). More reasonable aims are to suggest productive experiments and to contribute to the interpretation of the results. These are not unworthy goals.

Drug discovery and development

It is a sobering experience to ask a classroom full of students how many would be alive today without at least one course of drug therapy during a serious illness. (This ignores diseases escaped through vaccination.) Or to ask the students how many of their surviving grandparents would be leading lives of greatly reduced quality without regular treatment with drugs. The answers are eloquent. They engender fear of the new antibiotic-resistant strains of infectious micro-organisms. Other drugs target human proteins, to deal with protein dysfunction or to adjust regulatory controls. It is necessary to develop new drugs, which, in combination with genomic information that can improve their specificity, will extend and improve our lives.

However, it is not easy to be a drug. For a chemical compound to qualify as a drug, it must be:

1. safe
2. effective
3. stable—both chemically and metabolically
4. deliverable—the drug must be absorbed and make its way to its site of action
5. available—by isolation from natural sources or by synthesis
6. novel, that is, patentable

Steps in the development of new drug are summarized in the Box. The process involves scientific research, clinical testing to prove safety and efficacy, and very important economic and legal aspects involving patent protection and estimation of returns on the very high required investment.

Steps in the development of a new drug

1. Understanding the biological nature and symptoms of a disease. Is it caused by:
   - an infectious agent—bacterium, virus, other?
   - a poison of nonbiological origin?
   - a mutant protein in the patient?
2. Developing an assay. Given a candidate drug, can you test it by:
   - its effect on the growth of a micro-organism?
   - its effect on cells grown in tissue culture?
   - its effect on animals that suffer the disease or an analogue?
   - its binding to a known protein target?

→

→

Steps in the development of a new drug (*continued*)

3. Is an effective agent from a natural source known from folklore/tradition? If so, go to 6.
4. Identify a specific molecular target, usually a protein. Determine its structure experimentally or by model-building.
5. Get a general idea of what kind of molecule would fit the site on the target. Is there a known substrate or inhibitor?
6. Identification of a *lead compound:* a chemical that shows the desired biological activity to *any* measurable extent. A lead compound is a bridgehead; finding lead compounds and subsequently modifying them are quite different kinds of activities.
7. Development of the lead compound: Extensive study of variants of the compound, with the goal of building in all the desired properties and enhancing the biological activity.
8. Preclinical testing, *in vitro* and with animals, to prove effectiveness and safety. At this point the drug may be patented. (In principle, one wants to delay patenting as long as possible because of finite lifetime of the patent, and many lengthy steps still remain before the drug can be sold.)
9. In the USA: submission of an Investigational New Drug Application to the Federal Drug Administration (FDA). This is followed by three phases of clinical trials.
10. Phase I clinical trials. Test the compound for safety on healthy volunteers. Determine how the body deals with the drug—how it is absorbed, distributed, metabolized, excreted. The results suggest a safe dosage range.
11. Phase II clinical trials. Test the compound for efficacy against a disease on approximately 200 volunteer patients. Does it cure the disease or alleviate symptoms? Calibrate the dosage.
12. Phase III clinical trials. Test approximately 2000 patients, to demonstrate conclusively that the compound is better than the best known treatment. These are randomized double-blind tests, either against a placebo or against a currently-used drug. These trials are very expensive; it is not uncommon to kill a project before embarking on this step, if the phase II trials expose side effects or unsatisfactory efficacy.
13. File a New Drug Application with the FDA, containing supporting data proving safety and efficacy. FDA approval allows selling the drug. Only now can the drug generate income.
14. Phase IV studies, subsequent to FDA approval and marketing, involve continued monitoring of the effects of the drug, reflecting the wider experience in its use. New side effects may turn up in some classes of patients, leading to restrictions on the use of the drug, or even possibly its recall.

To develop a drug, first you must choose a target disease. You will want to study what is known about its possible causes, its symptoms, its genetics, its epidemiology, its relationship to other diseases—human and animal—and all known treatments. Assuming that the potential utility of a drug justifies the major time, expense, and effort required to develop one, you are now ready to begin.

From the target disease, you must select a target protein. About 500 proteins are the targets of known drugs.

You must develop a suitable assay with which to detect success in the initial phase. If a known protein is the target, binding can be measured directly. A potential anti-bacterial drug can be tested by its effect on growth of the pathogen. Some compounds might be tested for effects on eukaryotic cells grown in tissue culture. If a laboratory animal is susceptible to the disease, compounds can be tested on animal subjects. However, compounds may have different effects on animals and humans. For example, tamoxifen, now a drug used widely against breast cancer, was originally developed as a birth-control pill. In fact it is a fine contraceptive for rats but *promotes* ovulation in women.

The lead compound

A goal in the early stages of drug development is identification of one or more **lead compounds**. A lead compound is any substance that shows the biological activity you seek. It demonstrates that a compound exists that possesses at least some of the desired properties.

There are a number of ways to find lead compounds:

1. Serendipity: penicillin is the classic example.
2. Survey of natural sources. 'Grind and find' is the medicinal chemist's motto. Sometimes traditional remedies point to a source of active compounds. For example, digitalis was isolated from leaves of the foxglove, which had been used for congestive heart failure. (Why not just continue to use the traditional remedy? Isolation of the active principle makes it possible to regulate dosage, and to explore variants.) Approximately half the drugs in current use are based on natural products.
3. Study what is known about substrates, inhibitors, and the mechanism of action, and select potentially active compounds from these properties.
4. Consider drugs effective against similar diseases.
5. Large-scale screening. Techniques of combinatorial chemistry permit parallel testing of large sets of related compounds. A special technique applicable to polypeptides is phage display.
6. Occasionally, from side effects of existing drugs. Minoxidil (2,4-diamino-6-piperidino-pyrimidine-3-oxide), originally designed as an antihypertensive, was found to induce hair growth. Viagra, originally developed as heart medicine, is another example.

7. Experimental screening. The US National Cancer Institute has screened tens of thousands of compounds. (Screening of variants is also very important *after* a lead compound has been found.)
8. Computer screening and *ab initio* computer design.

Discovery of a lead compound triggers other kinds of research activities. Many variants of the lead compound must be tested to improve its effectiveness, and to build in other essential properties. For instance, a compound that binds to its target *in vitro* is no good as a drug unless it can get to the target *in vivo*. Deliverability of a drug to a target within the body requires the capacity to be absorbed and transported. It requires metabolic stability, and 'shelf' stability. It requires the proper solubility profile—a drug must be sufficiently water-soluble to be absorbed, but not so soluble that it is excreted immediately; it must (in most cases) be sufficiently lipid-soluble to get across membranes, but not so lipid-soluble that it is merely taken up by fat stores. There must be a reasonable synthetic route to produce the compound in quantity.

Improving on the lead compound: Quantitative Structure-Activity Relationships (QSAR)

For any compound with pharmacological activity, similar compounds typically exhibit related activity but vary in potency and specificity. Starting with a lead compound, chemists must survey large numbers of related molecules to optimize desired pharmacological properties. To search systematically, it would be very useful to understand how the variation in structural and physicochemical features in the family of molecules is correlated with pharmacological properties. The problem is that there are very many possible descriptors for characterizing molecules. These include structural features such as the nature and distribution of substituents; experimental features such as solubility in aqueous and organic solvents, or dipole moments; and computed features such as charges on individual atoms.

Quantitative Structure-Activity Relationships (QSAR) provide methods for predicting the pharmacological activity of a set of compounds from the relationship between molecular features and pharmacological activity, based on test cases. The method was developed by C. Hansch and colleagues in the 1960s, and has been of very widespread use.

C. Hansch, J. McClarin, T. Klein and R. Langridge applied QSAR methods to study inhibitors of carbonic anhydrase. Carbonic anhydrase is an enzyme that catalyses the reaction $CO_2 + H_2O \rightleftharpoons H^+ + HCO_3^-$. Clinical applications of carbonic anhydrase inhibitors include diuretics, treatment of high interocular pressure in glaucoma by supressing secretion of aqueous humour (the fluid within the eye), and anti-epileptic agents. High-altitude climbers take carbonic anhydrase inhibitors for relief of symptoms of acute mountain sickness.

Measurements of carbonic anhydrase binding of 29 phenylsulphonamides:

X
SO_2NH_2

(X stands for a set of substituents on the ring that are variable in both structure and position) showed that the binding constant was related to the Hammett electronic substituent constant σ, a measure of the electron-withdrawing or -donating strength of the substituent; the octanol-water partition coefficient P of the unionized form of the ligand; and the location (*ortho* or *meta*) of the substitution:

$$\log K \approx 1.55\sigma + 0.65 \log P - 2.07I_1 + 3.28I_2 + 6.94$$

in which K = binding constant, $I_1 = 1$ if X is *meta* and 0 otherwise, and $I_2 = 1$ if X is *ortho* and 0 otherwise. The substituents X were of the form –alkyl, or –COO-alkyl, or –CONH-alkyl.

This type of correlation has two implications:

1. A large number of compounds can be screened in the computer and those predicted to be the best tested experimentally.
2. It is possible to visualize the binding site from analysis of the parameters:

   - The positive coefficient of σ, implying that electron-withdrawing substituents are favoured, suggests that the ionized form of the $-SO_2NH_2$ moiety binds to the Zn ion in the carbonic anhydrase active site.
   - The positive coefficient of $\log P$ suggests a hydrophobic interaction between the protein and ligand.
   - The negative coefficients of I_1 and I_2 suggest steric clashes with substituents in the *meta* or *ortho* positions.

Structures of ligated carbonic anhydrase confirm these conclusions (see Weblem 5.9).

Bioinformatics in drug discovery and development

Computing and information retrieval contribute to several steps in drug discovery and development projects. These include: target identification; design, analysis, and enhancement of ligands; and selection and *in silicio* screening of libraries. Information systems are also important in the organization of the theoretical predictions, the experimental designs, and analysis of the data. D. Searls has called the intimate interplay between theory and experiment 'wet-dry cycles'.

Target selection

To develop a drug against a disease, it is necessary to select a protein linked to the disease in a way that suggests that it would be therapeutically useful to affect its function or expression. New high-throughput data sources, particularly of genome sequences and protein expression patterns, provide a rich source of material for identifying potential drug targets. **Differential genomics and proteomics**, the comparisons of healthy and diseased humans or animals, can pinpoint which particular protein is missing, dysfunctional, improperly regulated, or expressed only in affected cells. Information about protein-protein complexes make it possible to target not just a single protein, but a specific protein-protein interaction.

Knowledge of prokaryotic and viral genomes supports identification of targets for drugs against infectious disease. Of particular interest are metabolic pathways specific to micro-organisms, and the proteins that participate in them. A drug affecting such a target is less likely to interact with a human homologue, with consequent side effects. Proteins with sequences similar across bacterial clades offer the possibility of broad-spectrum antibiotics. Conversely, gene duplications warn of potential redundant functions, with concomitant insensitivity to inactivation of the target. Knowledge of the relative speed of evolution of different proteins, including horizontal gene transfer rates, indicates the expected stability of a therapy against development of resistant strains.

Commitment to a target by a large pharmaceutical company involves a very heavy investment of resources. The profit expected to flow from a successful drug exerts a very important influence on the choice of targets actively pursued. Analysis of the history of drugs that currently yield high profits suggests that prediction of economic returns is not a very precise science. Now, even generously-supported bioinformatics efforts are much less expensive than laboratory work. The possibility that calculations will improve predictions and enhance profit is behind the espousal of bioinformatics by the pharmaceutical industry, in addition to the purely scientific contributions of bioinformatics to drug discovery. This contribution to economic forecasting is especially important when a company considers high-risk projects, such as those aimed at developing a drug against a new class of targets. Such projects must compete with lower-risk activities such as trying to improve on a competitor's success.

Prediction of a lead compound

Methods for predicting ligands suitable as lead compounds for drug discovery can be divided into inductive and deductive approaches.

Inductive methods depend on correlations between known affinities of some test set of compounds, and molecular features characterizing entire libraries of potential ligands. These features include structural properties such as size, geometry, charge distributions and specific functional groups including hydrogen-bond donors and acceptors. They include general 'drug-like' qualities such as solubility in aqueous and organic solvents, easy route of administration, appropriate distribution in body tissues and metabolic turnover rate. Medicinal chemists apply an equivalent of the duck test: if it walks like a drug, swims like a drug and quacks like a drug, then maybe it will be a drug. The relevant characteristics of compounds are compiled into a **feature vector** used to compare the overall match between compounds of known affinity and a complete library. The requirements for organization, encoding, storage, and searching of information about small molecules has created a new field, **chemoinformatics**, which complements bioinformatics in applications to drug discovery.

Deductive methods are applicable if the binding site on the target protein is known or can be inferred. However, because binding affinity and specificity are only two requirements for a lead compound—admittedly essential ones—it is necessary to combine deductive methods with the correlation to desirable properties as in the purely inductive approach. Binding assays on purified systems give little idea of the behaviour of a compound as a drug in its biological context. Bioinformatics has a contribution to make in integrating the information available from molecular and cell biology, and physiology and pharmacology, to help bridge the gap between *in vitro* experiments and therapeutic activities.

Molecular modelling in drug discovery

A central problem in drug discovery is the identification of a compound that will bind tightly and specifically to a target protein. Tight binding is necessary for efficacy at low concentrations. Specificity is necessary to minimize side effects.

If the structure of the target is known from experiment, it is possible to apply molecular modelling directly to ligand design. If the structure of the target is unknown, a picture of the binding site must be created from indirect evidence, and ligand design is correspondingly more difficult. Ligand design without the target structure is like trying to catch a bank robber from eyewitness descriptions; ligand design to a target of known structure is like trying to catch the bank robber from a clear image on a CCTV videotape.

Goals of molecular modelling applied to drug design include:

- Ideally: suggestion of a lead compound that already shows reasonable affinity and specificity. This is a rare achievement.
- Analysis of compounds known to bind to the target. Understanding the important interactions serves as a guide to design and testing of potential ligands, and for selecting structural features to build into combinatorial synthesis of libraries. In the case of antibacterial or antiviral projects, a model of the protein-ligand complex can give some idea of how easy it would be for the pathogen to develop resistance by mutations that lower the affinity.
- **Pharmacophore** identification is the extraction of common substructures of many compounds that share a pharmacological activity, or at least that bind to the same site on a protein. The hypothesis is that there is some common constellation of atoms within the structures that is responsible. The computational problem of extracting the pharmacophore from a set of compounds is similar to that of structural alignment of a set of homologous proteins. However, although typical ligands are much smaller than proteins, the combinatorial problems are more severe because one has lost the linear ordering of the residues in proteins. (see page 235.) Inferred pharmacophore properties are integrated with QSAR methods to filter libraries of compounds for candidate ligands.
- *In silicio* screening: predicting of affinities, even qualitatively, suggests candidate ligands from a library of chemical structures. The results can be used either for setting priorities in experimental tests, or be integrated into broader approaches to computer screening of libraries on the basis of features correlated with favourable chemical and pharmacological properties. Many readers will be aware of the harnessing of screensavers worldwide to search for potential drugs.[†] At present, over 2.6 million computers have joined this project. They have contributed a cumulative total of over 320 000 years of CPU power.

see Box, Docking: prediction of ligand geometry and affinity

† http://www.chem.ox.ac.uk/curecancer.html

- **Lead compound improvement:** Once a compound is identified that binds to a target protein, albeit with low affinity and specificity, interactive modelling can suggest modifications that are expected to enhance the fit. Synthesis and testing of compounds predicted to show enhanced affinity, and even solution of crystal structures of their complexes, can guide the search for improved compounds. The modelling is usually coupled with combinatorial chemistry and experimental screening of libraries of compounds.

Docking: prediction of ligand geometry and affinity

Docking is prediction of ligand binding. In includes prediction both of binding of small molecules to proteins, and of protein-protein binding. The goals of docking are (1) to identify the binding site on the protein, and determine the position and orientation of the ligand, and (2) to estimate the affinity.

(1) Identification of mode of binding Docking of small molecules to proteins requires matching of the ligand to a site on a protein of known structure. The binding site may be known in advance, or it may be necessary to try many different modes of apposition of the ligand and protein to predict the optimal binding site.

The basis for docking is the identification of complementarity in size, shape, and distribution of charge, polarity, and potential for hydrophobic and hydrogen-bonding interactions. A complication is the possibility of flexibility in both partners. Small organic molecules containing many single bonds have a high degree of conformational flexibility. (Drug designers love structures with rings and bridges.) Many proteins show conformational changes upon binding ligands. Therefore the experimental structure of an unligated protein cannot be assumed to serve as a rigid target for docking. However, allowing for flexibility complicates docking calculations substantially.

Water molecules at interfaces present another difficulty. They can contribute to the surface complementarity, and provide bridging hydrogen bonds.

(2) Estimation of affinity It is difficult to estimate absolute affinities. However, comparative docking can provide useful information about *relative* affinities. A suitable scoring function, that can predict the ranking of different ligands in approximate order of affinity, allows selectivity, and setting of priorities, in experimental testing. Such scoring schemes can be *ab initio*—based on the kinds of force fields described on pages 257–258—or empirical. Conversely, comparative docking of one ligand to many proteins can predict the specificity of the interaction.

→

Compare:

Docking calculation	Information provided
1 ligand –1 protein	mode of binding, estimate of affinity
many ligands –1 protein	ranking of affinities of a series of potential ligands
1 ligand – many proteins	prediction of specificity

Docking and scoring are important steps in the filter between a full potential library and testing at the bench. A typical narrowing of the funnel might run as follows:

overall library size	10^{12} compounds
after general filters	10^5
docking	10^4
scoring	10^3
visual	10–100 for experimental testing

Two case studies illustrate the range of chemical and molecular-biological techniques involved in drug development, and show some interesting similarities and contrasts. They concern two well-known families of analgesic drugs—colloquially, 'pain-killers'—typified by morphine and aspirin. The two groups of compounds have different mechanisms of actions, different potencies, and different spectra of side effects.

Case Study 5.1: Development of analgesic drugs based on morphine‡

Morphine and codeine are natural alkaloids contained in the latex of the opium poppy (*Papaver somniferum*) (Fig. 5.17). The pharmacological effects have been known since antiquity. Modern chemistry has explored and developed many variants. Heroin was synthesized in 1874 (Fig. 5.17). More hydrophobic than the natural compounds, heroin traverses the blood-brain barrier more readily, giving it a more rapid onset of action.

Both codeine and heroin are metabolized to produce morphine, the active form. Codeine is therefore a natural example of a **prodrug**, an inactive agent that is converted to an active one. The conversion depends on a cytochrome, CYP2D6, which is absent in 5–10% of Caucasians and 1–3% of Afro-Americans and Asians, in whom codeine is ineffective.

Morphine and codeine have been applied in medicine and surgery as analgesics, drugs to relieve severe pain. Side effects include passivity and euphoria,

‡ Coop, A. & MacKerell, A. D. Jr. (2000), The future of opioid analgesics, *Amer. J. Pharm. Educ.*, **66**, 153–156.

Case Study 5.1 *(continued)*

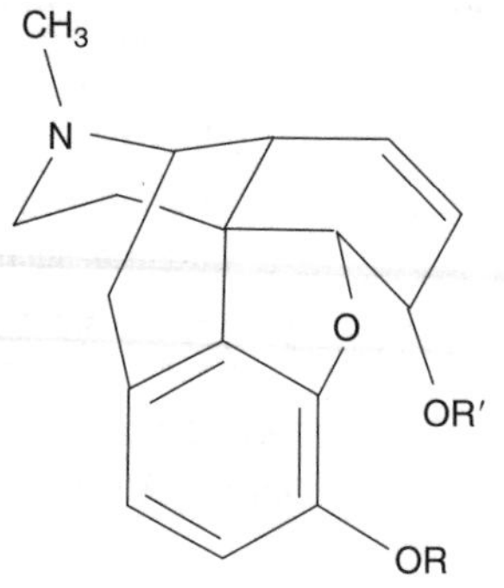

Fig. 5.17 Morphine, codeine and heroin have structures differing only in substituents at two positions:

Compound	**R**	**R′**
Morphine	–H	–H
Codeine	$-CH_3$	–H
Heroin	$-COCH_3$	$-COCH_3$

and physical dependence and addiction. Drug developers have therefore long sought a compound that would relieve pain without the harmful side effects. Of course there was no guarantee that this would be possible.

Synthetic variants of morphine allow correlation of biological effects with chemical structure.

One approach is to try to simplify the structure. The goals are (1) to infer the minimal pharmacophore required for activity, and (2) if possible, to dissect the parts of the structure that relieve pain from those causing addiction. Morphine, codeine and heroin are rigid compounds containing five fused rings. Levorphanol differs from morphine by loss of the bridging oxygen (removal of the tetrahydrofuran ring) and one of the hydroxyl groups (Fig. 5.18). It is a more potent analgesic than morphine but still addictive. Benzomorphan, cyclazocine and pentazocine break the cyclohexene ring (Fig. 5.19). The addictive effects of these compounds are lesser than those of morphine and levorphanol. Demerol, which opens the cyclohexene ring, and methadone, which has *no* fused rings, retain analgesic activity, sharing even smaller common substructures with morphine.

From these structures, one can infer the pharmacophore shown in Fig. 5.20.

In contrast to simplifying the molecule to identify a pharmacophore, attempts to enhance specificity have retained the pharmacophore but made the molecule more complex. Some success has been achieved. Etorphine and buprenorphine, discovered in the 1960s, are far more powerful analgesics than morphine (etorphine is used for sedation of large animals), and have lower addictive potential (see Fig. 5.21). Indeed, the most important clinical use of buprenorphine is in treatment of drug addiction, rather than in analgesia.

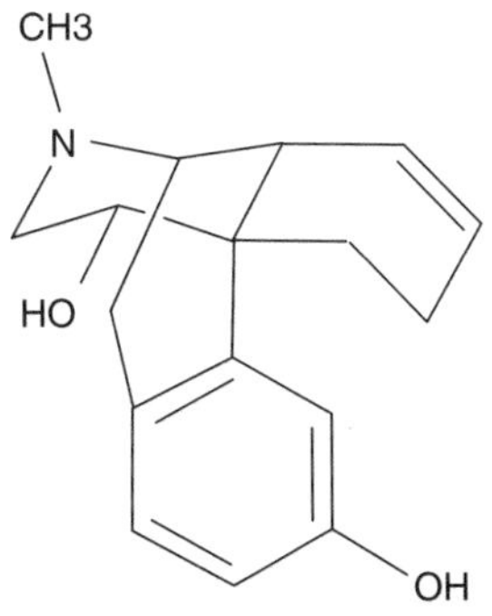

Fig. 5.18 The structure of levorphanol.

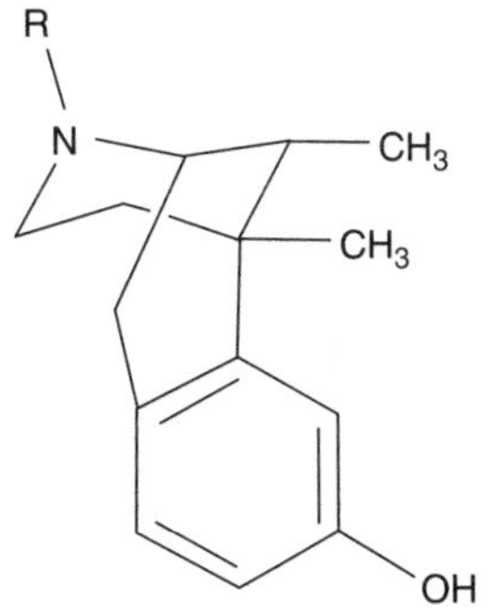

Fig. 5.19 The structures of benzomorphan: R = CH_3; cyclazocine: R = CH_2-cp (cp = cyclopropane); pentazocine: R = $CH_2CH{=}C(CH_3)_2$.

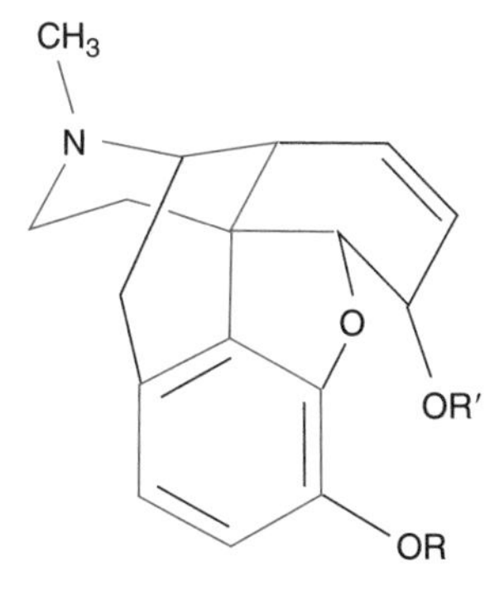

Fig. 5.20 Pharmacophore (red) derived from structural comparisons among morphine derivatives. (After A. D. MacKerell, Jr.)

Case Study 5.1 *(continued)*

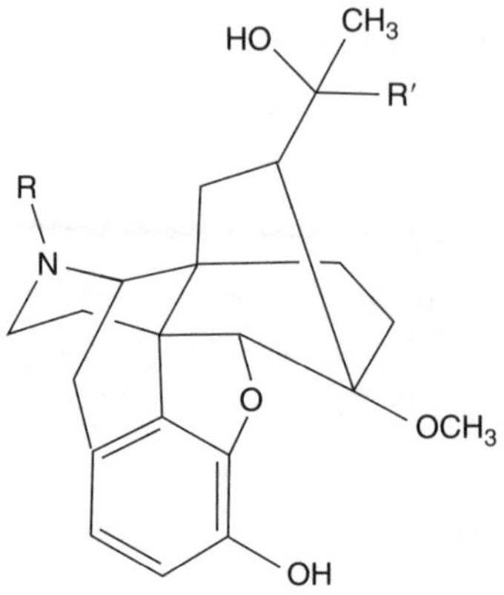

Fig. 5.21 The strucures of etorphine: R = CH_3, R′ = C_3H_7; buprenorphine: R = CH_2-cp (cp = cyclopropane), R′ = *t*-butyl.

This exploration of variants went on before the natural receptors were identified. We now know that the natural targets of action of morphine and related molecules are receptors for endogenous peptides called endorphins. These include:

β-endorphin	YGGFMTSEKSQTPLVTLFKNAIIKNAYKKGE
dynorphin	YGGFLRRIRPKLKWDNQ

and their cleavage products:

Met-enkephalin	YGGFM
Leu-enkephalin	YGGFL

Morphine is therefore a natural **peptidomimetic**, a non-peptide that shares a structure and activity with a peptide.

Several classes of receptors are known, including μ, κ, and δ types, and a recently-discovered fourth type, called ORL-1 (ORL = opiate-receptor like). Their sequences are about 50–70% identical at the residue level. They are G-protein-coupled receptors, similar in structure to bacteriorhodopsin (see Fig. 4.7). Different ligands—natural and synthetic—have different affinity to different receptors, and different kinetics of binding and dissociation. The natural targets of morphine are μ receptors. It is thought that μ receptors tend to be more involved in physical dependence and addiction than κ receptors, although this statement of the situation is extremely oversimplified. Nevertheless, it suggests that to produce a drug that provides analgesia with reduced side effects one should look at the *distribution* of affinities of compound for the different types of receptor.

Case Study 5.2: Computer-aided drug design: specific inhibitors of prostaglandin cyclooxygenase 2

Prostaglandins are a family of natural compounds that mediate a wide variety of physiological processes. Pharmacological applications include the use of prostaglandins themselves, and, conversely, drugs that block prostaglandin synthesis. Prostaglandin E_2 (dinoprostone) is used in obstetrics to induce labour. Aspirin, ibuprofen, acetaminophen (tylenol), and other **nonsteroidal anti-inflammatory drugs (NSAIDs)** are effective against arthritis and related diseases (see Box, page 283). They achieve this effect by inhibiting enzymes in the pathway of prostaglandin synthesis, specifically, prostaglandin cyclooxygenases. A well-known side effect of aspirin is bleeding from the walls of the stomach. This occurs because prostaglandins (the production of which aspirin inhibits) suppress acid secretions by the stomach and promote formation of a mucus coating protecting the stomach lining.

Aspirin and other NSAIDs inhibit *two* closely-related prostaglandin cyclooxygenases, called COX-1 and COX-2. (Unfortunately the same abbreviations are used for cytochrome oxidases 1 and 2.) COX-1 is expressed constitutively in the stomach lining. COX-2 is inducible, and up-regulated in response to inflammation. This suggests that a drug that would inhibit COX-2 but not COX-1 would retain the desired activity of NSAIDs but reduce unwanted side effects. [Note added at time of going to press: some COX-2 inhibitors have recently been implicated in increased risk of cardiovascular disease.]

The amino acid sequences and crystal structures of COX-1 and COX-2 are known. (These proteins have 65% sequence identity.) Figure 5.22 shows part of the structure of COX-1, acetylated by the aspirin analogue 2-bromoacetoxybenzoic acid (aspirin brominated on the methyl group of the acetyl moiety). The salicylate moiety binds nearby. The effect is to block the entrance to the active site. Most NSAIDs bind but do not covalently modify the enzyme.

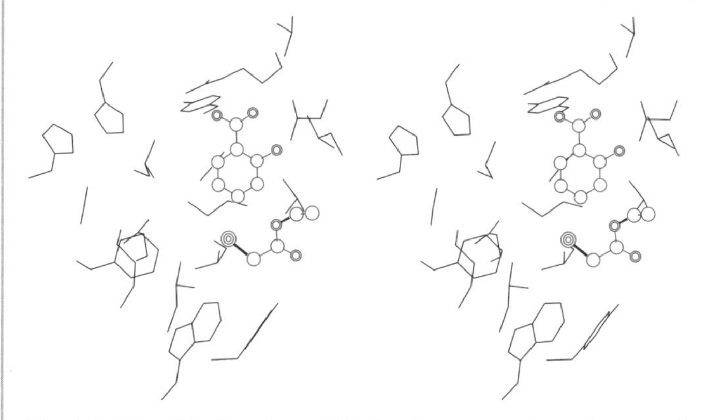

Fig. 5.22 The binding site in COX-1 for an aspirin analogue, 2-bromoacetoxybenzoic acid. The ligand has reacted with the protein, transferring the bromoacetyl group to the sidechain of serine 530. The protein is shown in skeletal representation, in black. The aspirin analogue is shown in ball-and-stick representation, in red.

→

Case Study 5.2 *(continued)*

Figure 5.23 shows the same figure with the corresponding region of COX–2 superposed. Can you see regions of structural difference, that could be clues to the design of selective drugs? Figure 5.24 shows the region of COX–2 with the selective inhibitor SC-558 (1-phenylsulphonamide-3-trifluoromethyl-5-parabromophenylpyrazole, made by Searle). From Fig. 5.25 we can see why SC-558 cannot inhibit COX–1. There would be steric clashes with the isoleucine sidechain, which corresponds to a valine in COX–2.

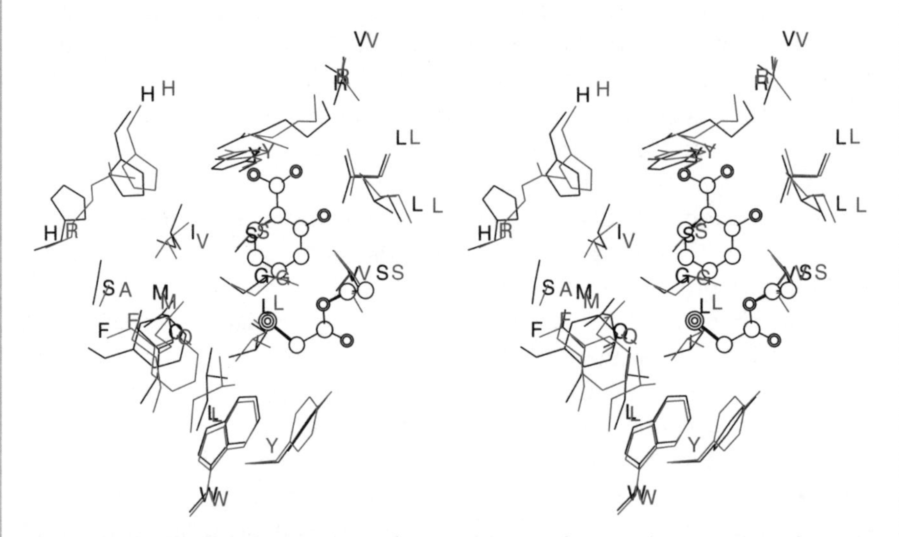

Fig. 5.23 The binding site in COX–1 for an aspirin analogue, 2-bromoacetoxybenzoic acid, in black, and the homologous residues of COX-2, in red. Can you see what unoccupied space exists in the site that could accommodate a larger ligand? Can you see any sequence differences that might be exploited to design an inhibitor that would bind to COX–2 (red) but *not* to COX–1 (black)?

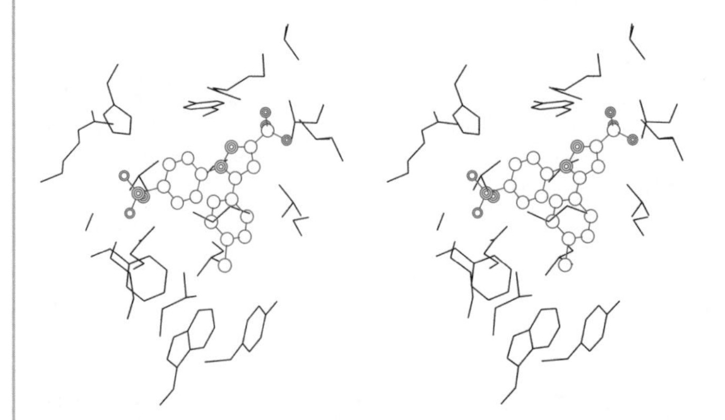

Fig. 5.24 The binding site in COX–2 (black) for a *selective* inhibitor of COX–2, SC-558 (1-phenylsulphonamide-3-trifluoromethyl-5-parabromophenylpyrazole) (red).

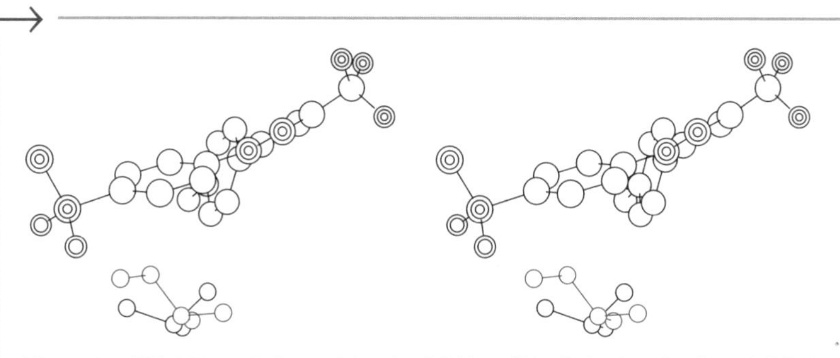

Fig. 5.25 SC-558 and the residue in COX–1 (black, isoleucine) and COX–2 (red, valine) that appears to produce the selectivity. SC-558 cannot bind to COX–1 because there would be steric contacts between it and the isoleucine.

Aspirin

Aspirin is one of the oldest of folk remedies and newest of scientific ones. At least 5000 years ago Hippocrates noted the effectiveness of preparations of willow leaves or bark to assuage pain and reduce fever. The active ingredient, salicin, was purified in 1828, and synthesized in 1859 by Kolbe. The mechanism of its action was unknown, and indeed remained unknown until, in the 1970s, J. Vane and colleagues discovered that aspirin acts by blocking prostaglandin synthesis. Not knowing the mechanism of action was never an impediment to its use.

A century ago, sodium salicylate was used in the treatment of arthritis. Because stomach irritation was a serious side effect, F. Hoffman sought to reduce the compound's acidity by forming acetylsalicylic acid, or aspirin. Aspirin was the first synthetic drug, which started the modern pharmaceutical industry. (The name salicin comes from the Latin name for willow, *salix*, and the name aspirin comes from 'a' for acetyl and 'spir' from the *spirea* plant, another natural source of salicin.)

Aspirin has the effect of reducing fever, and giving relief from aches and pains. In high doses it is effective against arthritis. Aspirin is also used for prevention and treatment of heart attacks and strokes. The applications to cardiovascular disease depend on inhibition of blood clotting by suppressing prostaglandin control over platelet clumping. The many applications of aspirin reflect the many physiological processes that involve prostaglandins.

Aspirin's many uses

Small doses	Medium doses	Large doses
Interferes with blood clotting	Fever/pain	Reduces pain and inflammation of arthritis and related diseases

Recommended reading

Protein folding

Baldwin, R. L. & Rose, G. D. (1999), Is protein folding hierarchic? I. Local structure and peptide folding. II. Folding intermediates and transition states, *Trends Biochem. Sci.*, **24**, 26–32; 77–83. [Introduction to current thinking about protein stability and folding.]

Structure alignment and sequence-structure relationships

Holm, L. & Sander, C. (1995), Dali: a network tool for protein structure comparison, *Trends Biochem Sci.*, **20**, 478–480. [Describes DALI and its applications to structural alignment.]

Smith, T. F. (1999), The art of matchmaking: sequence alignment methods and their structural implications, *Structure Fold. Des.*, **7**, R7–R12. [Description of work melding sequence and structure analysis.]

Connections among sequences, structures and functions

Das, R., Junker, J., Greenbaum, D. & Gerstein, M. B. (2001), Global perspectives on proteins: comparing genomes in terms of folds, pathways and beyond, *The Pharmacogenomics Journal*, **1**, 115–125.

Galperin, M. Y. & Koonin, E. V., Sequence–Evolution–Function / Computational Approaches in Comparative Genomics (Boston: Kluwer 2003).

State of the art in homology modelling and its application in structural genomics

Tramontano, A. (2004), Integral and differential form of the protein folding problem, *Physics of Life Revs.*, **1**, 103–127.

Guex, N., Diemand, A. & Peitsch, M. C. (1999), Protein modelling for all, *Trends. Biochem. Sci.*, **24**, 364–367.

Peitsch, M. C., Schwede, T. & Guex, N. (2000), Automated protein modelling–the proteome in 3D, *Pharmacogenomics*, **1**, 257–266. [What it will take to complete the structural genomics problem.]

Schwede, T., Kopp, J., Guex, N. & Peitsch, M. C. (2003), SWISS-MODEL: An automated protein homology-modeling server, *Nucl. Acids Res.*, **31**, 3381–3385. [Descriptions of SWISS-MODEL.]

Martí-Renom, M. A., Stuart, A. C., Fiser, A., Sánchez, R., Melo, F., Šali, A. (2000), Comparative protein structure modeling of genes and genomes, *Annu. Rev. Biophys. Biomol. Struct.*, **29**, 291–325.

Pieper, U., Eswar, N., Braberg, H., Madhusudhan, M. S., Davis, F. P., Stuart, A. C., Mirkovic, N., Rossi, A., Martí-Renom, M. A., Fiser, A., Webb, B., Greenblatt, D., Huang, C. C., Ferrin, T. E. & Šali, A. (2004), MODBASE, a database of annotated comparative protein structure models, and associated resources, *Nucl. Acids Res.*, **32**, D217–D222.

Other protein structure prediction methods

Bonneau, R. & Baker, D. (2001), Ab initio protein structure prediction: Progress and Prospects, *Annu. Rev. Biophys. Biomol. Struct.*, **30**, 173–189. [Recent review of structure prediction methods by authors of the most successful of them.]

Classics still well worth reading

Kauzmann, W. (1959), Some factors in the interpretation of protein denaturation, *Adv. Protein Chem.*, **14**, 1–63.

Richards, F. M. (1977), Areas, volumes, packing and protein structure, *Annu. Rev. Biophys. Bioeng.*, **6**, 151–176.

Chothia, C. (1984), Principles that determine the structure of proteins, *Annu. Rev. Biochem.*, **53**, 537–572.

Richards, F. M. (1991), The protein folding problem, *Scientific Amer.*, **264(1)**, 54–57, 60–63.

Exercises, Problems, and Weblems

Exercises

5.1 The heat of sublimation of ice = 51 kJ mol^{-1} at the freezing point. In the solid state, each molecule of H_2O makes two hydrogen bonds. What is the energy of a single water-water hydrogen bond?

5.2 Which pairs are orthologues, which are paralogues, and which are neither?

(a) Human haemoglobin α and human haemoglobin β.

(b) Human haemoglobin α and horse haemoglobin α.

(c) Human haemoglobin α and horse haemoglobin β.

(d) Human haemoglobin α and human haemoglobin ζ

(e) The proteinases human chymotrypsin and human thrombin.

(f) The proteinases human chymotrypsin and kiwi fruit actinidin.

5.3 On a photocopy of Plate VI, indicate the locations in the structure that correspond to X, Y and Z in the following diagram.

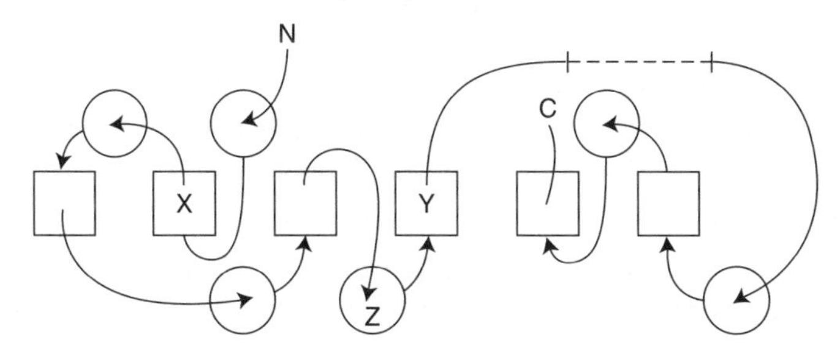

5.4 On a photocopy of Fig. 5.8b, highlight the region of 3_{10} helix that was not predicted to be helical.

5.5 Which of the following shows the correct topology—correct strand order in the sequence and orientation—of the β-sheet in Fig. 5.8b?

(a) ↑ ↑ ↑ ↑   (b) ↑ ↓ ↑ ↓   (c) ↑ ↑ ↓ ↑
     1 2 3 4   3 4 2 1   1 3 2 4

5.6 In the structure prediction of the *H. influenzae* hypothetical protein, Fig. 5.11: (a) What are the differences in folding pattern between the target protein and the experimental parent or template? (b) What are the differences in folding pattern between the prediction by A. G. Murzin and the target? (c) What are the differences in folding pattern between the prediction by A. G. Murzin and the experimental parent? In what respects is Murzin's prediction a better representation of the folding pattern than the experimental parent?

5.7 Draw the chemical structures of aspirin and 2-bromoacetoxy-benzoic acid.

5.8 Many proteins from pathogens have human homologues. Suppose you had a method for comparing the determinants of specificity in the binding sites of two homologous proteins. How could you use this method to select propitious targets for drug design?

5.9 In the neural network illustrated on page 246 (lower figure), how many parameters—variable weights and thresholds—are available to adjust, assuming a linear decision procedure?

5.10 What is the geometrical interpretation of a neuron that accepts two inputs x and y and 'fires' if and only if $x + 2y \geq 2$?

5.11 Sketch a neuron with two inputs x and y each of which may have any numerical value, that will emit 1 if and only if the value of the first input is greater than or equal to that of the second. What is the geometric interpretation of this neuron?

Problems

5.1 In the table of aligned sequences of ETS domains (see Problem 1.1): (a) Which are the most similar and which the most distant members of the family? (b) Suppose that an experimental structure is known only for the first sequence. For which others would you expect to be able to build a model with an overall r.m.s. deviation of $\leq$1.0 Å for 90% or more of the residues?

5.2 Sketch a neural network that accepts 8 inputs, each of which has value 0 or 1, with the interpretation that the 8 inputs correspond to the residues in a sequence of 8 amino acids, and that the value of the ith input is 0 if the ith residue is hydrophilic and 1 if the ith residue is hydrophobic. The network should output 1 if the pattern appears helical—for simplicity demand that it be PPHHPPHH where H = hydrophobic (uncharged) and P = polar or charged—and 0 otherwise.

5.3 Write a more reasonable set of patterns to identify helices from the hydrophobic/hydrophilic character of the residues in a 10-residue sequence. Your patterns might include 'wild cards'—positions that could be either hydrophobic or hydrophilic, or correlations between different positions. Generalize the previous problem by sketching neural networks to detect these more complex patterns.

5.4 We, and computers, can do logic with arithmetic. Define: 1 = TRUE and 0 = FALSE. Sketch simulated neurons with two inputs, each of which can have only the values 0 or 1, and a linear decision process for firing, for which (a) the output is the logical AND of the two inputs and (b) the output is the logical OR of the two inputs. (c) What is the simplest neural network, with each neuron having a linear decision process for firing, that produces as its output the EXCLUSIVE OR of the two inputs (the EXCLUSIVE OR is TRUE if either one of the inputs is TRUE, FALSE if neither or both of the inputs are TRUE.) Can this be done with a single layer? If not, what is the minimum number of layers in the network required?

5.5 Modify the PERL program for drawing helical wheels (pages 232–233) so that different amino acids are all represented in the same font, but appear in different colours, as follows: GAST, cyan; CVILFYPMW, green; HNQ, magenta, DE, red; and KR blue.

5.6 Hydrophobic cluster analysis. Suppose a region of a protein forms an α-helix. To represent its surface, imagine winding the sequence into an α-helix (even if in fact it forms a strand of sheet or loop in the native structure). Then 'ink' the surface of this helix, and roll it onto a sheet of paper, to print the names of the residues. By rolling the helix over twice, all faces are simultaneously visible.

From such a diagram, hydrophobic patches on surfaces of helices can be identified. In this way it is possible to try to predict which regions of the sequence actually form helices in the native structure. Comparisons of hydrophobic clusters can also be used to detect distant relationships.

Write a PERL program to produce such diagrams.

5.7 In the 2000 Critical Assessment of Structure Prediction (CASP4), one of the targets in the category for which no similar fold was known was the N-terminal domain of the human DNA end-joining protein XRCC4, residues 1–116.

The secondary structure prediction by B. Rost, using the method PROF: profile-based neural network prediction, is as follows (An H under a residue means that that residue is predicted to be in a Helix, an E means that that residue is predicted to be in an Extended conformation, or strand, and – means Other):

```
                    1         2         3         4         5         6
                    0         0         0         0         0         0
Sequence   MERKISRIHLVSEPSITHFLQVSWEKTLESGFVITLTDGHSAWTGTVSESEISQEADDMA
Prediction ---EEEEEEE-----HHHHHH-HHHHHHH--EEEEEE------EE---HHHHHHHHHHHH

                                                  1         1
                    7         8         9         0         1
                    0         0         0         0         0
Sequence   MEKGKYVGELRKALLSGAGPADVYTFNFSKESCYFFFEKNLKDVSFRLGSFNLEKV
Prediction HHH-HHHHHHHHHHHH-----EEEEEE-----EEEEE------EEEE-----HHHH
```

The experimental structure of this domain, released after the predictions were submitted (PDB entry [1FU1]), is shown here:

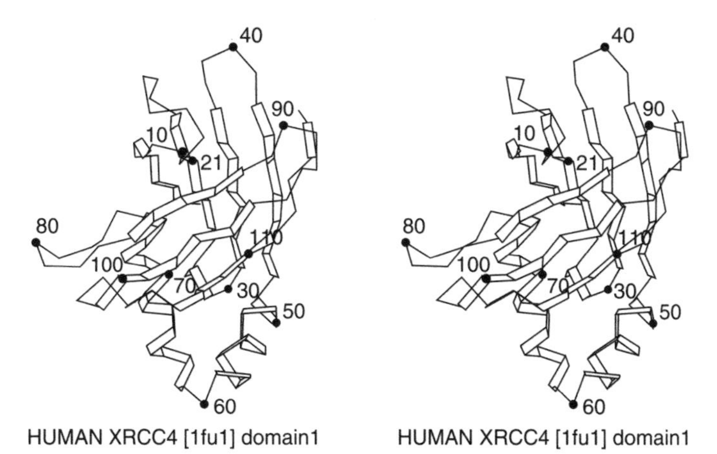

HUMAN XRCC4 [1fu1] domain1 HUMAN XRCC4 [1fu1] domain1

The secondary structure assignments from the PDB entry are:

Secondary Structure	Residue ranges
Helix	27–29, 49–59, 62–75
Sheet 1	2–8, 18–24, 31–37, 42–48, 114–115
Sheet 2	84–88, 95–101, 104–111

(a) Calculate the value of Q3, the percentage of residues correctly assigned to helix (H), strand (E) and other (-).

(b) On a photocopy of the picture of XRCC4, highlight, in separate colours, the regions *predicted* to be in helices and strands.

(c) From the result of (b): How many predicted helices overlap with helices in the experimental structure? How many strands overlap with strands in the experimental structure?

5.8 In CASP4, the group of Bonneau, Tsai, Ruczinski and Baker made a prediction of the full three-dimensional structure of protein XRCC4, residues 1–116. The secondary structure prediction derived from their model is as follows (H = helix, E = strand (extended), - = other):

```
                     1         2         3         4         5         6
                     0         0         0         0         0         0
Sequence    MERKISRIHLVSEPSITHFLQVSWEKTLESGFVITLTDGHSAWTGTVSESEISQEADDMA
Prediction  ----E--EEEE---EEEE--EHHHHHHHH----EEEE--EEEE-----HHHHHHHHHHHH

                     7         8         9         0         1
                     0         0         0         0         0
Sequence    MEKGKYVGELRKALLSGAGPADVYTFNFSKESCYFFFEKNLKDVSFRLGSFNLEKV
Prediction  HHH---HHHHHHHHHHH-----EEEEEEE--EEEEEEE------HHHH----HHHH
```

(a) What is the value of Q3 for this prediction? (b) In this case, which method gives the better results, as measured by Q3, for the prediction of secondary structure: the neural network that produces only a secondary structure prediction, or a prediction of the full three-dimensional structure?

5.9 Write and test PERL programs that implements the neural networks shown on page 247.

5.10 Suppose that you are trying to evaluate, using a threading approach, whether a sequence of length M is likely to have the folding pattern of a protein of known structure of length $N > M$. (a) How many different alignments of the sequences are possible. (b) Suppose that half the residues of the known protein form α-helices, and no gaps within helical regions are permitted. How many different alignments of the sequences are now possible? (c) How many alignments are there, under each of these assumptions, if $N = 200$, and $M = 150$?

5.11 Write a PERL program to calculate approximate values of π by a Monte Carlo method, as follows: The square in the plane with corners at (0, 0), (1,0), (0, 1), and (1,1) has area 1. Compute a series of *pairs* of random numbers (x, y) in the range [0, 1] to generate points distributed at random in this square. Count the number of points that lie within a circle of radius 0.5 inscribed in the square. The ratio of the number of points that fall within the circle to the total number of points = the ratio of the area of the circle to the area of the square = $\pi/4$.

Determine the average relationship between the number of points chosen and the number of correct digits in the calculated value of π. Estimate the number of points required to determine π correctly to 50 decimal places.

5.12 To convert the output of a neuron from a step function to a smooth function (see page 248), one can replace a statement of the form 'Let X be some weighted sum of the inputs; then output 1 if $X > 0$ else output 0' with 'Let X be some weighted sum of the inputs; then output $1/(1 + e^{-X})$.' (a) Verify that as $X \rightarrow -\infty$, $1/(1 + e^{-X}) \rightarrow 0$, as $X \rightarrow +\infty$, $1/(1 + e^{-X}) \rightarrow 1$, and that if $X = 0$, $1/(1 + e^{-X}) = 0.5$. (b) Suppose the network for determining whether a point lies within a triangle (page 247, bottom) is so altered, so that the output of each neuron is described by the smooth function $1/(1 + e^{-X})$ rather than a step function, and that a point is considered inside the accepted area if the output of the network is >0.5. Write a PERL program to determine what area is then defined.

Weblems

5.1 The bacterium *Pseudomonas fluorescens* and the fungus *Curvularia inaequalis* each possess a chloroperoxidase, an enzyme that catalyses halogenation reactions. Do these enzymes have the same folding pattern?

5.2 Check the prediction of transmembrane helical segments in bacteriorhodopsin from the secondary structure assignments in the experimental structure in the Protein Data Bank.

5.3 Plate VI showed the structure of a thiamin-binding domain, identified by M. Gerstein as one of the five most common folding patterns appearing in archaea, bacteria and eukarya. Using facilities available in SCOP, draw pictures of the four other structures.

5.4 Using either the results of Weblem 5.3, or pictures available in *Introduction to Protein Architecture: The Structural Biology of Proteins*, draw simplified topology diagrams analogous to the one in Plate VI for the other four structures.

5.5 Does the human θ_1 globin gene encode an active globin? Or is it really a pseudogene? Send the amino acid sequence of human θ_1 globin to SWISS-MODEL, including a request for a WhatCheck report on the result. What can you conclude from the result about the status of human θ_1 globin?

5.6 Compare the number of entries in SCOP, in different categories, listed on page 240, with the number that SCOP contains now.

5.7 Align the sequences of γ-chymotrypsin and *S. aureus* epidermolytic toxin A using pairwise sequence alignment methods. Compare the result with the structural alignment shown in the text.

5.8 Align the sequences of human neutrophil elastase and *C. elegans* elastase. (a) In the optimal alignment, how many identical resdues are there? (b) Would it be reasonable to build a model of *C. elegans* elastase starting from the structure of human neutrophil elastase?

5.9 S. Chakravarty and K. K. Kannan solved the structures of carbonic anhydrase with a benzenesulphonamide ligand (Protein Data Bank entry 1CZM.) Draw pictures of the binding site showing the nature of the interactions between the protein and ligands. Describe the nature of the interactions in terms of the conclusions drawn from the QSAR analysis.

CHAPTER 6

Proteomics and systems biology

Chapter contents

DNA microarrays *293*
Analysis of microarray data *295*
Mass spectrometry *301*
Identification of components of a complex mixture *301*
Protein sequencing by mass spectrometry *304*
Genome sequence analysis by mass spectrometry *306*
Systems biology *311*
Networks and graphs *313*
Network structure and dynamics *318*
Protein complexes and aggregates *320*
Properties of protein-protein complexes *321*
Protein interaction networks *324*
Regulatory networks *329*
Structures of regulatory networks *330*
Structural biology of regulatory networks *336*
Recommended reading 339
Exercises, Problems, and Weblems 339

Learning goals

1. To understand the goals of proteomics—the measurement of amounts and distributions of proteins within a cell or organism.
2. To be familiar with the data derivable from microarrays and their application to inferring and interpreting similarities and differences in gene expression patterns. To understand the relationship between typical 'raw' microarray data (see for instance Plate VII) and the gene expression table.
3. To be familiar with the data derivable from mass spectrometry, and their application to analysis of mixtures of proteins, to partial protein sequencing, and to high-throughput nucleic acid sequencing and searching for variant genetic sequences.
4. To appreciate a trend towards a new basic point of view: The theme of systems biology is integration.
5. To recognize the distinction between physical and logical networks.
6. To understand the general features of graphs, including the distinction between undirected, directed, and labelled graphs. To understand the representation of networks by graphs.
7. To appreciate the different possible kinds of dynamic states of networks.
8. To understand the characteristics of protein-protein and protein-nucleic acid complexes.
9. To understand the structure and some of the building blocks of regulatory networks.

Proteomics is the study of the distribution and interactions of proteins in time and space in a cell or organism. High-throughput experimental methods of data analysis, including microarray analysis and mass spectrometry, are giving us a large-scale picture of the protein economy in living things, and of gene and protein variation in populations.

The goal of systems biology is the synthesis of proteomic, genomic, and other data into an integrated picture of the structure, dynamics, logistics, and ultimately the logic of living things. For any protein in a cell, a systems biologist will combine study of the protein, its gene, the molecules that control its expression or its activity once expressed, and the set of other proteins with which it interacts. A systems biologist will assemble into a metabolic network the chemical reactions catalysed by the enzymes of a cell, and assemble into control networks the mechanisms that regulate their activities and expression.

DNA microarrays

DNA microarrays analyse (1) the mRNAs in a cell, to reveal the expression patterns of proteins; or (2) genomic DNAs, to reveal absent or mutated genes.

See Box: Applications of DNA microarrays, Chapter 1, page 53

1. **For an integrated characterization of cellular activity, we want to determine what proteins are present, where, and in what amounts.** To find the expression pattern of a cell's genes, we measure the relative amounts of many different mRNAs. Hybridization is an accurate and sensitive way to detect whether any particular nucleic acid sequence is present. The key to high-throughput analysis is to run many hybridization experiments in parallel.
2. **Knowing the human genome sequence can help to identify genes associated with propensities to diseases.** Some diseases, such as cystic fibrosis, arise from mutations in single genes. For these, isolating a region by classical genetic mapping can lead to pinpointing the lesion. Other diseases, such as asthma, depend on interactions among many genes, with environmental factors as complications. To understand the aetiology of multifactorial diseases requires the ability to determine and analyse multiple expression patterns of genes, which may be distributed around different chromosomes.

DNA microarrays, or DNA chips, are devices for checking a sample *simultaneously* for the presence of many sequences.

The basic idea is this: To detect whether one oligonucleotide has a particular known sequence, test whether it can bind to an oligo with the complementary sequence ('one-to-one'). To detect the presence or absence of a query oligo in a mixture, spread the mixture out, and test each component of the mixture for binding to the oligo complementary to the query ('many-to-one'). This is a Northern or Southern blot. To detect the presence or absence of *many* oligos in a mixture, synthesize a set of oligos, one complementary to each sequence of the query list, and test each component of the mixture for binding to each member of the set of complementary oligos ('many-to-many'). Microarrays provide an efficient, high-throughput way of carrying out these tests in parallel. They also permit measuring the expression levels of thousands or even tens of thousands of genes in a single sample.

To achieve parallel hybridization analysis, a large number of DNA oligomers are affixed to known locations on a rigid support, in a regular two-dimensional array. The mixture to be analysed is prepared with fluorescent tags, to permit detection of the hybrids. After exposing the array to the mixture, each element of the array to which some component of the mixture has become attached bears the tag. Because we know the sequence of the oligomeric probe in each spot in the array, measurement of the *positions* of the probes identifies their sequences. This analyses the components present in the sample.

Such a DNA microarray is based on a small wafer of glass or nylon, typically 2 cm square. Oligonucleotides are attached to the chip in a square array, at densities between 10 000 and 250 000 positions per cm^2. The spot size may be as small as ~150 μm in diameter. The grid is typically a few cm across. A *yeast chip* contains over

6000 oligos, covering all known genes of *Saccharomyces cerevisiae*. A DNA array, or DNA chip, may contain 400 000 probe oligomers. Note that this is larger than the total number of genes even in higher organisms (excluding immunoglobulin genes).

To analyse a mixture, expose it to the microarray under conditions that promote hybridization, then wash away any unbound sample. To compare two sets of oligos, tag the samples with differently-coloured fluorophores (Plate VII). Scanning the array collects the data in computer-readable form.

Different types of chips designed for different investigations differ in the types of DNA immobilized.

1. In an **expression chip**, the immobilized oligos are cDNA samples, typically 20–80 base pairs long, derived from mRNAs of known genes. The goal of such an experiment is to determine the expression patterns of genes that correspond to the CDNA samples. This is by far the most common application of microarrays. The target sample might be a mixture of mRNAs from normal or diseased tissue.
2. In **mutation microarray analysis**, one looks for patterns of single-nucleotide polymorphisms (SNPs).
3. In **genomic hybridization**, one looks for gains or losses of genes, or changes in copy number. The probe sequences, fixed on the chip, are large pieces of genomic DNA, from known chromosomal locations, typically 500–5000 base pairs long. The probe mixtures contain genomic DNA from normal or disease states. For instance, some types of cancer arise from chromosome deletions, which can be identified by microarrays.

Microarrays are capable of comparing concentrations of components of the sample. This allows quantitative investigation of responses to changed conditions. However, the precision is low. Moreover, mRNA levels, detected by the array, do not always accurately reflect protein levels. Indeed, usually mRNAs are reverse-transcribed into more stable cDNAs for microarray analysis; the yields in this step may also be nonuniform. Microarray data are therefore semiquantitative, in that a distinction between presence and absence is possible, determination of relative levels of expression in a controlled experiment is more difficult, and measurement of absolute expression levels are beyond the capacity of current microarray techniques.

Web resources: Microarray databases

Microarrays provide another high-throughput stream of data production in bioinformatics. A standard called MIAME (Minimum Information About a Microarray Experiment) describes the contents and format of the information to be recorded in the experiment, and deposited.

Major publicly-available microarray databases include:

The European Bioinformatics Institute hosts a database, ArrayExpress:
http://www.ebi.ac.uk/arrayexpress/

The US National Center for Biotechnology Information hosts the Gene Expression Omnibus database:
http://www.ncbi.nlm.nih.gov/geo/

The Stanford Microarray Database:
http://genome-www5.stanford.edu/MicroArray/SMD/

A listing of microarray databases for plants appears in:
http://www.univ-montp2.fr/~plant_arrays/databases.html

For additional lists, see:

Butte, A. (2002), The use and analysis of microarray data, *Nat. Revs. Drug Discov.*, **1**, 951–960.

Penkett, C. J. & Bähler, J. (2004), Getting the most from public microarray data, *European Pharmaceutical Review*, **1**, 8–17.

Analysis of microarray data

The raw data of a microarray experiment is an image, in which the colour and intensity of the fluorescence reflect the extent of hybridization to alternative probes. The two sets of probes are tagged with red and green fluorophores. If only one probe hybridizes, the spot appears red; if only the other probe hybridizes, the spot appears green. If both hybridize, the colour of the corresponding spot appears red + green = yellow (see Plate VII).

The initial goal of data processing is a **gene expression table**. This is a matrix in which the rows correspond to different genes, and the columns to different samples. Different spots in a microarray pattern such as that shown in Plate VII correspond to different genes. For each gene, results from different sets of samples appear in the red or green channel, respectively (or neither, or both). There is extensive redundancy in the oligos in a microarray—each gene may be represented by several spots, corresponding to different regions of the gene sequence; inclusion of controls with a deliberate mismatch allows data verification. Typically one gene may correspond to ~30–40 spots.

The samples may vary according to experimental conditions and/or physiological states, or they may be extracted from different individuals, different tissues or different developmental stages.

The process of data reduction to produce the gene expression matrix involves many technical details of image processing, checking internal controls, dealing with missing data, selecting reliable measurements, and putting the results of different arrays on consistent scales. The derived gene expression table indicates *relative* expression levels. A change in expression levels of a gene between two samples by a factor $\geq$1.5–2 is generally considered significant.

Extraction of reliable biological information from a gene expression table is not straightforward. Despite extensive internal controls, there is considerable noise in the experimental technique. In many cases, variability is inherent within the samples themselves. Micro-organisms can be cloned; animals can be inbred to a comparable degree of homogeneity. However, experiments using RNA from human sources—for example, a set of patients suffering from a disease and a corresponding set of healthy controls—are at the mercy of the large individual variations that humans present. Indeed, inbred animals, and even apparently identical eukaryotic tissue-culture samples, show extensive variability.

Another intrinsic disadvantage—and a severe one—in interpreting gene expression data, is the fact that the number of genes is much larger than the number of samples. Computationally, we are trying to understand the relationship of a space of very many variables (the genes) to a space of observations (the phenotype), from only a few measured points (the samples). The sparsity of the observations does not give us anywhere near adequate coverage. Statistical methods bear a heavy burden in the analysis to give us confidence in the significance of our conclusions.

Two general approaches to the analysis of a gene expression matrix involve (a) *comparisons focussed on the genes*; that is, comparing distributions of expression patterns of different genes by comparing rows in the expression matrix, or (b) *comparisons focussed on samples*; that is, comparing expression profiles of different samples by comparing columns of the expression matrix:

(a) **Comparisons focussed on genes: How do gene expression patterns vary among the different samples?** Suppose a gene is known to be involved in a disease, or to a change in physiological state in response to changed conditions. Other genes coexpressed with the known gene may participate in related processes contributing to the disease or change in state. More generally, if two rows (two genes) of the gene expression matrix show similar expression patterns across the samples, this suggests a common pattern of regulation, and possibly some relationship between their functions, including but not limited to a possible physical interaction.

(b) **Comparisons focussed on samples: How do samples differ in their gene expression patterns?** A consistent set of differences among the samples may characterize the classes from which the samples originate. If the samples are from different controlled groups (for instance, diseased and healthy animals), do samples from different groups show consistently different expression patterns? If so, given a novel sample, we can assign it to its proper class on the basis of its observed gene expression pattern.

How then do we measure the similarity of different rows or columns? Each row or column of the expression matrix can be considered as a vector, in a space of many dimensions. The row-vectors (a row corresponds to a gene), each entry of which refers to the same gene in different samples, has as many elements as there are samples. The column-vectors (a column corresponds to a sample), each entry of which refers to a different gene in a single sample, has as many elements as there are genes reported. It is possible to calculate the 'angle' between different row-vectors, or between different column-vectors, to provide a measure of their similarities. It is then natural to ask whether subsets of the points form natural clusters—points with high mutual similarity—characterizing either sets of genes or sets of samples.

Depending on the origin of the samples, what is already known about them, and what we want to learn, data analysis can proceed in different ways.

(1) The simplest case is a carefully controlled study, using two different sets of samples *of known characteristics*. For instance, the samples might be taken from

bacteria grown in the presence or absence of a drug, from juvenile or adult fruit flies, or from healthy humans and patients with a disease. We can focus on the question, what differences in gene expression pattern characterize the two states? Can we design a classification rule such that, given another sample, we can assign it to its proper class? This would be applicable in diagnosis of disease. For instance, determination of the subtype of a leukaemia permits more accurate treatment and prognosis. Subject to the availability of adequate data, such an approach can be extended to systems of more than two classes.

Computationally, training such a classification algorithm is called 'supervised learning'. The expression pattern of each sample is given by a vector corresponding to a single column of the matrix. This corresponds to a point in a many-dimensional space—as many dimensions as there are genes. In favourable cases, the points may fall in separated regions of space. Then a scientist, or a computer program, will be able to draw a boundary between them. In other cases, separation of classes may be more difficult. Consider the distribution of football players during a match. At the start of play, a line drawn across the midfield separates the teams; that is, the midfield line divides the field into two regions, each region containing exclusively the players of one of the teams. During play, the teams become commingled, and it is very difficult to divide the field into regions that separate the teams.

(2) In a different experimental situation, we might not be able to *preassign* different samples to different categories. Instead, we hope to extract the classification of samples from the analysis. The goal is to cluster the data to *identify* classes of samples and the differences between the genes that characterize them.

Many clustering algorithms have been applied to microarray data, including those that try to work out simultaneously both the *number* of clusters and the *boundaries* between them. All algorithms must face the difficulty arising from the sparsity of sampling of the very high-dimensional space of the measurement. Sometimes it is possible to simplify the problem by identifying a small number of combinations of genes that account for a large portion of the variability. This is called **reduction of dimensionality** (see Box).

Reduction of dimensionality

The distribution of gene expression data in a space of a large number of dimensions means that (1) coverage of the space with a limited number of samples is sparse and (2) it is difficult to visualize the distribution of sample points. In some cases, the distribution may depend primarily on fewer equivalent variables, and it is very advantageous to find them and transform the data accordingly.

A simple example illustrates the basic idea. Consider a distribution of groups of people picnicking on a beach. Represent the position of each person by the *x*, *y*, and *z* coordinates of the tip of his or her nose. Take the *x*-axis to be parallel

→

→

Reduction of dimensionality (*continued*)

to the shoreline, the *y*-axis perpendicular to the shoreline, and the *z*-axis vertical. Obviously height is irrelevant: this is really a two-dimensional, not a three-dimensional, distribution. To cluster the people into groups (perhaps families, or surfing clubs) the *x* and *y* coordinates carry all the significant data, and the *z*-coordinate carries only irrelevant information, such as the heights of the people and whether or not they are standing up or sitting on the sand. In this case, to reduce the dimensionality from three to two we need only ignore the *z*-coordinate. (Indeed, if the tide comes in and the beach area becomes narrower, the dimension along the shoreline carries the bulk of the information and the dimensionality could be further reduced from two to one.)

Now suppose the people are not distributed on the beach, but climbing a vertical rock face rising parallel to the shoreline. This also is really a two-dimensional, not a three-dimensional, distribution, but in this case it is the *x* and *z* coordinates that carry the information.

In more complex cases, reduction in dimension requires more than simply picking coordinates to ignore. Suppose the people are distributed on a ski slope. To reduce the distribution from three to two dimensions, we could not simply ignore a coordinate, but would have to *project* the data onto the oblique plane parallel to the slope. (The plane parallel to the sloping ground is oblique to a coordinate system oriented along horizontal and vertical directions.) This idea of *projection* of the data onto a lower-dimensional space, which contains the important components of the variation, is the key to the methods.

Practical problems of data analysis are harder than these simple illustrations. For one thing, the starting dimensions are much higher than three and the reduction in dimensionality is potentially much greater. For another, it is not obvious how to achieve the dimensionality reduction because we don't have the easily visualizable picture of the physical space and the distribution of people on a beach, rock face, or ski slope.

Nevertheless, the questions to be answered remain: Along what directions should we project the data to retain the largest discrimination using the fewest dimensions? Mathematical methods known as **Principal-Component Analysis (PCA)** using the **Singular Value Decomposition (SVD)** can solve this problem. These methods automatically select a new coordinate system that best represents the variability of the data along the fewest axes, and, for each new coordinate axis, the calculation gives a measure of the contribution of that coordinate to accounting for the variability of the data.

Although two dimensions may well not contain all important components of the variation, we can always pick the best two-dimensional projection and plot the result on a graph; this has the immense advantage of allowing scientists to stare at the data and think about them. (Three-dimensional distributions can also be represented visually, with somewhat greater difficulty.)

Case Study 6.1: Interpretation of microarray data: Regulation of genes by BRCA1 and implications for the role of BRCA1 dysfunction or silencing in carcinogenesis

The *BRCA1* gene encodes a tumour suppressor. It is mutated in approximately 50% of patients with familial predisposition to breast and ovarian cancer. A single defective *BRCA1* allele is sufficient to increase risk, for in any cell the normal copy of the gene may be lost, or, in a small fraction of cases, rendered inactive by promotor methylation.

BRCA1 is an 1863-residue protein. It has an N-terminal ring finger domain, followed by a predicted helical coiled-coil region, followed by two tandem BRCT domains, that bind other proteins, and also regulate transcription. (BRCT abbreviates BRCA C-terminal domain.)

BRCA1 interacts with many other proteins to form functional complexes and is thereby involved in several different activities, including:

- **Sensing and signalling of lesions in DNA.** *BRCA1* responds to several types of DNA damage—for instance double-strand breaks—and activates repair mechanisms appropriate to each.
- **Preserving chromosome structure.** As chromosome integrity may suffer *as a consequence of* inaccurate repair of DNA damage, these functions are related.
- **Mediating checkpoint tests at points in the cell cycle,** in part at least by regulating transcription of genes encoding proteins involved in checkpoint enforcement.

A unifying idea about *BRCA1* is that the protein encoded mediates responses to DNA damage by eliciting repair mechanisms and, in case repair is unsuccessful, checkpoint mechanisms that stop cells with unrepaired damage from propagating. Loss of BRCA1 function leads to the accumulation of damaged DNA in cells, enhancing the chances of transition to a cancerous state.

The variety and complexity of the processes involving BRCA1 make it difficult to sort out the detailed mechanism of its relation to cancer:

1. Is tumour formation a direct consequence of loss of one or more functions of BRCA1 and its interacting partners? If so, which one(s)?
2. What is the importance of transcriptional regulation—of *BRCA1* by products of other genes, of other genes by BRCA1, or both? To what extent do changing expression patterns involving *BRCA1* lead *indirectly* to tumorigenesis? We shall see that the distinction between direct and indirect effects is not really a hard and fast one: BRCA1 binds directly to some of the proteins the expression of which it regulates.
3. DNA repair mechanisms are common to many types of cells. Why does BRCA1 dysfunction or silencing specifically lead to increased risk of cancers of the breast and ovary (and other epithelial tissues, including pancreas and prostate)?

→

Case Study 6.1. *(continued)*

One function of BRCA1 is control over transcription. In order to investigate the regulatory context of the relationship of *BRCA1* to cancer risk, Welcsh et al. used microarray analysis to compare the expression patterns of genes in cells producing high and low levels of BRCA1, using a cell line in which BRCA1 expression was selectively inducible. (See Plate VIII.) The chip used for detection of the response contained oligonucleotides representing ~6800 human genes. (Note that this is a relatively small fraction of the total human proteome.)

The results implicated 373 genes, differentially expressed by significant and reproducible amounts in response to higher levels of BRCA1 expression. Standing out among these were 57 up-regulated genes and 15 down-regulated genes, for which expression levels changed by factors ≥2. These candidates for involvement in functions of BRCA1 relevant to tumourigenesis were checked for differential expression in cancer tissues from patients and normal controls.

Clustering the gene expression matrix shows the clear distinction between up- and down-regulated genes, and gives an impression of the variability among replicates (Plate VIII). Many of the proteins encoded by up-regulated genes are hormone receptors and structural proteins. Many of the proteins encoded by down-regulated genes are involved in DNA replication and translation.

Notable among the genes identified in the study are the following:

Consistent with the tissue-specific appearance of tumours as a result of BRCA1 dysfunction, some of the genes with altered expression patterns are involved in oestrogen-mediated control pathways, suggesting a possible link to the tissue-specificity enigma. The set of proteins implicated includes cyclin D1 and myc, which are up-regulated by lower levels of *BRCA1*. Cyclin D1 and myc are observed to be overexpressed in 20% of breast cancers, consistent with their repression by functional BRCA1. (For comparison with the clinical setting, low levels of BRCA1 expression correspond to patients with reduced or absent BRCA1 function, that is, the high-risk group; and high levels are analogous to normal controls. However, the experiments of Welcsh et al. did not try to reproduce actual endogenous BRCA1 expression levels observed in patients and normal counterparts.)

Conversely, JAK and STAT proteins are down-regulated by decreased levels of BRCA1. These proteins are implicated as growth inhibitors in control pathways that govern proliferation, differentiation, apoptosis and transformation. Loss of BRCA1 activity would be expected to reduce JAK1 and STAT1 levels, promoting cellular proliferation and reducing apoptosis. This is consistent with the observation that *Stat1*-null mice develop tumours more readily than normals.

The relationships detected by Welcsh et al. are part of the cell's control network. However, some of the products of genes regulated by BRCA1 are also known to be involved in formation of functional complexes

→

with BRCA1. For instance, the product of *myc*—a potent oncogene—binds to BRCA1, suggesting a direct inhibition of myc function by BRCA1. Thus reduced BRCA1 levels would have the dual effect of reducing the inhibition of *myc* through binding, and increasing its expression through loss of transcriptional repression.

Thus, *myc* is linked to BRCA1 through both physical and regulatory interactions. We shall see in a later section that the idea of two parallel interaction networks in cells—physical interactions and regulatory interactions—is an attractive distinction. However, it is one that is difficult to maintain in a system such as BRCA1 function in which the two are so closely intertwined.

Mass spectrometry

Mass spectrometry is a physical technique that characterizes molecules by measurements of the masses of their ions. Investigations of large-scale expression patterns of proteins require methods that give high throughput rates as well as fine accuracy and precision. Mass spectrometry achieves this, which has stimulated its development into a mature technology in widespread use. Applications to molecular biology include:

- Rapid identification of the components of a complex mixture of proteins.
- Sequencing of proteins and nucleic acids, including high-throughput genomic sequencing, and surveying populations for genetic variability.
- Analysis of post-translational modifications, or substitutions relative to an expected sequence.

Identification of components of a complex mixture

First the components are separated by electrophoresis, then the isolated proteins digested by trypsin to produce peptide fragments with r.m.m. about 800–4000 (see Fig. 6.1). Trypsin cleaves proteins after Lys and Arg residues. Given a typical amino acid composition, a protein of 500 residues yields about 50 tryptic fragments. The spectrometer measures the masses of the fragments with very high accuracy. The list of fragment masses, called the **peptide mass fingerprint**, characterizes the protein (see Fig. 6.2). Searching a database of fragment masses identifies the unknown sample.

Construction of a database of fragment masses is a simple calculation from the amino acid sequences of known proteins, translations of open reading frames in genomes, or (in a pinch) of segments from EST libraries. The fragments correspond to segments cut by trypsin at lysine and arginine residues, and the masses of the amino acids are known. (Note that trypsin doesn't cleave Lys-Pro peptide bonds, and may also fail to cleave Arg-Pro peptide bonds.)

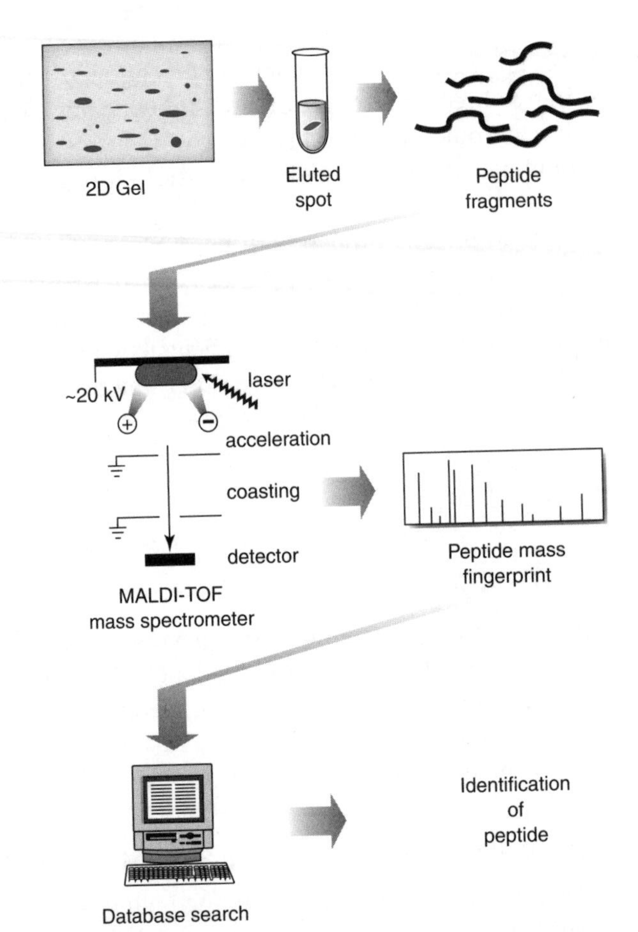

Fig. 6.1 Identification of components of a mixture of proteins by elution of individual spots, digestion and fingerprinting of the peptide fragments by MALDI–TOF (Matrix-Assisted Laser Desorption Ionization–Time of Flight) mass spectrometry, followed by looking up the set of fragment masses in a database.

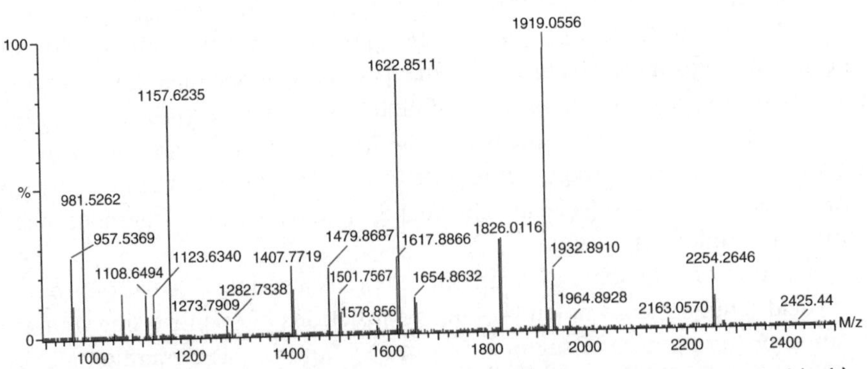

Fig. 6.2 Mass spectrum of a tryptic digest. Of the 21 highest peaks (shown in black), 15 match expected tryptic peptides of the 39 kDa subunit of Cow mitochondrial complex I. This easily suffices for a positive identification. (Figure courtesy of Dr. I. Fearnley.).

Web resources: Identification of proteins from peptide mass fingerprints

Two commonly-used databases compiling predicted peptide mass fingerprints of proteins are:

www.matrixscience.com

prospector.ucsf.edu/ucsf.html3.4/msfit.htm

Mass spectrometry is sensitive and fast. Peptide mass fingerprinting can identify proteins in sub-picomole quantities. Measurement of fragment masses to better than 0.1 mass units is quite good enough to resolve isotopic mixtures. It is a high-throughput method, capable of processing 100 spots/day (though sample preparation time is longer). However, there are limitations. Only proteins of known sequence can be identified from peptide mass fingerprints, because only their predicted fragment masses are included in the databases. (As with other fingerprinting methods, it would be possible to show that two proteins from different samples are likely to be the same, even if no identification is possible.) Also, post-translational modifications interfere with the method by altering the masses of the fragments.

The results shown in Fig. 6.2 are from an experiment in which the molecular masses of the ions were determined from their time-of-flight (TOF) over a known distance, as illustrated in Fig. 6.1. The operation of the spectrometer involves these steps:

1. Production of the sample in an ionized form in the vapour phase.
2. Acceleration of the ions in an electric field. Each ion emerges with a velocity proportional to its charge/mass ratio.
3. Passage of the ions into a field-free region, where they 'coast'.
4. Detection of the times of arrival of the ions. The time-of-flight (TOF) indicates the mass-to-charge ratio of the ions.
5. The result of the measurements is a trace showing the flux as a function of the mass-to-charge ratio of the ions detected.

Because proteins are fairly delicate objects, it has been challenging to vaporize and ionize them without damage. Two 'soft-ionization' methods that solve this problem are:

1. **Matrix-Assisted Laser Desorption Ionization (MALDI)** The sample is introduced into the spectrometer in dry form, mixed with a substrate or matrix that moderates the delivery of energy. A laser pulse, absorbed initially by the matrix, vaporizes and ionizes the protein. The MALDI-TOF combination, that produced the results shown in Fig. 6.2, is a common experimental configuration.
2. **Electrospray Ionization (ESI)** This method starts with the sample in liquid form. Spraying it through a small capillary with an electric field at the tip creates an aerosol of highly-charged droplets. As the solvent evaporates, the

droplets contract, bringing the charges closer together and increasing the repulsive forces between them. Eventually the droplets explode into smaller droplets, each with less total charge. This process repeats, ultimately creating ions, which may be multiply-charged, devoid of solvent. These ions are transferred into the high vacuum region of the mass spectrometer.

Because the sample is initially in liquid form, ESI lends itself to automation in which a mixture of tryptic peptides passes through a high-performance liquid chromatograph (HPLC) into the mass spectrometer directly.

Protein sequencing by mass spectrometry

Fragmentation of a peptide produces a mixture of ions. Conditions under which cleavage occurs primarily at peptide bonds yield series of ions differing by the masses of single amino acids (Fig. 6.3). The amino acid sequence of the peptide is therefore deducible from analysis of the mass spectrum (Fig. 6.4), subject to ambiguities: for instance, Leu and Ile have the same mass and cannot be distinguished in fragments cleaved only at peptide bonds. Discrepancies from the masses of standard amino acids signal post-translational modifications. In practice, the sequence of about 5–10 amino acids can be determined from a peptide of length < 20–30 residues.

In current practice, the fragments are produced *in situ*: First the peptide is vaporized, then it is fragmented by Collision-Induced Dissociation (CID) with Argon gas. This approach requires two mass analysers, operating in tandem in the same instrument (called MS/MS). The vaporized sample first passes through

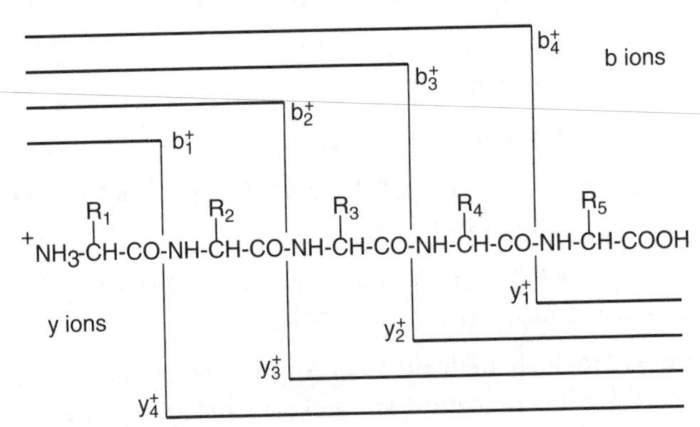

Fig. 6.3 Fragments produced by peptide bond cleavage of a short peptide. b ions contain the N-terminus; y ions contain the C-terminus. The difference in mass between successive b ions or successive y ions is the mass of a single residue, from which the peptide sequence can be determined. Two ambiguities remain: Leu and Ile have the same mass and cannot be distinguished, and Lys and Gln have almost the same mass and usually cannot be distinguished. In Collision-Induced Dissociation, bond breakage can be largely limited to peptide linkages by keeping to low-energy impacts. Higher energy collisions can fragment sidechains, occasionally useful to distinguish Leu/Ile and Lys/Gln.

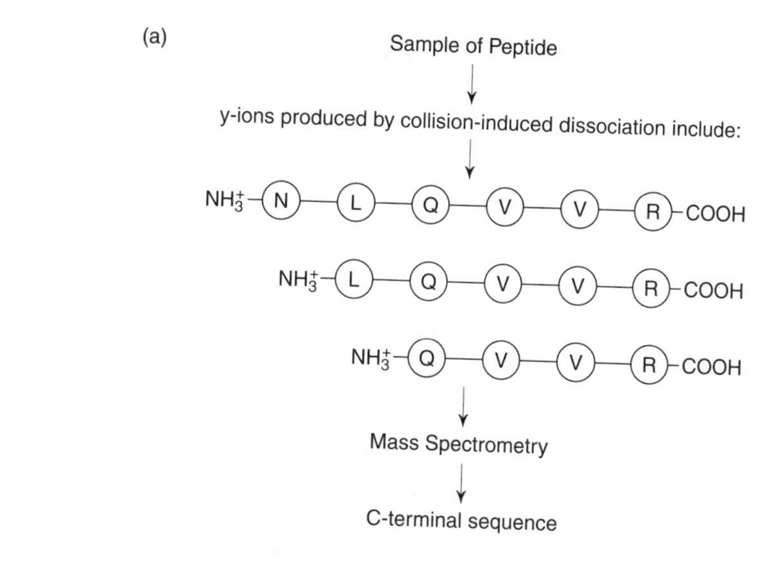

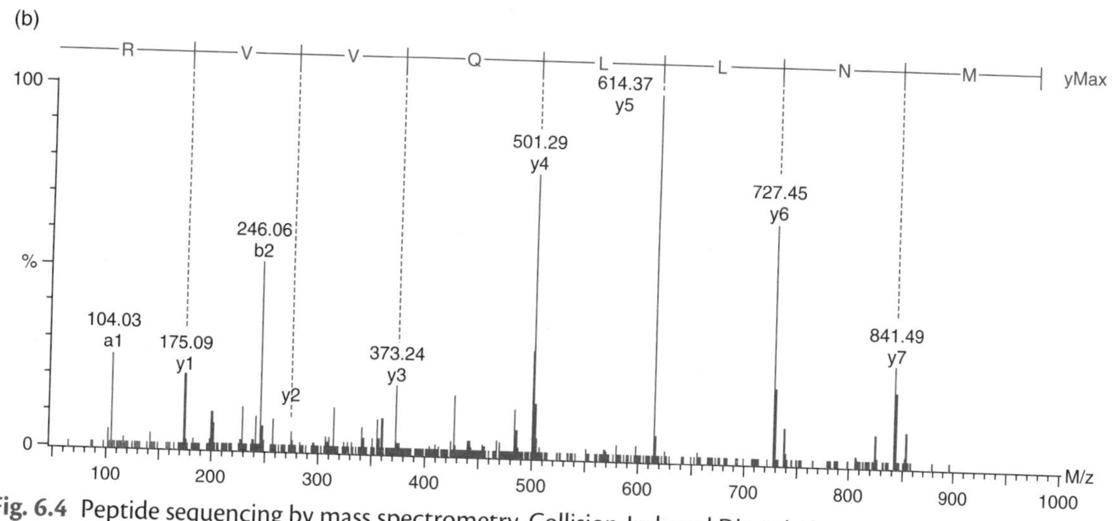

Fig. 6.4 Peptide sequencing by mass spectrometry. Collision-Induced Dissociation (CID) produces a mixture of ions. (a) The mixture contains a series of ions, differing by the masses of successive amino acids in the sequence. In CID the ions are not *produced* in sequence as suggested by this list, but the mass-spectral measurement automatically sorts them in order of their mass/charge ratio. (b) Mass spectrum of fragments suitable for C-terminal sequence determination. The greater stability of y ions over b ions in fragments produced from tryptic digests simplifies the interpretation of the spectrum. The mass differences between successive y-ion peaks are equal to the individual residue masses of successive amino acids in the sequence. Because y ions contain the C-terminus, the y-ion peak of smallest mass contains the C-terminal residue, etc. and therefore the sequence comes out 'in reverse'. The two leucine residues in this sequence could not be distinguished from isoleucine in this experiment. From Carroll, J., Fearnley, I. M., Shannon, R. J., Hirst, J. & Walker, J. E. (2003), Analysis of the subunit composition of complex I from bovine heart mitochondria, *Mol. Cell. Proteomics*, **2**, 117–126 (Supplementary figure S138).

one mass analyser, to separate an ion of interest. The selected ion passes into the collision cell where impact with Argon atoms excite and fragment it. By keeping the energy of impact low, the fragmentation can be limited largely to peptide bond breakage (Fig. 6.3). The second mass analyser determines the masses of the fragments.

Masses of amino acid residues, standard isotopes

Gly	57.02146	Ala	71.03711	Ser	87.03203
Pro	97.05276	Val	99.06881	Thr	101.04768
Cys	103.00919	Leu	113.08406	Ile	113.08406
Asn	114.04293	Asp	115.02694	Gln	128.05858
Lys	128.09496	Glu	129.04259	Met	131.04049
His	137.05891	Phe	147.06841	Arg	156.10111
Tyr	163.06333	Trp	186.07931		

Genome sequence analysis by mass spectrometry

Mass spectrometry of nucleic acids provides a very precise and high-throughput technique for quantitative analysis of DNA and RNA sequences in individuals and in populations. Its advantages include:

- **High precision** The standard deviation of typical mass spectral concentration measurement replicates is ~3% compared with ~200% for microarray measurements.
- **More data per sample** A mass spectrum contains many peaks rather than a single value. This allows analysis of mixtures; and permits 'multiplexing', or simultaneous analysis of several features of a set of mixed samples.
- **High specificity and sensitivity** Very small sample sizes are required. PCR amplification can be pushed to very high gain, as there is little risk of mistaking a contaminant for a true sample amplicon. In fact, it is possible to determine sequences from individual cells or even single DNA strands.

To prepare for the measurement, samples are treated by gene-specific PCR amplification by allele-specific primer extension, to produce single-stranded oligonucleotides. Products are purified and embedded in a matrix suitable for MALDI vaporization and mass analysis. No hybridization step is required for detection. Assembly of many subjects on an array allows for automation of data collection. (Throughput rates can reach 10^5 spectra per instrument per day.)

The typical relative molecular mass of an oligonucleotide measured is ~6000, corresponding to about 20 bases. Under conditions where the amplified products of different alleles contain different numbers of bases, the mass difference is ≥300, a very large difference relative to the accuracy of mass spectrometry. In fact, it is feasible to pick up a single-base substitution in oligos of the same length, or even the methylation of a gPc site. For nucleotide substitutions, the mass differences between bases range from ±9 for t↔a to ±40 for c↔g.

Applications include:

(1) **Measurement of allele frequencies in populations, or detection of alleles in individuals, by identification of single nucleotide polymorphisms (SNPs)** For population studies, samples from several individuals in the selected groups can be pooled, and genotype frequencies measured to about 3% accuracy. Several SNPs can be determined from a single spectrum. Such studies have impact on a wide variety of fields, including anthropology, agriculture, and forensics, but medical applications are the major driving force. For example, controlled comparisons between healthy populations and those predisposed to a disease can identify genetic factors of clinical importance.

(2) **Characterization of individual genotypes** A selection of 100 000 SNPs offers about 3 polymorphisms per gene, enough for fairly thorough characterization of the protein-coding portion of an individual person's genome. Determination of one individual's SNP profile is achievable using one instrument for one day. Clinical applications include: (a) diagnosis, based on systematic differences, between healthy individuals and those with a disease, previously established from controlled population studies, and (b) pharmacogenomics, to distinguish patients who will benefit from treatment with a drug from those who will not benefit or even risk severe side effects.

(3) **Measurement of individual haplotypes** Haplotypes are local combinations of genetic polymorphisms that tend to be co-inherited. (See Box, Haplotype distributions.) Haplotypes simplify the search for phenotype-genotype correlations, because they reduce the number of variables with which to characterize the genotype. Mass-spectrometric methods based on amplifying regions around SNPs in a sample containing a single DNA molecule provide an accurate and high-throughput method of individual haplotype determination.

Haplotype distributions

Our individual genomes are characterized by a distribution of genetic markers. Single-nucleotide polymorphisms are convenient features to observe, and to study within and across populations. Although the overall density of SNPs in our genomes is ~1 SNP/5 kbp, many 100 kbp regions show only a few (typically 2–4) of the possible combinations of SNPs, suggesting that recombination is rare within the region. These segments, which remain intact, are separated by intervals in which recombination is more frequent.

The few discrete combinations of SNPs define the **haplotype** of an individual. The International HapMap project collects and curates haplotype distributions from several human populations.

→

Haplotype distributions (*continued*)

Haplotypes are difficult to measure, because it is essential to determine which SNPs appear in the *same* DNA strand. Clearly, study of mixed samples from several individuals can determine the frequencies of individual SNPs but not their correlation into haplotypes. Even a sample containing both chromosomes from a single diploid cell mixes the contributions of both copies of the region. However, mass spectral studies of amplified single-copy DNA molecules, produced by dilution, can identify the *combination* of SNPs appearing together on the same chromosome, allowing unambiguous haplotyping.

(4) **Measurement of gene expression levels on an absolute scale, with a precision of ~3%** This is achieved by spiking the RNA extracted from a sample with a known amount of a related oligoribonucleotide, and measuring the relative amounts of signal from the calibrating oligo and the natural ones.

(5) **Noninvasive prenatal diagnosis based on the small amount of foetal DNA that leaks into maternal blood** Because of the 95–99% maternal DNA background, only paternal contributions to the foetus can be identified. However, the technique is sensitive enough to detect the *SRY* gene, demonstrating that the foetus is male, or other paternal alleles that may be useful in diagnosing genetic abnormalities. It should be emphasized that the use of only a *maternal blood sample* avoids the significant risks of an invasive procedure to sample amniotic fluid.

(6) **Genomic sequencing** Mass spectrometry has the potential to compete in accuracy and throughput with gel-based methods for large-scale DNA sequence determination.

Case Study 6.2: Application of combined genomic, proteomic, and structural methods to antibiotic resistance in tuberculosis

Tuberculosis is an infectious disease caused by *Mycobacterium tuberculosis*. Despite development of vaccines and drugs, it remains a potent killer. Tuberculosis, and AIDS, are the most common cause of death from infectious disease, claiming about three million victims per year. The World Health Organization estimates that there are eight million new cases per year. HIV infection exacerbates the mortality of tuberculosis infection by lowering resistance.

Our front-line defences against most bacterial infections include macrophages, cells of the immune system that engulf bacteria and attack them with a variety of chemical and biochemical agents. *M. tuberculosis*, exceptionally, is adapted to survive *within* the macrophage. Part of its adaptation is structural: cells of *M. tuberculosis* and close relatives surround themselves with a waxy coat. The low permeability of the coat shields them from the inhospitable environment within the macrophage, including low pH and oxidative stress. The bacteria also make substantial changes to gene expression patterns, to adapt their physiological state to these surroundings.

After several decades of decline following the development of effective drugs, the incidence of tuberculosis began to increase in the mid-1980s. One reason is emergence of resistant strains.

A primary drug used in prevention and therapy of tuberculosis is isoniazid (isonicotinic acid hydrazide). Isoniazid attacks *M. tuberculosis* by interfering with the synthesis of its cell wall, without which the bacterium cannot survive. Targets of isoniazid include an NADH-dependent enoyl acyl carrier protein reductase (InhA), and a β-keto-acyl ACP synthase (KasA). These enzymes participate in synthesis of mycolic acids, major components of the cell wall.

Isoniazid must be converted to an active form after absorption by the bacterial cell. The enzyme that effects the conversion, KatG, is a natural suspect for involvement in resistance. Its natural function is to detoxify peroxides.

Several methods have been applied to elucidate the adaptations responsible for isoniazid resistance:

(1) Changes in gene expression patterns were detected using microarrays.

(2) Genes that change expression were sequenced in susceptible and resistant strains, and mutations observed.

(3) The crystal structure of isoniazid bound to InhA has been determined.

(1) Changes in gene expression patterns

Wilson and colleagues* examined susceptible and resistant strains of *M. tuberculosis* at times up to 8 hours of exposure to isoniazid. (Plate IX shows the results after 4 hours exposure.) The array included almost all ORFs identified in the *M. tuberculosis* genome. Although biochemical studies had already implicated some proteins in resistance, a general screen was carried out in order to identify as many potential drug targets as possible.

The genome of *M. tuberculosis* is about 4.4 Mbp long and contains about 4000 genes.

Exposure to isoniazid greatly enhanced the transcription of two classes of genes. One set is involved in cell-wall synthesis, including an operon-like cluster encoding components of a fatty-acid synthase complex (FAS-II). Additional genes, including a subunit of alkyl hydroxyperoxide reductase (AhpC), that handles oxidative stress, were also up-regulated. The logic of the experiment is that the treated cells are recognizing the effects of the drug, and feedback mechanisms are acting to try to compensate for reduced activities, by enhanced expression.

(2) Mutations conferring resistance to isoniazid

On the basis of the changed expression profiles, Ramaswamy et al. (2003) sequenced a total of 2.6 Mbp from 124 *M. tuberculosis* isolates.† These include

* Wilson, M., DeRisi, J., Kristensen, H. H., Imboden, P., Rane, S., Brown, P. O. & Schoolnik, G. K. (1999), Exploring drug-induced alterations in gene expression in *Mycobacterium tuberculosis* by microarray hybridization, *Proc. Natl. Acad. Sci. USA*, **96**, 12833–12838.

† Ramaswamy, S. V., Reich, R., Dou, S. J., Jasperse, L., Pan, X., Wanger, A., Quitugua, T. & Graviss, E. A. (2003), Single nucleotide polymorphisms in genes associated with isoniazid, *Antimicrob. Agents Chemother.*, **47**, 1241–1250.

Case Study 6.2. *(continued)*

mutations in KatG that impede activation of isoniazid, and mutations in InhA to escape inhibition by the activated form.

Note that because oxidative stress is part of the host's natural defence to infection, simple knockout of KatG could be a dangerous strategy for the bacterium. Ideally, to achieve resistance, the bacterium would reduce the activity of the enzyme in isoniazid activation but retain activity against small peroxides. In this way it would reduce susceptibility to the drug while maintaining its general fitness in the environment within the macrophage. Precisely this balance is attained by the most common KatG mutation in resistant strains, S315T.

The most common mutation in InhA is S94A. The inhibitory effectiveness of activated isoniazid is reduced in this modified protein.

(3) Crystallography

Rozwarski et al. (1998) solved the structure of the complex between the activated form of isoniazid and InhA (Fig. 6.5). The drug is covalently attached to the nicotinamide ring of NAD, bound to the active site of InhA. The sidechain of S94 is also shown. In the inhibitory complex, the protein binds the NAD-activated isoniazid adduct. The coupling of these molecules can occur *only* on the enzyme (in solution activated isoniazid and NADH do not react).

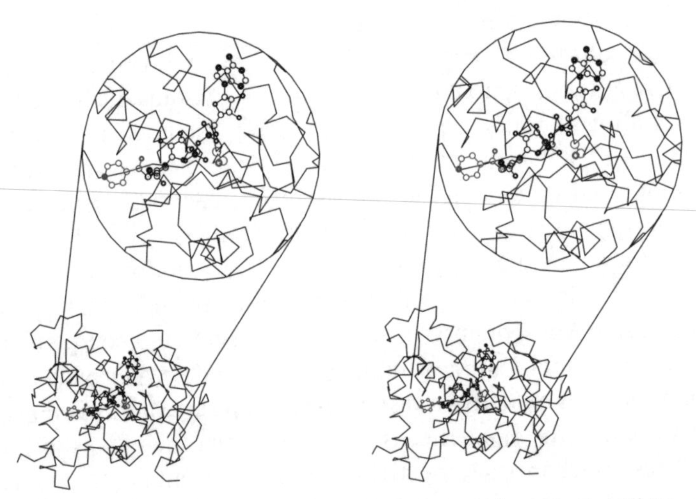

Fig. 6.5 Structure of long fatty acid chain enoyl-acp reductase (InhA) in complex with inhibitor [1ZID]. The ligand is an adduct of activated isoniazid and NADH. Shown in red are the isoniazid moiety of the inhibitor (at left in blown-up circle), and the sidechain of Ser94 (centre of blown-up circle). The mutation S94A contributes to isoniazid resistance. (See Rozwarski, D. A., Grant, G. A., Barton, D. H., Jacobs, W. R., Jr. & Sacchettini, J. C. (1998), Modification of the NADH of the isoniazid target (InhA) from *Mycobacterium tuberculosis*, *Science*, **279**, 98–102.)

How does the S94A mutant achieve resistance? In the absence of inhibitor, the enzyme can bind either substrate first and then cofactor, or cofactor first and then substrate. Because the substrate occupies the same site on the enzyme as the inhibitor, only if cofactor is bound first can an inhibitory complex form. Two pathways are possible, the first leading exclusively to product, the other producing an inhibitory complex (E = enzyme, S = substrate, C = cofactor and I = inhibitor):

```
If substrate binds first:
(1a) E + S → ES   + C ↛ ESC → E + C + P

(1b) E + S → ES   + I →

If cofactor binds first:
(2a) E + C → EC   + S → ESC → E + C + P

(2b) E + C → EC   + I → ESI = inhibitory
                              complex
```

If substrate binds first (1a and 1b), the inhibitory complex cannot form. If cofactor binds first (2a and 2b), a stable inhibitory complex may form, taking the enzyme out of the game permanently.

The S94A mutation reduces the affinity of the enzyme for NADH. This enhances the substrate-bound-first pathway (1a and 1b), lowering the amount of inhibitory complex produced (2b), and also enhancing the dissociation rate of the inhibitory complex.

It is also possible that S94A and other mutations reduce the affinity of the adduct.

Research and development of anti-tuberculosis drugs is a continuing challenge. This example shows the effectiveness of coordinated application of many different techniques.

Systems biology

Systems biology focusses on the integration and control of gene and protein activity.

Proteins are social animals, and life depends on their interactions. Because individual proteins have specialized functions, control mechanisms are required to integrate their activities. The right amount of the right protein must function in the right place at the right time. Failure of control mechanisms can lead to disease and even death.

Under unchanging environmental conditions, an organism's biochemical systems must be stable. Under changing conditions, the system must be robust, accommodating both neutral and stressful perturbations. Over longer periods of time, processes must have their rates altered, or even be switched on and off. This regulation includes short-term adjustments, for instance in the stages of the cell cycle; or responses to external stimuli such as changes in the composition or levels of nutrients or of oxygen. Longer-term regulatory activities include control over developmental stages during the entire lifetime of an organism.

Metabolism is the flow of molecules and energy through pathways of chemical reactions. Of course many metabolic reactions involve proteins and nucleic acids as well as small compounds such as amino acids and sugars. The full panoply of metabolic reactions forms complex traffic patterns. Some patterns are linear pathways, such as the multistep synthesis of tryptophan from chorismate. Others form closed loops, such as the tricarboxylic acid (Krebs) cycle. Moreover, the pathways interlock densely. The structure of the totality of metabolic pathways—its connectivity or topology—and its activity patterns, can be analysed in terms of a mathematical apparatus dealing with graphs and flows and throughputs.

To control metabolic flow patterns, **regulatory pathways** connect proteins and metabolite concentrations. The structure and dynamics of the regulatory pathways are different from those of the metabolic pathways. Corresponding to the succession of enzymatic transformations in metabolism, a regulatory pathway is an assembly of signalling cascades.

Systems biology describes metabolic and regulatory interactions in terms of **interaction networks.**

Two parallel networks: physical and logical

In cells, the two interaction networks operate in parallel: (1) a **physical network** of protein-protein and protein-nucleic acid complexes, and (2) a **logical network** of control cascades. Metabolic pathways partake of both: many but not all metabolic pathways are mediated by physical protein-protein interactions and regulated by logical interactions.

Examples of purely physical interactions include the assembly of photosynthetic reaction centres, complexes of proteins and cofactors that convert light to chemical energy; and assemblies of collagen in connective tissue. Examples of logical interactions, *not* mediated entirely by direct physical interaction between proteins, include feedback loops in which the increase in concentration of a product of a metabolic pathway inhibits an enzyme catalysing one of the early steps in the pathway, or the secretion of a small molecule as a signal to other cells ('fire and forget' mode). In these cases the logical interaction is transmitted by a diffusing small molecule. Many other examples appear in the very common theme of regulation of gene expression. A transcription factor, binding to DNA, may never interact physically with the proteins the expression of which it controls.

See Box: Cell-cell communication in micro-organisms: Quorum sensing

The allosteric change in haemoglobin is an example of simultaneous physical and logical interaction: The subunits of haemoglobin respond to changes in oxygen levels by a conformational change that alters oxygen affinity. Another example is the transmission of a signal from the surface of a cell across the membrane to the interior by dimerization of a receptor. This can be the initial trigger of a cascade that ultimately affects gene expression. Not all links of this process need involve protein-protein interactions; some may be mediated by diffusion of small molecules such as cyclic AMP.

Even though certain protein-protein and protein-nucleic acid complexes participate in both physical and logical networks, the two networks remain distinct in terms of their logic and their biological function, and it is useful to keep the distinction in mind, *especially* when considering proteins that participate in both.

Cell-cell communication in micro-organisms: Quorum sensing

Control mechanisms *not* involving direct protein-protein interactions mediate intercellular signalling in micro-organisms. *Vibrio fischeri* is a marine bacterium that can adopt alternative physiological states in which bioluminescence is active or inactive. (Literally a 'light switch'.) The organism can live free in seawater, or colonize the light organs of certain species of fish or squid. It is bioluminescent only when growing within the animal.

The bacteria respond to the local density of cells; a form of communication called **quorum sensing**. In *V. fischeri*, quorum sensing is mediated by secretion and detection of a small signalling molecule, N-(3-oxohexanoyl)-homoserine lactone. Related species use other N-acyl homoserine lactones, abbreviated to AHL. AHL can diffuse freely out of the cells in which it is synthesized. Within the light organs, culture densities can reach 10^{10}–10^{11} cells/ml, and the AHL concentration can exceed the threshold of about 5–10 nM for triggering the physiological switch.

Bacterial genes *LuxI* and *LuxR* govern the regulation. The product of *LuxI* is involved in the synthesis of AHL. The *LuxR* gene product contains a membrane-bound domain, that detects the AHL signal; and a transcriptional activator domain. *LuxR* activates an operon that includes (1) genes for synthesis of luciferase (the enzyme responsible for the bioluminescence), and (2) *LuxI*, expression of which synthesizes additional AHL, amplifying the signal and sharpening the transition.

The host also senses the bacteria: the light organs of squid grown in sterile salt water do not develop properly. This appears to be a reaction to the luminescence, rather than to the AHL. For the animal, the luminescence contributes to camouflage: disguise from predators living at lower depths, by blending with illumination from the sky. The masking of shadows is a natural form of 'make-up'. (The bioluminescence also regularly surprises diners in seafood restaurants, who jump to the conclusion that their glowing dinner is of extraterrestrial origin. However, most bioluminescent bacteria are harmless, although some strains of the related *Vibrio* species *V. cholerae*, the causative agent of cholera, are weakly bioluminescent. In fact the virulence of *V. cholerae* is also under the control of quorum sensing, by a related mechanism.)

Networks and graphs

In the abstract, networks have the form of graphs.

The routes between cities on the map of Sweden in Fig. 4.3 (page 177) is a network represented by a graph, similar to those appearing in systems biology. Each city is a node. The thick lines joining them indicate roads. Other examples

See Box: The idea of a graph (page 315).

familiar to many readers are the map of the London Underground,‡ and maps of the subway systems of other cities. Each station is a node of the graph, and edges correspond to tracks connecting the stations. The modern London Underground map shows the *topology* of the network; it does not quantitatively represent the geography of the area. An early map, from 1925, did maintain geographic accuracy.§ This was possible when the system was simpler than it is now. Some of the maps now posted in the Paris Métro are fairly accurate geographically. *Considered as networks, a geographically accurate map and a simplified map with the same topology correspond to the same graph.*

The London Underground network is fully connected, in that there is a path between any two stations. Many questions familiar to commuters are shared in the analysis of biological networks; for example: What are the paths connecting station A and station B? Regarding different lines as subnetworks, how easy is it to transfer from one to another; that is, what is the nature of the patterns of connectivity? In case of failure of one or more links, is the network still robust (that is, does it remain fully connected?).

Biological systems need to be robust, both for survival of individuals under stress and for the plasticity required for evolution. In yeast, for example, single gene knockouts of over 80% of the ~6200 open reading frames are survivable injuries.

In principle, networks can achieve robustness through redundancy. The most direct mechanism is simple **substitutional redundancy**: if two proteins are each capable of doing a job, knock out one and the other takes over. In the London Underground this would correspond to a second line running over the same route. For instance, when the Circle Line is not running, passengers travelling between Paddington and King's Cross stations can travel by the Hammersmith & City line running on the same tracks.

In cells, some genes have closely-related homologues resulting from gene duplication, and some of these contribute to substitutional redundancy. For example, in investigating mouse models for diabetes it appeared that mice and rats (but not humans) have two similar but non-allelic insulin genes. However, substitutional redundancy requires equivalence not only of function but of control of expression. In the mouse, knocking out either insulin gene leads to compensatory increased expression of the other and normal phenotype. Equivalent expression patterns are more probable among duplicated genes than among unrelated ones. For example, *E. coli* contains two fructose-1,6-bisphosphate aldolases. One, expressed only in the presence of special nutrients, is nonessential under normal growth conditions. However, the other is essential. In this case functional redundancy does *not* provide robustness. These two enzymes are probably homologous, but they are distant relatives, not the product of a recent gene duplication. One is a member of a family of fructose-1,6-bisphosphate aldolases typical of bacteria and eukaryotes, and the other is a member of another family that occurs in archaea. *E. coli* is unusual in containing both.

‡ http://www.transportforlondon.gov.uk/tfl/tube_map.shtml or http://www.afn.org/~alplatt/tube.html Exercises 6.11 and 6.12, Problem 6.5 and Weblem 6.4 also make use of this map.

§ http://www.ltmuseum.co.uk/collections/posters_b.html

An alternative mechanism of network robustness is **distributed redundancy**: the same effect achieved through different routes. In normal *E. coli* approximately two-thirds of the NADPH produced in metabolism arises via the pentose phosphate shunt, which requires the enzyme glucose-6-phosphate dehydrogenase. Knocking out the gene for this enzyme leads to metabolic shifts, after which increased NADH produced by the tricarboxylic acid cycle is converted to NADPH by a transhydrogenase reaction. The growth rate of the knockout strain is comparable to that of the parent.

The idea of a graph

- Mathematically, a graph consists of a set of vertices *V* and a set of edges *E*.
- Each edge is specified by a pair of vertices.
- In a **directed graph** the edges are **ordered** pairs of vertices.
- In a **labelled graph** there is a value associated with each edge. (A directed graph is a special case of a labelled graph: consider the arrowheads as labels.)

An undirected unlabelled graph specifies the connectivity of a network but not the distances between vertices (the topology but not the geometry, as in the modern London Underground map). Labels on the edges can indicate distances. For example, some phylogenetic trees indicate only the topology of the ancestry. Others indicate quantitatively the amount of divergence between species. Phylogenetic trees are often drawn with the lengths of the branches indicating the time since the last common ancestor. This is a pictorial device for labelling the edges.

Some graphs do not correspond to physical structures, and in any event edge labels need not reflect geometry in the usual sense. For example, the links in a network of metabolic pathways might be labelled to reflect flow patterns.

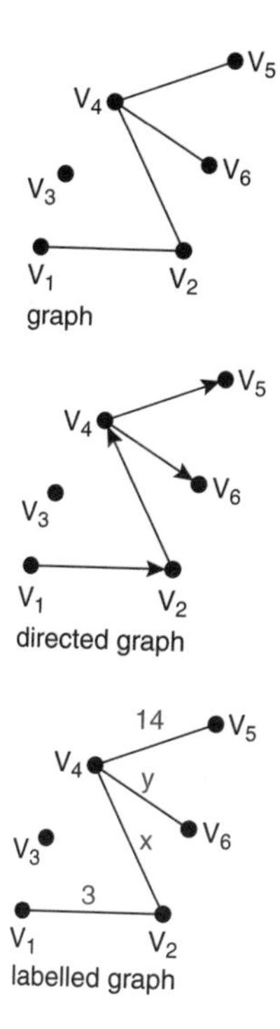

Examples of graphs

- Sets of people who have met each other
- Electricity distribution systems
- Phylogenetic trees
- Metabolic pathways
- Chemical bonding patterns in molecules
- Citation patterns in the scientific literature
- The World Wide Web

A fundamental property of a network is its **connectivity**.

If V_A and V_Z are vertices in a graph, a **path** from V_A to V_Z is a series of vertices: V_A, V_B, V_C, ... V_Z, such that an edge in the graph connects each successive pair of vertices. The number of vertices in the chain is called the **length** of the path. A **cycle** is a path of length >2 in a nondirected graph for which the initial and final endpoints are the same, but in which no intermediate link is repeated.

A graph that contains a path between any two vertices is called **connected.** Alternatively, a graph may split into several connected components, called **cliques**. The graph in the margin on page 315 contains two cliques, one containing five vertices and one containing only one vertex. (In the extreme, a graph could contain many vertices but no edges at all.) It is often useful to determine the *shortest path* between any two nodes, and to characterize a network by the distribution of path lengths. The well-known assertion that any pair of people are connected by 'six degrees of separation' means that: if the people in the world are vertices of a graph and the graph contains an edge whenever two people know each other, then the graph is fully connected, and there is a path between any two vertices with length ≤6.

A **tree** is a special form of graph. A tree is a connected graph containing only one path between each pair of vertices. A hierarchy is a tree: examples include military chains of command, and Linnean taxonomy. Note that some family trees are not trees in the mathematical sense; examples are plentiful in the royal families of Europe. A tree cannot contain a cycle: if it did, there would be two paths from the initial point (= the final point) to each intermediate point. In the graph in the margin on page 315, the subgraph consisting of vertices V_1, V_2, V_4, V_5, and V_6, is a tree. Adding an edge from V_1 to V_5 would create an alternative path from V_1 to V_5, and the cycle $V_1 \rightarrow V_2 \rightarrow V_4 \rightarrow V_5 \rightarrow V_1$; the graph would no longer be a tree.

The **density of connections**, that is, the mean number of edges per vertex, characterizes the structure of a graph. A fully-connected graph of N vertices has $N - 1$ connections per vertex; a graph with no edges has 0. Nervous systems of higher animals achieve their power not only by containing large numbers of neurons but also by high connectivities.

In some systems there are limits on numbers of connections: For many human societies, in the graph in which individuals are the vertices, and edges link people married to each other, each node has connectivity 0 or 1. In hydrocarbons, the graphs in which carbon and hydrogen atoms are the vertices and edges link atoms bonded to each other, each node has ≤4 connections. In other networks, connectivities follow observable regularities. (See Box: 'Small-world' networks.) For instance, the World Wide Web can be considered as a directed graph. Individual documents are the nodes, and hyperlinks are the edges. It is observed that the distribution of incoming and outgoing links follow power laws: $P(k)$ = probability of k edges is proportional to k^{-q}, where $q = 2.1$ for incoming links, and $q = 2.45$ for outgoing links.

'Small-world' networks

Many observed networks, including biological networks, the World Wide Web, and electric power distribution grids, have the characteristics of high clustering and short path lengths. They include relatively few nodes with very large numbers of connections, called 'hubs', and many that contain few connections. These combine to produce short path lengths between all nodes. From this feature they are called 'small-world' networks. Such networks tend to be fairly robust—staying connected after failure of random nodes. Failure of a hub would be disastrous but is unlikely, because there are few hubs.

Many networks, notably the World Wide Web, are continuously adding nodes. The connectivity distribution tends to remain fairly constant as the network grows. These are called **'scale-free' networks**.

The density of connections is very important in defining the properties of a network. For instance, the interactions that spread disease among humans and/or animals form a network. Whether a disease will cause an epidemic depends not only on the ease of transmission in any particular interaction, but on the density of connections. As the density of connections—the rate of interactions—increases, the system can exhibit a *qualitative* change in behaviour, analogous to a phase change in physical chemistry, from a situation in which the disease remains under control to an epidemic spreading through an entire population. The classic approach of 'quarantine'—isolating people for forty days—works by cutting down the degree of connectivity of the disease-transmission network. Note that a carrier who shows no symptoms—'Typhoid Mary¶' was a classic case—serves as a hub of the disease transmission network.

Two historical epidemics associated with wars demonstrate the distinction between topology and geometry in network connectivity. (1) In the early years of the Peloponnesian War, Athens suffered a severe epidemic. (Despite Thucydides' detailed description of the symptoms, the disease has not been definitively identified, but was probably bubonic plague.) A factor contributing to its transmission was the crowding of people into the city from the more vulnerable surrounding countryside. (2) After the First World War, an epidemic of influenza killed an estimated 20 million people, more than died in the war itself. Long-distance travel by soldiers returning from the war helped spread the disease. Any epidemic needs an infectious agent, and a high density of routes of transmission. The controlling factor is the density of the *connections* and not the density of the people.

A change in behaviour analogous to the transition to an epidemic appears in nuclear fission. In a sample of Uranium-235, decaying nuclei produce neutrons that can trigger fission of other nuclei. If the sample is small, so many secondary neutrons are lost through the surface that the sample remains stable. Above a critical mass, enough neutrons are captured within the sample to create a chain

¶ Mary Mallon (1869–1938) presented the following unfortunate combination of features: (1) she was infected with typhoid, (2) she did not show symptoms, and (3) she worked for many families as a cook.

reaction. If the atoms are vertices of a graph, and the edges are the trajectories of neutrons from one atom to another, the change in behaviour can be seen as the effect of increasing the connectivity of a network. (The background of Michael Frayn's recent popular play, Copenhagen, involves the attempts, before and during the Second World War, to estimate the size of the critical mass, in order to determine whether nuclear weapons would be feasible.)

Network structure and dynamics

An unlabelled, undirected graph gives a *static* structure of the topology of a network. For our molecular interaction networks, this may be an adequate description of many of the physical interactions.

For some networks, such as metabolic pathways or patterns of traffic in cities, the *dynamics* of the system depend on the transmission capacities of the individual links. These capacities can be indicated as labels on the edges of the graph. This allows modelling of patterns of flow through the network. Examples include route planning, in travel or deliveries. Note that the shortest path may well not give optimal throughput. Taxi drivers are exquisitely sensitive—and, in some cities, insensitively voluble—about optimal traffic paths.

In molecular biology, metabolic pathways and signal transduction cascades are networks that lend themselves to pathway and flow analysis. Even optimal sequence alignment by dynamic programming (see Chapter 4) involves determining the optimal path through an edit graph.

Although much is known about the mechanisms of individual elements of control and signalling pathways, understanding their integration is a subject of current research. For instance, the idea that healthy cells and organisms are in stable states is certainly no more than an approximation (and in most cases an idealization). The description of the actual dynamic state of the metabolic and regulatory networks is a very delicate problem. Understanding *how* cells achieve even an apparent approximation to stability is also quite tricky. It is likely that great redundancy of control processes lies at its basis. Regulation is based on the resultant of many individual control mechanisms—here a short feedback loop, there a multistep cascade. Somehow the independent actions of all the individual signals combine to achieve an overall, integrated result. It is like the operation of the 'invisible hand' that, according to Adam Smith, coordinates individual behaviour into the regulation of national economies.

Several types of dynamic states of a network are possible (see Box):

- Equilibrium
- Steady-state
- States that vary periodically
- Unfolding of developmental programs
- Chaotic states
- Runaway or Divergence
- Shutdown

States of a network of processes

- At **equilibrium** one or more forward and reverse processes occur at compensating rates, to leave the amounts of different substances unchanging:

$$A \rightleftarrows B$$

Chemical equilibria are generally self-adjusting upon changes in conditions, or in concentrations of reactants or products.

- A **steady state** will exist if the total rate of processes that produce a substance is the same as the total rate of processes that consume it. For instance, the two-step conversion:

$$A \rightarrow B \rightarrow C$$

could maintain the amount of B constant, provided that the rate of production of B (the process A → B) is the same as the rate of its consumption (the process B → C). The net effect would be to convert A to C.

A cyclic process could maintain a steady state in all its components:

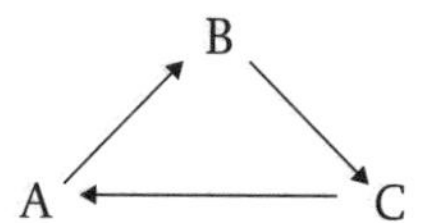

A steady state in such a cyclic process with all reactions proceeding in one direction is very different from an equilibrium state. Nevertheless, in some cases, it is still true that altering external conditions produces a shift to another, neighbouring, steady state.

- **States that vary periodically** appear in the regulation of the cell cycle, e.g. circadian rhythms, and seasonal changes such as annual patterns of breeding in animals and flowering in plants. Circadian and seasonal cycles have their origins in the regular progressions of the day and year, but have evolved a certain degree of internalization.

- Many equilibrium and some steady-state conditions are **stable**, in the sense that concentrations of most metabolites are changing slowly if at all, and the system is robust to small changes in external conditions. The alternative is a **chaotic state**, in which small changes in conditions can cause very large responses. Weather is a chaotic system: the meteorologist Lorenz asked, 'Does the flap of a butterfly's wings in Brazil set off a tornado in Texas?' In a carefully regulated system, chaos is usually well worth avoiding, and it is likely that life has evolved to damp down the responses to the kinds of fluctuations that might give rise to it. Chaotic dynamics does sometimes produce the approximations to stable states—these are called **strange attractors**. Understanding stability in dynamical systems subject to changing environmental stimuli is an important topic, but beyond the scope of this book.

→

→

States of a network of processes (*continued*)

- **Unfolding of developmental programs** occurs over the course of the lifetime of the cell or organism. Many developmental events are relatively independent of external conditions, and are controlled primarily by regulation of gene expression patterns.
- **Runaway** or **Divergence**. Breakdown in control over cellular proliferation leads to unconstrained growth, in cancer.
- **Shutdown** is part of the picture. Apoptosis is the programmed death of a cell, as part of normal developmental processes, or in response to damage that could threaten the organism, such as DNA strand breaks. Breakdown of mechanisms of apoptosis—for instance, mutations in protein p53—is an important cause of cancer.

Protein complexes and aggregates

The basis of our understanding of how life within a cell is organized and regulated is the set of protein-protein and protein-nucleic acid interactions. The development of high-throughput methods for detecting interactions has been a focus of recent interest.

Interacting proteins and nucleic acids span a range of structures and functions:

- Simple dimers or oligomers in which the monomers appear to function independently.
- Oligomers with functional 'cross-talk', including ligand-induced dimerization of receptors, and allosteric proteins such as haemoglobin, phosphofructokinase and asparate carbamoyltransferase.
- Large fibrous proteins such as actin or keratin.
- Non-fibrous structural aggregates such as viral capsids.
- Large aggregates with dynamic properties such as F1-ATPase, pyruvate kinase, the GroEL-GroES chaperonin, and the proteasome.
- Protein-nucleic acid complexes, including ribosomes, nucleosomes, transcription regulation complexes, splicing and repair particles, and viruses. In many cases initial binding is followed by recruitment of additional proteins to form large complexes.
- Many proteins, whether monomeric or oligomeric, function by interacting with other proteins. These include all enzymes with protein substrates, and many antibodies, inhibitors, and regulatory proteins.
- Protein interactions are frequently associated with disease, as misfolded or mutant proteins are prone to aggregation. (See Box: Diseases associated with protein aggregates.)

Diseases associated with protein aggregates

Disease	Aggregating protein	Comment
Sickle-cell anaemia	Deoxyhaemoglobin–S	Mutation creates hydrophobic patch on surface
Classical amyloidoses	Immunoglobulin light chains, transthyretin, and many others	Extracellular fibrillar deposits
Emphysema associated with Z-antitrypsin	Mutant α_1-antitrypsin	Destabilization of structure facilitates aggregation
Huntington	Altered huntingtin	One of several polyglutamine repeat diseases
Parkinson	α-synuclein	Found in Lewy bodies
Alzheimer	$A\beta,\tau$	$A\beta$ = 40–42 residue fragment
Spongiform encephalopathies	Prion proteins	Infectious, despite containing no nucleic acid

Properties of protein-protein complexes

Stoichiometry—what is the composition of the complex?

Stable oligomeric proteins may contain many copies of one protein, or combine different ones. Among aggregates of a single protein, complexes containing *odd* numbers of molecules are less common than those containing even numbers. Oligomers (complexes containing a few copies of the same protein—dimers, trimers, . . .) usually show symmetry. For instance, insulin is a hexamer with three-fold and two-fold axes.

Some prokaryotic proteins containing identical subunits are homologous to eukaryotic proteins containing related but nonidentical subunits, arising by gene duplication and divergence. The proteasome is an example. Some viruses achieve diversity *without* duplication, by combining proteins with the same sequence but different conformations.

Protein complexes vary widely in the numbers and variety of molecules they contain. Some complexes contain only a few proteins, but others are very large: for example, pyruvate dehydrogenase contains hundreds of subunits, and some viral capsids contain thousands.

Many very large aggregrates have clinical importance, including Bovine spongiform encephalopathy (BSE, or 'mad cow disease'), Alzheimer and Huntington disease. Amyloidoses are diseases characterized by extracellular fibrillar deposits, usually with a common crossed-β-sheet structure. They arise from a variety of causes, including destabilizing mutations, overproduction of a protein, and inadequate clearance in renal failure. Misfolded proteins are more prone to aggregate, and mutated proteins are more prone to misfold. Large local concentrations, such as can occur in myelomas that overproduce immunoglobulin light chains, also heighten the threat of aggregation.

See Box, Diseases associated with protein aggregation

Affinity—how stable is the complex?

A common index of the affinity of a complex is the **dissociation constant**, K_D, the equilibrium constant for the *reverse* of the binding reaction:

$$\text{Protein} \cdot \text{Ligand} = \text{Protein} + \text{Ligand} \qquad K_D = \frac{[p][l]}{[pl]}$$

The Michaelis constant of an enzyme is the dissociation constant of the Enzyme-Substrate complex.

[P], [L] and [PL] denote the numerical values of the concentrations of Protein, Ligand, and Protein-Ligand complex, respectively, expressed in mol ℓ^{-1}. The lower the K_D, the tighter the binding. K_D corresponds to the concentration of free ligand at which half the proteins bind ligand and half are free: [P] = [PL].

The K_D is related to the Gibbs free energy change of dissociation by the relationship:

$$\text{PL} = \text{P} + \text{L} \qquad \Delta G^{\circ} = \Delta H^{\circ} - T\Delta S^{\circ} = -RT \ln K_D$$

Dissociation constants of protein-ligand complexes span a wide range:

Biological context	Ligand	Typical K_D
Allosteric activator	Monovalent ion	$10^{-4} - 10^{-2}$
Coenzyme binding	NAD, for instance	$10^{-7} - 10^{-4}$
Antigen-antibody complexes	Various	$10^{-4} - 10^{-11}$
Thrombin inhibitor	Hirudin	5×10^{-14}
Trypsin inhibitor	Bovine pancreatic trypsin inhibitor	10^{-14}
Streptavidin	Biotin	10^{-15}

Structural studies have elucidated several important features of the interactions between soluble proteins, that contribute to affinity:

- **What holds the proteins together?** Burial of hydrophobic surface, hydrogen bonds and salt bridges.
- **Do proteins change conformation upon formation of complexes?** In some cases they do. In these cases the interaction energy has to 'pay for' the conformational change, and the interface tends to be larger.
- **What determines specificity?** Complementarity of the occluding surfaces—in shape, hydrogen-bonding potential, and charge distribution. (See Box: Features of protein-protein interfaces.) Prediction of protein complexes from the structures of the partners is the **docking problem.** Reliable solution of this problem, together with progress in structural genomics, would permit *in silicio* screening of proteomes for interacting partners.

Kinetics of formation and breakup, average lifetime

The dissociation constant of a complex indicates the fraction of time that the components spend in the bound state and the fraction of time in which they are unbound. But the **average lifetime** of the bound state can vary without affecting K_D. Defining individual rate constants for association and dissociation, k_{on} and k_{off}, the dissociation constant is equal to their ratio:

$$P + L \underset{k_{off}}{\overset{k_{on}}{\rightleftarrows}} [PL] \quad K_D = k_{off}/k_{on}$$

A short average lifetime, corresponding to large values of both k_{off} and k_{on}; or a long average lifetime, corresponding to small values of both k_{off} and k_{on}, can produce the same K_D. Lifetimes are important: if you want to purify a complex, it is important that its average lifetime be longer than the duration of the isolation procedure! Conversely, if a protein-protein complex is to mediate transmission of

Features of protein-protein interfaces

- **Burial of protein surface.** The accessible surface area (ASA) of a protein is calculated by rolling a probe sphere the size of a water molecule (radius 1.4 Å) over the protein surface. The surface buried by formation of a complex is the difference between the ASA of the complex and the sum of the ASAs of the components separately.

  A typical protein-protein interface might involve 22 residues, and 90 atoms of which 20% would be mainchain atoms, and an occasional water molecule. A histogram of surface area buried in binary protein complexes shows a peak centred at 1600 Å^2.

  The minimum buried surface for stability of a protein-protein complex is about 1000 Å^2. Complexes that bury > 2000 Å^2 tend to involve conformational changes upon complex formation.

  Each Å^2 of hydrophobic surface buried contributes about 105 J mol^{-1} to the free energy of stabilization.
- **The composition of the interface.** The chemical character of protein-protein interfaces is intermediate between that of the surfaces and interiors of monomeric globular proteins. Interfaces are enriched in neutral polar atoms at the expense of charged atoms. The amino acid compositions of interfaces are enriched in aromatic residues—His, Phe, Tyr, Trp—relative to remaining exposed surface. There is a lesser degree of enrichment in aliphatic sidechains—Leu, Ile, Val, Met—and Arg (but surprisingly, not Lys).
- **Complementarity** of interfaces is responsible for specificity. Complementarity involves both good packing at the occluding surfaces and proper juxtaposition of hydrogen-bonded and charged atoms. Typically there is one hydrogen bond per 170 Å^2 of interface area. Isolated water molecules occupy sites in many interfaces. Typically there is one fixed water molecule per 100 Å^2 of interface.

a signal, a short lifetime provides a natural 'reset mechanism' to preclude the signal's being locked 'on' for too long.

The 'on rate' is limited by diffusion rates. Under ordinary conditions $k_{on} \leq 10^{-9}$ M s^{-1}. If a conformational change is required for binding k_{on} may be considerably smaller. Typical k_{on} rates are 10^{-6}–10^{-7} M^{-1} s^{-1}, and typical lifetimes ~1 s.

How are complexes organized in three dimensions?

When two proteins form a complex, each leaves a 'footprint' on the surface of the other, defining the portion of the surface involved in the interaction. If two proteins interact using the *same* surface on both, the complex is **closed.** If two proteins interact through *different* surfaces, the complex is **open.** The significance is that a closed complex does not allow additional proteins to bind with the same interaction. An open complex, in which the surface of potential interaction is not occluded, can grow by accretion of additional subunits. Thus, open but not closed complexes are compatible with formation of aggregates by replication of the interaction.

Do proteins change conformation upon complex formation?

Some protein complexes form by the coming together of rigid subunits. The subunits in these complexes have the same structure in the complex that they have separately. Other protein complexes involve structural changes upon complex formation. These include complexes of subunits that are not independently stable separately.

Protein interaction networks

The units from which interaction networks are assembled are:

- For physical networks, a protein-protein or protein-nucleic acid complex.
- For logical networks, a dynamic connection in which the activity of a process is affected by a change in external conditions, or by the activity of another process.

Most experiments reveal only pairwise interactions. The challenges are to integrate pairwise interactions into a network and then to study the structure and dynamics of the system.

Many techniques detect physical interactions directly. These include:

- **X-ray and NMR structure determinations** cannot only identify the components of the complex, but reveal how they interact, and whether conformational changes occur upon binding.
- **Two-hybrid screening systems.** Transcriptional activators such as Gal4 contain a DNA-binding domain and an activation domain. Suppose these two domains are separated, and one test protein is fused to the DNA-binding domain and a

second test protein is fused to the activation domain. Then a reporter protein will be expressed only if the components of the activator are brought together by formation of a complex between two test proteins. High-throughput methods allow parallel screening of a 'bait' protein for interaction with a large number of potential 'prey' proteins. (See Box: Protein interaction networks determined by two-hybrid screening systems.)

Protein interactions detected by two-hybrid screening systems*

	H. pylori	*S. cerevisiae*	*C. elegans*	*D. melanogaster*
Total proteome size	1 576	5 585	33 469	13 843
Proteins tested	732	987 / 790	1 415	4 685
Interactions detected	1 465	936 / 800	2 131	4 876

The two sets of numbers for yeast are the results of independent investigations.

* From: Aloy, P. & Russell, R. B. (2004), Ten thousand interactions for the molecular biologist, *Nature Biotechnology*, **22**, 1317–1321.

- **Chemical crosslinking** fixes complexes so that they can be isolated. Subsequent proteolytic digestion and mass spectrometry permits identification of the components.
- **Coimmunoprecipitation.** An antibody raised to a 'bait' protein binds the bait together with any other 'prey' proteins that interact with it. The interacting proteins can be purified and analysed, for instance by Western blotting, or mass spectrometry.
- **Chromatin immunoprecipitation** identifies DNA sequences that bind proteins. Treatment with formaldehyde crosslinks proteins and DNA, fixing the complexes that exist within a cell. Then, isolation of the chromatin and breaking the DNA into small fragments allows separation of proteins by binding to specific antibodies, carrying the DNA sequences along with them. Reversal of the crosslink followed by sequencing of the DNA identifies the specific DNA sequence to which each protein binds.
- **Phage display.** Genes for a large number of proteins are individually fused to the gene for a phage coat protein, to create a population of phages each of which carries copies of one of the extra proteins exposed on its surface. Affinity purification against an immobilized 'bait' protein selects phages displaying potential 'prey' proteins. DNA extracted from the interacting phages reveals the amino acid sequences of these proteins.
- **Surface plasmon resonance** analyses the reflection of light from a gold surface to which a protein has been attached. The signal changes if a ligand binds to the immobilized protein. (The method detects localized changes in the

refractive index of the medium adjacent to the gold surface. This is related to the mass being immobilized.)

- **Fluorescence Resonance Energy Transfer.** If two proteins are tagged by different chromophores, transfer of excitation energy can be observed over distances up to about 60 Å.

Other methods provide complementary information:

- **Domain recombination networks.** Many eukaryotic proteins contain multiple domains. A feature of eukaryotic evolution is that a domain may appear in different proteins with different partners. In some cases proteins in a bacterial operon catalysing successive steps in a metabolic pathway are fused into a single multidomain protein in eukaryotes. The domains of the eukaryotic protein are individually homologous to the separate bacterial proteins. (Examples of proteins fused in prokaryotes and separate in eukaryotes are also known.)

  It is possible to create a network by defining an interaction between two protein domains whenever homologues of the two domains appear in the same protein. This is evidence for some functional link between the domains, even in species where the domains appear in separate proteins.

- **Coexpression patterns.** Clustering of microarray data identifies proteins with common expression patterns. They may have the same tissue distribution, or be up- or down-regulated in parallel in different physiological states. This is also suggestive evidence that they share some functional link. In the response of *M. tuberculosis* to isoniazid (page 308), genes for the Fatty Acid Synthesis complex are coordinately up-regulated. They are on an operon-like gene cluster, and in fact these proteins do form a physical complex. On the other hand, alkyl hydroperoxidase (AHPC) is also up-regulated in response to isoniazid. AHPC acts to relieve oxidative stress. There is no evidence that it physically interacts with the Fatty Acid Synthesis complex, or that it mediates a metabolic transformation coupled to fatty acid synthesis. It is a second component of the response to isoniazid.

- **Phylogenetic distribution patterns.** The **phylogenetic profile** of a protein is the set of organisms in which it and its homologues appear. Proteins in a common structural complex or pathway are functionally linked and expected to coevolve. Therefore proteins that share a phylogenetic profile are likely to have a functional link, or at least to have a common subcellular origin. There need be no sequence or structural similarity between the proteins that share a phylogenetic distribution pattern. A welcome feature of this method is that it derives information about the function of a protein from its relationship to *nonhomologous* proteins.

There are many ways to link proteins, including direct physical protein-protein interactions, two-hybrid complementarity, domain recombination, coexpression patterns, and phylogenetic profiles. Each provides a basis for a protein interaction network. The networks formed by combining each set of interactions are different, although they overlap, to a greater or lesser extent. They give different views of the kinds of relationships between proteins that exist in cells. It is possible to form a more comprehensive network by combining different types of interactions. For instance, the DIP database http://dip.doe-mbi.ucla.edu/ is a curated collection of experimentally-determined protein-protein interactions. It contains data about 44 349 interactions between 17 048 proteins from 107 organisms.

Plate X shows a portion of an interaction network of yeast proteins, based on sets of proteins that have been found together in solved structures.

Web resources: Interaction databases

Intact: An open source molecular interaction database
http://www.ebi.ac.uk/intact/

DIP: Database of Interacting Proteins
http://dip.doe-mbi.ucla.edu/

MIPS Comprehensive Yeast genome database
http://mips.gsf.de/

BIND: Biomolecular Interaction Network Database
http://www.bind.ca/

MINT: A Molecular Interactions database
http://cbm.bio.uniroma2.it/mint/

SPiD: *Subtilis* Protein interaction Database
http://genome.jouy.inra.fr/cgi-bin/spid/index.cgi

GRID: General Repository for Interaction Datasets
http://biodata.mshri.on.ca/grid/servlet/Index

PathCalling: Protein-protein interactions in *S. cerevisiae*
http://portal.curagen.com/extpc/com.curagen.portal.servlet.Yeast

HPID: The Human Protein Interaction Database
http://www.hpid.org/

Case Study 6.3: Components of the primosome assembly in *Bacillus subtilis*

The first step in DNA replication in *Bacillus subtilis* is the binding of initiator proteins to specific DNA sequences that serve as origins of replication. These then recruit a nucleoprotein complex called the primosome. A major component of the primosome is DnaC, a hexameric replicative helicase.‖

It is believed that steps in the process include:

1. Binding of an initiator protein, DnaA or PriA, to an appropriate single-stranded DNA sequence.

‖ Be aware that the nomenclature of these proteins differs between *E. coli* and *B. subtilis*.

→

Case Study 6.3 (*continued*)

2. Other proteins—DnaB, DnaC and DnaI—are recruited. DnaB and DnaI are regulators of DnaC activity.
3. DnaC is loaded onto the single-stranded DNA, forming a hexameric assembly.
4. DnaG is recruited to prime DNA synthesis.

Scientists at the *Institut National de la Récherche Agronomique* maintain a database of the protein interaction network of *B. subtilis.*** (See http://genome.jouy.inra.fr/cgi-bin/spid/index.cgi)

Figure 6.6 shows a small fragment of the network, limited to immediate neighbours of DnaC.

The web site is active: Clicking on a node either *adds* the interaction partners of the node to the graph, or *replaces* the graph with another centred on the selected protein. By adding partners, one can look at more extended neighbourhoods of DnaC. By replacing the graph, one can walk through the network. (See Weblem 6.5.)

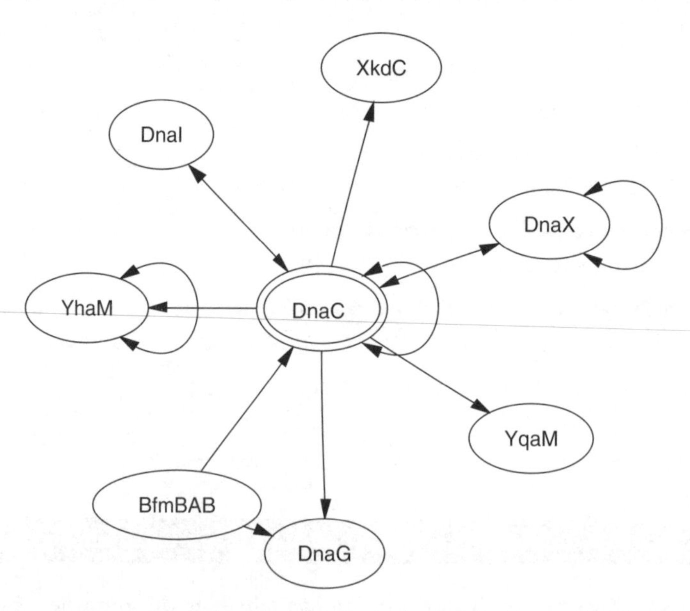

Fig. 6.6 DnaC and proteins that interact directly with it. Arrows linking partners point from 'bait' to 'prey'; bidirectional arrows indicate cases where the interaction was detected in reciprocal experiments. In the original web site, the arrows are colour coded according to the nature of the evidence for the interaction. Reproduced by permission.

** Hoebeke, M., Chiapello, H., Noirot, P. & Bessiéres, P. (2001), SPiD: a subtilis protein interaction database, *Bioinformatics*, **17**, 1209–1212; Noirot-Gros, M. F., Dervyn, E. Wu, L. J., Mervelet, P., Errington, J., Erlich, S. D. & Noirot, P. (2002), An expanded view of bacterial DNA replication, *Proc. Natl. Acad. Sci. USA*, **99**, 8342–8347.

Regulatory networks

Individual control interactions are organized into linear signal transduction cascades, and reticulated into control networks. Regulatory networks pervade living processes.

Any regulatory action requires (1) a stimulus, (2) transmission of a signal to a target, (3) a response, and (4) a 'reset' mechanism to restore the resting state (see Fig. 6.7). Many regulatory actions are mediated by protein-protein complexes. Transient complexes are common in regulation, as dissociation provides a natural reset mechanism.

Some stimuli arise from genetic programs. Some regulatory events are responses to current internal metabolite concentrations. Others originate outside the cell; the signal is detected by surface receptors, and transmitted across the membrane to an intracellular target.

Two components of regulatory networks are (1) the **signal transduction network**, and (2) the **transcriptional control network**. The signal transduction network exerts control '*in the field*', by a variety of mechanisms such as: inhibitors; dimerization, ligand-induced conformational changes including but not limited to allosteric effects; GDP-GTP exchange or kinase-phosphorylase switches; and differential turnover rates. This component acts fast, on sub-second timescales. The transcriptional regulatory network exerts control '*at headquarters*', through control over gene expression. This component is slower, acting on a timescale of minutes.

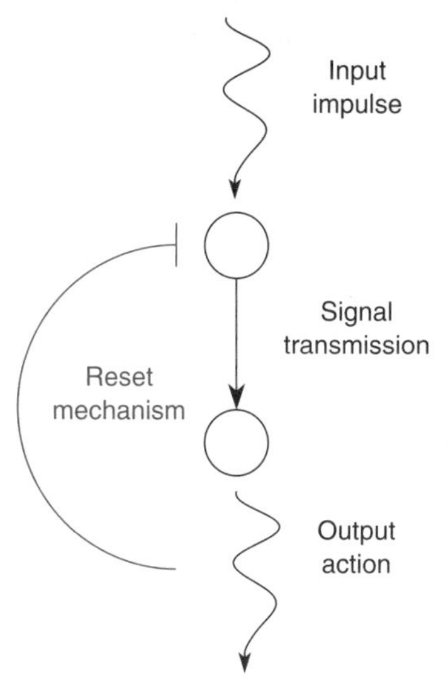

Fig. 6.7 The elementary step in a regulatory network. An input impulse is received by a node, which transmits a signal to a downstream node, causing an output action. This is followed by reset of the upstream node to its inactive state. Combination of such elementary diagrams gives rise to the complex regulatory networks in biology.

General characteristics of all control pathways include:

- a single signal can trigger a single response or many responses
- a single response can be controlled by a single signal or influenced by many signals
- each response may be stimulatory—increasing an activity—or inhibitory—decreasing an activity
- transmission of signals may damp out stimuli or amplify them

There are ample opportunities for complexity, and cells have taken extensive advantage of these.

Structures of regulatory networks

Think of control, or regulatory, networks as assemblies of *activities*. Although mediated in part by physical assemblies of macromolecules—protein-protein and protein-nucleic acid complexes—regulatory networks:

(1) **Tend to be unidirectional**. A transcription activator may stimulate the expression of a metabolic enzyme, but the enzyme may not be involved directly in regulating the expression of the transcription factor. (See page 82 for a discussion of control of the tryptophan synthase pathway in *E. coli*.)

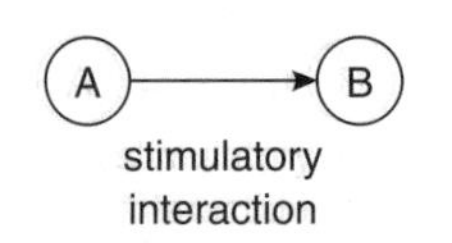

stimulatory interaction

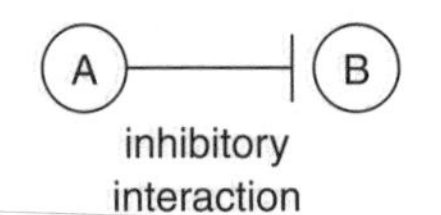

inhibitory interaction

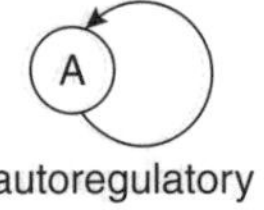

autoregulatory interaction

reciprocal interaction

(2) **Have a logical component**. It is not enough to describe the connectivity of a regulatory network. Any regulatory action may stimulate or repress the activity of its target. If two interactions combine to activate a target, activation may require *both* stimuli (logical AND), or *either* stimulus may suffice (logical OR).

(3) **Produce dynamic patterns**. Signals may produce combinations of effects with specified time courses. Cell-cycle regulation is a classic example.

A regulatory network can be described by a graph, in which edges indicate steps in pathways of control. Regulatory networks are directed graphs (see page 315): the influence of vertex A on vertex B is expressed by a directed edge connecting A and B. An edge directed from vertex A to vertex B is called an **outgoing connection** from A and an **incoming connection** to B. Often, an arrow indicates a stimulatory interaction, and a T-symbol indicates an inhibitory interaction. An edge connecting

Case Study 6.4: Architecture and dynamics of the genetic regulatory network of *Saccharomyces cerevisiae*††

A recent study of transcription regulation in yeast treated a network containing 3562 genes, corresponding to approximately half the known proteome of *S. cerevisiae*. The genes included 142 that encode transcription regulators, and 3420 that encode target genes exclusive of transcription regulators. There are

†† Luscombe, N. M., Babu, M. M., Yu, H., Snyder, M., Teichmann, S. A. & Gerstein, M. B. (2004), Genomic analysis of regulatory network dynamics reveals large topological changes, *Nature*, **431**, 308–312.

→

7074 known regulatory interactions among these genes, including effects of regulators on one another, and of regulators on non-regulatory targets.

Analysis of the overall network architecture reveals that:

- The distribution of incoming connections to target genes has a mean value of 2.1, and is distributed exponentially. Most target genes receive direct input from about two transcription regulators. The probability that a gene is controlled by k transcription regulators, $k = 1, 2, \ldots$, is proportional to $e^{-\alpha k}$, with $\alpha = 0.8$.
- The distribution of outgoing connections has a mean value of 49.8, and obeys a power law. The probability that a given transcription regulator controls k genes is proportional to $k^{-\beta}$, with $\beta = 0.6$. Power-law behaviour is common in networks, and characterizes topologies in which a few nodes—the 'hubs'—have many connections, and many nodes have few. (See page 317.) In regulatory networks, hubs tend to be fairly far upstream, forming important foci of regulation with far-reaching control.
- The average number of intermediate nodes in a minimal path between a transcription regulator and a target gene is 4.7. The maximum number of intermediate nodes in a path between two nodes is 12.
- The clustering coefficient of a node is a measure of the degree of local connectivity within a network. If all neighbours of a node are connected to one another, the clustering coefficient of the node is 1. If no pair of neighbours of a node is connected to each other, the clustering coefficient of the node is 0. The mean clustering coefficient, averaged over all nodes, is a measure of the overall density of the network. For the yeast transcriptional regulatory network, the mean clustering coefficient is 0.11.

Figure 6.8 is a cartoon-like sketch of a fragment of such a network, indicating rather loosely some of the general features. Nodes are divided into **transcription regulators**, shown as circles, and **target genes**, shown as squares. Target genes are distinguished by having no output connections. There is extensive interregulation among the transcription factors, to a much higher density of interconnections than can intelligibly be shown in this diagram. Think of a seething broth of transcription factors, within the shaded area, sending out signals to target genes. The shaded area indicates only the *logical* clustering of the transcription regulators. There is no suggestion about physical localization; indeed, transcription regulators interact with DNA, and almost never interact physically with the proteins the expression of which they control.

Each transcription regulator directly influences approximately 50 genes on average, although, as with other 'small-world' networks following power-law distributions of connectivities, the distribution is very skewed—some 'hubs' have very many output connections, but most nodes have very few. A few of the interregulatory connections between transcription factors are shown in red. In about 10% of the cases, two neighbours of the same transcription factor

Case Study 6.4 *(continued)*

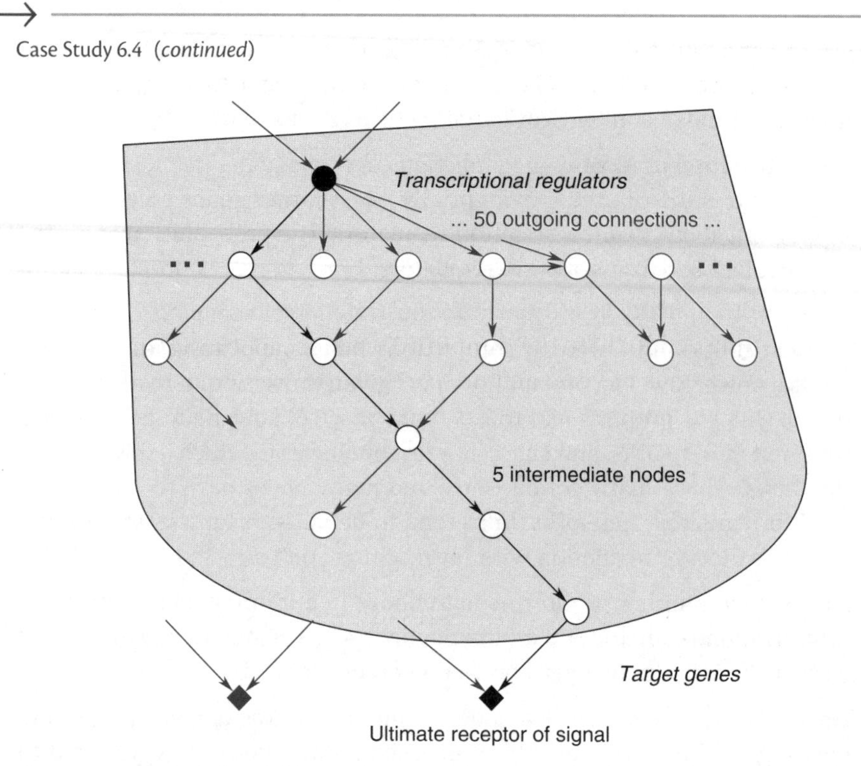

Fig. 6.8 Simplified sketch illustrating some features of an 'average' segment of the pathways in the yeast interaction network. Transcription regulators appear as circles. Target genes appear as squares. A transcription regulator typically has direct influence over about 50 genes, indicated by multiple connections from the filled black circle to the circles on the line below it. Roughly one in ten of the neighbours of any node is connected to another neighbour, indicated by the horizontal arrow on the second row. The ultimate receptor of the signal lies at the end of a pathway typically containing about five intermediate nodes (shown in black). This ultimate target gene receives on average about two inputs. This diagram shows only a small fragment of a network that is in fact quite dense.

interact with each other. A pathway from one regulator (filled black circle) to one ultimate receptor (filled black square), through five intermediate nodes, is shown in black. The intermediate nodes are other transcription regulators, connected both within the path drawn in black, and off this path. Even the transcription factor used as the origin of the path receives input connections. Although it is possible to identify target genes from the absence of outgoing connections, it is more difficult to identify ultimate initiators of signal cascades.

The ultimate receptor is a target gene that receives regulatory input but itself has no output links. This target is expected to receive (on average) a second control input. The black target node receives input via a black arrow, along the selected path, and via a red arrow suggesting the second input. Of course the second input may arrive via a path that shares common nodes with the black path, including other routes from the filled black circle.

The dense forest of additional pathways, from which this fragment is extracted, is not shown. Some 'back-of-the-envelope' calculations: There are ~3500 nodes, each receiving on average 2 input connections. There are ~140 transcription factors, making an average of 50 output connections. The number of input connections must equal the number of output connections, and indeed $3500 \times 2 = 140 \times 50 = 7000$.

Given the complexity, it is difficult to illustrate larger segments of the network in more detail than the simplified version appearing in Fig. 6.8. However, dissections of yeast and other regulatory networks have defined certain recurrent motifs that serve as building blocks. These might be considered the 'secondary structures' of network architectures. (See Box, page 335: Common motifs in biological control networks.)

The high ratio of interactions to transcription regulators implies that we cannot expect to associate individual regulatory molecules with single, dedicated, activities (as we can, for the most part, with metabolic enzymes). Instead, the activity of the network involves the coordinated activities of many individual regulatory molecules.

The network achieves versatility and responsiveness by reconfiguring its activities. This is seen by comparing the changes in the activities of networks controlling yeast gene expression patterns in different physiological regimes of the organism: cell cycle, sporulation, diauxic shift (the change from anaerobic fermentative metabolism to aerobic respiration as O_2 levels increase), DNA damage, and stress response. Cell cycling and sporulation involve the unfolding of endogenous gene expression programs; the others are responses to environmental changes.

Different states are characterized both by similarities and differences in gene expression patterns, and by the components of the regulatory network that are active. There is considerable shift in expression of *target genes*. About a quarter of the target genes are specialized to individual physiological states. That is, of the total of 3420 target genes, the expression of almost half (1514) do not show major changes in the different states. Of the 1906 that show altered expression levels in different states, almost half (803) are specialized to a single physiological state.

In contrast, different states show much more overlap in the usage of *transcription regulators*. For instance, for cell-cycle control, 280 target genes (8%) are differentially regulated by 70 (49%) of the transcription regulators. Clearly there is a much lower degree of specialization than of the target genes. In general, half the transcription factors are active in at least three out of the five physiological regimes. However, contrasting with the high overlap of usage of the transcription regulators (the nodes), the overlap of the activities within the network (the connections) is relatively low. Different components

Case Study 6.4 *(continued)*

of the interaction network organize the different gene expression patterns in different states.

Whereas different physiological states are characterized by substitutions of different sets of synthesized proteins, the regulatory network uses much of the same structure but reconfigures the pattern of activity. Think of the transcription factors as 'hardware' and the connections as reprogrammable 'software'. The molecules do not change but the interactions do: in different states, many transcription regulators change most, or a substantial part, of their interactions. In particular, the set of transcription regulators that form the hubs of the network—those with many outgoing nodes that form foci of control—are not a constant feature of the system. Some hubs are common to all states, but others step forward to take control in different physiological regimes. The result of the reconfiguration of activity is that over half of the regulatory interactions are *unique* to the different states.

The effect of the changes in the active interaction patterns is to alter the topological characteristics of the network in different states. For instance, under panic conditions—DNA damage and stress—the average number of genes under control of individual transcription regulators increases, the average minimal path length between regulator and target decreases, and the clustering becomes less dense (that is, there is less interregulation among transcription factors). This can be understood in terms of a need for fast and general mobilization—the equivalent of shouting 'Go! Go! Go!' over the radio. Normal circumstances—cell-cycle control for instance—allow for a more dignified and precise regulatory state, which permits finer control over the temporal course of expression patterns. In cell-cycle control and sporulation, there is a much denser interregulation among transcription factors, and longer minimal path lengths between transcription regulators and target genes.

Different physiological states also differ in their usage of the common motifs—fork, scatter, and 'one-two punch'. (See Box: Common motifs in biological control networks). Forks are more used in conditions of stress, diauxic shift, and DNA damage. They are appropriate to the need for quick action. Requirements for buildup of intermediates would delay the response. Conversely, the 'one-two punch' motif is more common in cell-cycle control. This is consistent with the need for a signal from one stage to be stabilized before the cell enters the next stage.

Much of evolution proceeds towards greater specialization. The human eye is a classic example. It is an intricate and fine-tuned structure, (features that were once adduced as evidence *against* Darwin's theory). Many evolutionary pathways show a trade-off between specialized adaptation and generalized adaptability.

Regulatory networks are an exception. Evolution has produced structures that are both specialized *and* versatile. The reconfigurability of regulatory networks allows them to respond robustly to changes in conditions, by creating many different structures, all specialized to the conditions that elicit them.

Common motifs in biological control networks

Within the high complexity of typical regulatory networks, certain common patterns appear frequently. In the architecture of networks, these form building blocks which contribute to higher levels of organization. Shen-Orr, Milo, Mangan and Alon* have described examples including: the *fork*, the *scatter*, and the '*one-two punch*' (a phrase from the boxing ring):

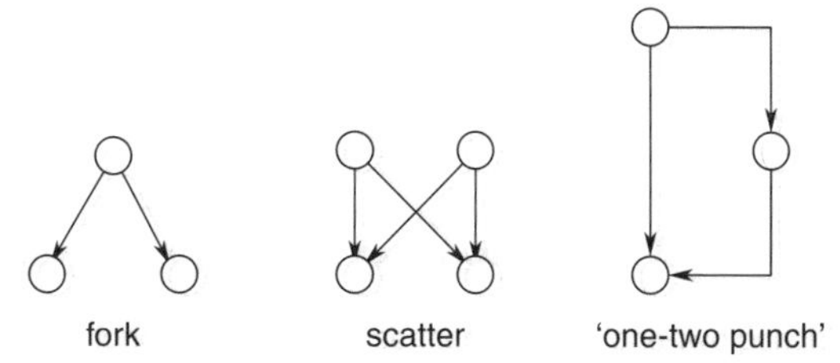

The **fork**, also called the single-input motif, transmits a single incoming signal to two outputs. Successive forks, or forks with higher branching degrees, are an effective way to activate large sets of genes from a single impulse. Generalizations of the binary fork include more downstream genes under common control (more tines to the fork), and autoregulation of the control node. Forks can achieve general mobilization. Moreover, if the regulated genes have different thresholds for activation, the dynamics of building up the signal can produce a temporal pattern of successive initiation of the expression of different genes.

The **scatter** configuration, also called the multiple input motif, can function as a logical OR operation: both downstream targets become active if *either* of the input impulses is active. Generalizations of the square scatter pattern shown may contain different numbers of nodes on both layers. Note that scatter patterns are superpositions of forks.

The '**one-two punch**', also called the 'feed-forward loop', affects the output both directly through the vertical link; and indirectly and subsequently, through the intermediate link. This motif can show interesting temporal behaviour if activation of the target requires simultaneous input from both direct and indirect paths (logical AND). Because buildup of the intermediate requires time, the direct signal will arrive before the indirect one. Therefore a short pulsed input to the complex will not activate the output—by the time the intermediate signal builds up, the direct signal is no longer active. The system can thereby filter out transient stimuli in noisy inputs. Conversely, the active state of the system can shut down quickly upon withdrawal of the external trigger.

A driver's action at a traffic light is an example of this control mechanism: The response to an amber light followed by a green light should be a cautious acceleration. The response to a red light should be an immediate stop. Control over drinking unfortunately *fails* to show this behaviour: the first drink itself up-regulates additional drinking, and there is no quick way to sober up!

* Shen-Orr, S. S., Milo, R., Mangan, S. & Alon, U. (2002), Network motifs in the transcriptional regulation network of *Escherichia coli*, *Nat. Genet.* **31**, 64–68.

a vertex to itself indicates autoregulation. A double-headed arrow indicates reciprocal stimulation of two nodes; note that this is *not* the same as an undirected edge.

Structural biology of regulatory networks

Any regulatory interaction involves one or more proteins and nucleic acids. Examples of regulatory mechanisms include a protein binding a ligand, undergoing chemical modification such as phosphorylation/dephosphorylation, changing conformation, or all of the above. X-ray crystallography and NMR spectroscopy have elucidated some of the general mechanisms underlying control processes.

Many molecules involved in regulation are multidomain proteins. A domain is a segment of a protein that has independent stability and can appear in conjunction with different partners through evolutionary recombination. Most multidomain proteins contain a linear sequence of domains each of which is relatively free to interact with other molecules. Assembly of a protein from domains therefore permits the joining into one molecule of a set of functions. 'Mixing and matching' of domains gives evolution access to a wide variety of functional combinations. (See Fig. 2.3, page 111.)

One important feature of regulatory proteins is recognition. An **interaction domain** is a part of a protein that confers specificity in ligation of a partner. Regulatory proteins contain a limited number of types of interaction domains, which have diverged to form large families with different individual specificities. For instance, the human genome contains 115 SH2 domains, and 253 SH3 domains. (*S*rc-*H*omology domains SH2 and SH3 are named for their homologies to domains of the src family of cytoplasmic tyrosine kinases.) Many individual interaction domains even interact with different partners as they participate in successive steps of a control cascade. Initial interactions may also trigger recruitment of additional proteins to form large regulatory complexes.

Many interaction domains are sensitive to the state of post-translational modification of their ligands, for instance binding preferentially to states of a ligand in which specific tyrosines, serines, or threonines are phosphorylated. These and other post-translational modifications function as switches, turning on or interrupting/resetting a signalling cascade.

Protein-protein complex formation allows a cell to detect a signal molecule in the external medium, and report its arrival to the cell interior, without the signal molecule itself ever needing to enter the cell. Many receptors use an ingenious dimerization mechanism: The receptor has external, transmembrane, and internal segments. An external ligand binds to *two* molecules of receptor. The juxtaposition of the external portions brings the internal portions together also, because they are tethered to the external regions by the transmembrane segments. Interaction between the interior segments triggers a conformational change that activates a process such as phosphorylation of a protein. This may initiate a signal transduction cascade that can amplify the original stimulus.

Figure 6.9 shows types of interaction domain complexes with ligands, including binding of peptides (which may be attached to proteins), protein-protein complexes, extracellular dimer formation upon binding a hormone, and a protein-nucleic acid complex.

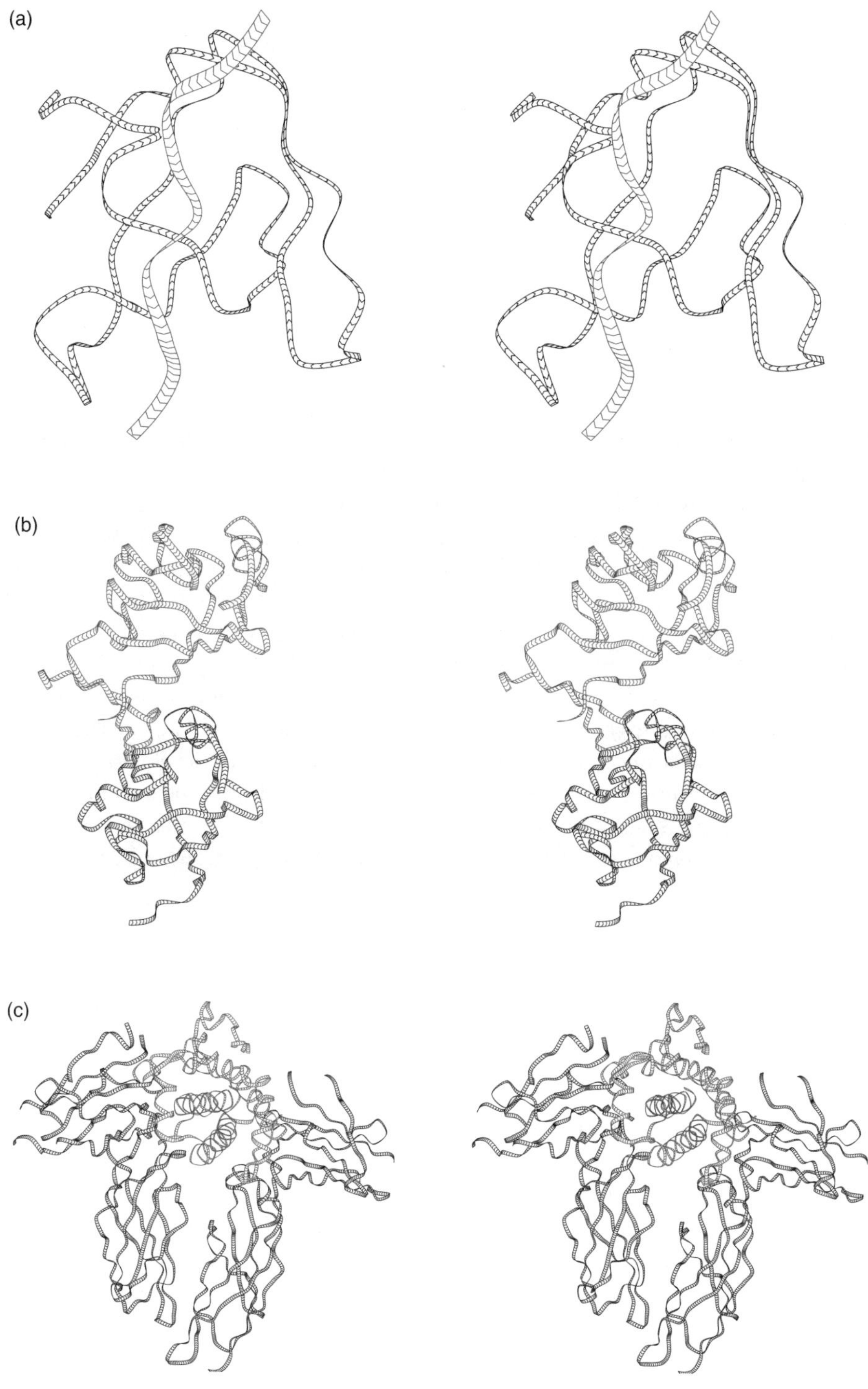

Fig. 6.9 (*Continued*)

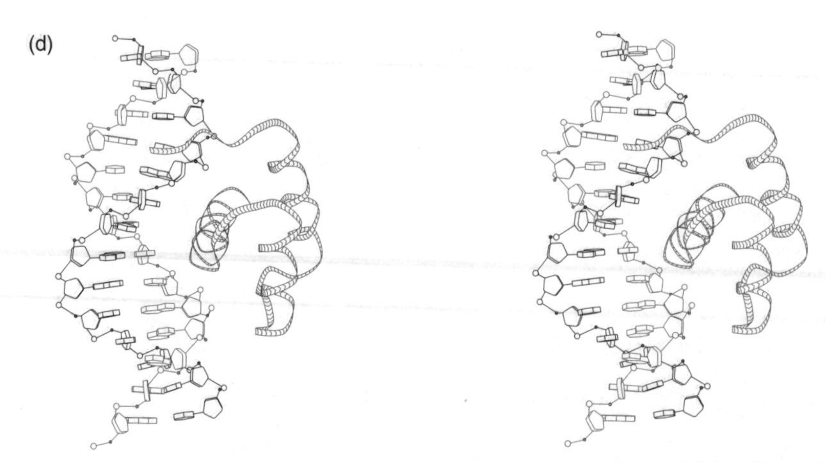

Fig. 6.9 Types of interactions involved in regulatory signalling. (a) Binding of a peptide by an SH3 domain [1CKA]. SH3 domains are common constituents of regulatory proteins. Functions of SH3 domains include signal transduction, protein and vesicle trafficking, cytoskeletal organization, cell polarization, and organelle biosynthesis. (b) domain-domain interaction: PDZ domains in syntrophin (black) and neuronal nitric oxide synthase (red) [1QAV]. (c) Binding of a molecule of human growth hormone (red) to two molecules of the external segment of the human growth hormone receptor (black). (d) The homeodomain antennapedia-DNA complex [9ANT]. Homeodomains are highly-conserved eukaryotic proteins, active in control of animal development. They regulate homeotic genes; that is, genes that specify locations of body parts. Antennapedia is a *Drosophila* protein responsible for initiating leg development. The earliest mutations found in antennapedia produced ectopic legs at the positions of, and instead of, antennae. Loss-of-function mutations convert legs into antennae. As with many DNA-binding proteins, an α-helix binds in the major groove of DNA.

Understanding the mechanism of regulation will require the structures of large protein and protein-nucleic acid complexes. The sizes of many of the large complexes challenges the limits of NMR spectroscopy. X-ray diffraction has had some major successes, but is at the mercy of being able to grow adequate crystals. Cryo-electron microscopy is another approach to structure determination of larger assemblies.

Electron microscopy of specimens at liquid nitrogen temperatures has revealed structures in the range r.m.m. = 5×10^5 to 4×10^8, 100–1500 Å in diameter. These results do not achieve atomic resolution. However, if the structures of individual components of a complex are known to high resolution from X-ray diffraction or NMR spectroscopy, the component structures can be fitted into the low-resolution structure determined by electron microscopy, to produce a detailed model of the entire assembly.

A limitation that remains is the difficulty of determining structures of transient complexes, or of systems showing substantial conformational changes upon assembly. The situation is shared with much of current molecular biology: we are coming to grips with static structures, but awaiting the development of methods for treating the dynamics.

Recommended reading

Albert, R. & Barabàsi, A. -L. (2002), Statistical mechanics of complex networks, *Rev. Mod. Phys.*, **74**, 47–97.
Barabàsi, A. -L., *Linked: How Everything Is Connected to Everything Else and What It means* (New York: Plume Books, 2003)
Ideker. T. (2004), A systems approach to discovering signaling and regulatory pathways—or, how to digest large interaction networks into relevant pieces, *Adv. Exp. Med. Biol.*, **547**, 21–30.
Babu, M. M., Luscombe, N. M, Aravind, L., Gcrstein, M. & Teichmann, S. A., (2004), Structure and evolution of transcriptional regulatory networks, *Curr. Opin. Struct. Biol.*, **14**, 283–291.
Tyers M. & Mann M., From genomics to proteomics, *Nature*, **422**, 193–197.

Exercises, Problems, and Weblems

Exercises

6.1 Hen egg white lysozyme has a relative molecular mass of about 14 300. If mass spectroscopy can measure mass to within 0.01%, could the following be confidently distinguished from the unmodified protein: (a) N-terminal acetylation? (b) phosphorylation of a single serine residue? (c) a single Lys→Gln substitution?

6.2 On photocopies of Fig. 6.4b, indicate the positions of the peaks if the sequence were: (a) MNLVQVR, (b) GNLQVVR, (c) MNLQVVG.

6.3 (a) What is the sequence of the fragment y_6 in Fig. 6.4b? (b) To which peak in Fig. 6.4b does the fragment NH_3^+-LQVVR-COOH correspond?

6.4 Oligonucleotide samples may vary by the binding of a Na^+ or K^+ ion to a phosphate, instead of a proton.

(a) What is the difference in mass between an oligonucleotide binding a proton or a Na^+ ion at a single site?

(b) What base change has the closest mass difference to the H^+–Na^+ mass difference?

(c) Would measuring mass to within 1 D be sufficiently accurate to distinguish this base change from the binding of a Na^+ ion instead of a proton, at a single site?

(d) In a mass spectrum of an oligonucleotide, what is the difference in mass between an oligonucleotide with a proton or a Mg^{2+} ion at a single site?

(e) What base change has the closest mass difference to the H^+–Mg^{2+} mass difference?

(f) Would measuring mass to within 1 D be sufficient accuracy to distinguish this base change from the binding of a Mg^{2+} ion instead of a proton, at a single site?

6.5 Assuming a typical single-nucleotide polymorphism (SNP) density of 1 SNP/5 kbp in a human genome, and only two possible bases observed at the position of any SNP, how many sequences could you expect to find throughout a population, within a 100 kbp region, if recombination were common at every position in the region? If only three of the possible combinations of SNPs—that is, three haplotypes—are observed, what fraction of possible sequences does this represent?

6.6 For which of the methods for determining interacting proteins (pages 325–326) (a) must one of the proteins be purified before testing for the interaction? (b) both of the proteins be purified before testing for the interaction?

6.7 In the top graph in the margin on page 315,

(a) Name two vertices such that if you add an edge between them at least one vertex has exactly two neighbours. (Note that two edges may cross without making a new vertex at their point of intersection.)

(b) Name two vertices such that if you add an edge between them to the original graph, the graph remains a tree.

(c) Name two vertices (neither of them V_1) such that if you add an edge between them to the original graph, the graph does not remain a tree.

(d) Name two vertices such that if you add an edge between them to the original graph, there are alternative paths, of lengths 3 and 4, between V_1 and V_5, with no vertices repeated. (In determining the length of a path, you have to count the initial and final vertices. A path of length 3 between V_1 and V_5 contains one intermediate vertex.)

(e) Name two vertices such that if you add an edge between them to the original graph, there is exactly one path between V_1 and V_3, with no vertices repeated, and it has length 4.

6.8 Of the examples of graphs at the bottom of page 315, (a) Which are directed graphs? (b) Which are labelled graphs? (c) In each example, what is the set of nodes? (d) In each example, what is the set of edges?

6.9 In a typical protein-protein interface of area 1700 Å^2: (a) How many intermolecular hydrogen bonds would you expect to be formed? (b) How many fixed water molecules would you expect to find in the interface? (c) If the entire buried area were hydrophobic, what contribution to the free energy of stabilization would you estimate it to make?

6.10 From the fragment of the *B. subtilis* protein interaction network shown in Fig. 6.6, what is the clustering coefficient of DnaC? The clustering coefficient of a node in a graph is defined as follows: Suppose the node has k neighbours. Then the total possible connections between the neighbours is $k(k-1)/2$. The clustering coefficient is the observed number of connections between neighbours divided by this maximum potential number of connections between neighbours.

6.11 In the London Underground: (a) What is the shortest path between Moorgate and Embankment stations. Note that, considered as a graph, the shortest path between two nodes is the path with the fewest intervening nodes, not the path that would take the minimal time or fewest interchanges. (b) What is the shortest cycle containing King's Cross, Holborn, and Oxford Circus stations? (c) If the neighbours of a station are the other stations that can be reached without passing through any intervening stations, what is the clustering coefficient of the Oxford Circus station? (See Exercise 6.10 for the definition of the clustering coefficient.)

6.12 In the London Underground: (a) What is the maximum path length between any two stations? That is, for which two stations does the shortest trip between them involve the maximum number of intervening stops? (b) If the District Line were not active, what stations if any would be inaccessible by underground? (c) If the Jubilee line were not active, what stations if any would be inaccessible by underground?

6.13 On a photocopy of the three common combinations of network control motifs (Box, page 335) (a) indicate which nodes are controlled by only *one* upstream node; (b) indicate which node exerts control over only *one* downstream node.

6.14 On a photocopy of the simplified fragment of the yeast regulatory network (Fig. 6.8) indicate examples of the network control motifs (a) fork (b) 'one-two punch'. (c) Add one arrow to create a scatter motif.

6.15 In the dimer between syntrophin and neuronal nitric oxide synthase (Fig. 6.9b), (a) is the dimer structure open or closed? (b) What secondary structure element is shared between the two domains?

6.16 In the overall yeast transcriptional regulatory network the number of incoming connections to target genes follows an exponential distribution. That is, the probability that a gene is controlled by k transcription regulators is proportional to $e^{-\alpha k}$, with $\alpha = 0.8$, $k = 1, 2, \ldots$. What is the ratio of the number of target genes receiving four input connections to the number receiving two input connections?

Problems

6.1 (a) How many positions in all are there in the microarray in Plate VII? (b) How many are complementary to RNAs from liver? (c) How many are complementary to RNAs from brain? (d) How many are complementary to RNAs from liver and brain? (e) How many are complementary to neither?

6.2 For avidin-biotin, $K_D = 10^{-15}$. Suppose k_{on} were as fast as the diffusion limit, $\sim 10^{-9}$ M s^{-1}. (a) What is the value of k_{off}? (b) What would be the half-life of the avidin-biotin complex? (c) Suppose k_{on} for avidin-biotin were 10^{-7} M^{-1} s^{-1}. What would be the half-life of the complex?

6.3 The anti-tuberculosis drug isoniazid requires activation by the *M. tuberculosis* enzyme KatG (a catalase-peroxidase), but the related drug ethionamide does not require activation. Suppose expression profiles were measured for the following:

(a) a strain with active KatG, not exposed to either drug,

(b) a strain with active KatG, exposed to isoniazid,

(c) a strain without active KatG, exposed to isoniazid,

(d) a strain with active KatG, exposed to ethionamide,

(e) a strain without active KatG, exposed to ethionamide.

Which two of (b), (c), (d), and (e) would show a similar pattern of enhancement of gene expression relative to (a)? Why whould you expect the enhancement pattern to be similar in: (b), (d) and (e) but not (c)?

6.4 J. Foote and G. Winter compared the dissociation constants of a natural mouse antilysozyme antibody (D1.3), an engineered 'humanized' antibody in which the antigen-binding site was grafted onto a human framework (Human-original) and several mutants of the 'humanized form', including Human-mutated. The antigen was hen egg white lysozyme.

Antibody	**Number of sequence differences to D1.3**	k_{on} $M^{-1}s^{-1}$	K_D
D1.3	0	1.4×10^{-6}	3.7×10^{-9}
Human-original	48	0.7×10^{-6}	260×10^{-9}
Human-mutated	44	1.3×10^{-6}	14×10^{-9}

(a) Calculate the 'off-rate' k_{off} for each antibody. (b) Which has the major effect on the dissociation constant: differences in 'on-rate' or differences in 'off-rate'?

6.5 Analyse the map of the London Underground by counting the number of connections made from each station in zone 1 (the central portion). Count connections *to* stations inside and outside zone 1, as long as they originate within zone 1. Count only one connection if two stations are connected by more than one line; in other words, for each station, the question is: How many other stations can be reached without passing through any intermediate stops?

(a) What is the maximum number of connections of any station?

(b) For each integer k from 1 to this maximum number, how many stations have k connections?

(c) Plot these data on a log-log plot. Does the relationship appear reasonably linear?

(d) If so, fit a straight line to the log-log plot and determine the exponent.

Results of network analysis of this sort are more significant if the data cover several orders of magnitude, but this is not possible for this example.

6.6 In the overall yeast transcriptional regulatory network the number of incoming connections to nodes follows an exponential distribution. That is, the probability P_k that a gene is controlled by k transcription regulators is given by $P_k = Ce^{-\alpha k}$, $k = 1, 2, \ldots$, with $\alpha = 0.8$.

(a) Determine the constant of proportionality C in terms of α, by summing the series $\sum_{k=1}^{\infty} Ce^{-\alpha k} = 1$.

(b) If $\alpha = 0.8$, what is the maximum value of k for which at least 1% of the nodes would be expected to have at least k incoming connections?

(c) If $\alpha = 0.8$, plot the expected histogram for $1 \leq k \leq 7$.

(d) Determine the mean value of k in terms of α. (Hint: in the solution of (a) you expressed $\sum_{k=1}^{\infty} e^{-\alpha k}$ as a function $f(\alpha)$. Differentiate this relationship with respect to α to produce the equation: $\sum_{k=1}^{\infty} - ke^{-\alpha k} = f'(\alpha)$. Then the mean value of k is given by $-f'(\alpha)/f(\alpha)$.)

(e) What is the mean value $\langle k \rangle$ corresponding to $\alpha = 0.8$?

(f) What is the median value of k? This is the value κ such that half the nodes have $\leq\kappa$ incoming connections, and half the nodes have $\geq\kappa$ incoming connections. Find κ in terms of α. (Hint: if $\sum_{k=1}^{\infty} Ce^{-\alpha k} = 1$, then $\sum_{k=\kappa+1}^{\infty} Ce^{-\alpha k} = \frac{1}{2}$. But $\sum_{k=\kappa+1}^{\infty} Ce^{-\alpha k} = e^{-\alpha\kappa}\sum_{k=\kappa+1}^{\infty} Ce^{-\alpha k}$. In general, this approach will provide a non-integral estimate of κ, just round this result to the nearest integer.)

(g) If $\alpha = 0.8$, what is the median value κ? How does it compare with the average value $\langle k \rangle$? Are the two values approximately equal?

6.7 Indicate how to connect a selection of the three common network control motifs so that a single input node can influence three output nodes.

Weblems

6.1 Define the following terms: (a) interactome (b) metabolome (c) signalome. (d) More difficult: can you think of, and define, a reasonable 'ome' that has not yet been proposed?

6.2 The catalase-peroxidase KatG of *M. tuberculosis* activates the drug isoniazid. The most common mutation in KatG associated with resistant strains is S315T. From a model of *M. tuberculosis* KatG based on the structure of the homologue from *Burkholderia pseudomallei* (PDB entry [1mwv]), suggest a mechanism for the reduced activity of the mutant to activate isoniazid while retaining activity against smaller substrates.

6.3 Find a fragment of the genealogy of the royal families of Europe containing a family tree that is not a graph.

6.4 One model for the growth of a scale-free network suggests that if new nodes and edges are added according to the rule that the probability of adding a new edge is proportional to the number of edges that a node already has, the network will remain scale-free and retain the same exponent. Using historical maps of the London Underground, check whether earlier networks are scale-free. Test whether addition of edges to the network has followed the rule that edges have been added preferentially to more highly-connected nodes.

6.5 From the *B. subtilis* protein interaction database SPiD (see Fig. 6.6), (a) what type of experimental evidence links DnaC to DnaG? (b) What type of experimental evidence links DnaC to DnaI?

6.6 From the *B. subtilis* protein interaction database SPiD (see Fig. 6.6), print a graph showing not only the immediate neighbours of DnaC, but the second neighbours (the neighbours of the neighbours).

Conclusions

How can we extrapolate from the current state of play to the bioinformatics of the future? Clearly, data collection will proceed and continue to accelerate. New high-throughput techniques will provide additional types of data, including information about the integration and control of life processes. Computing facilities of increasing power will be applied to the storage, distribution and analysis of the results. New databases will appear on the Web, and links between databases will become more effective. Improved algorithms will be devised to analyse and interpret the information given us and to transmute it from data to knowledge to wisdom.

One threshold will be reached when our knowledge of sequences and structures becomes more nearly complete, in the sense that a fairly dense subset of the available data from contemporary living forms has been collected. (Of course there is no question of being able to know everything.) This will be recognized operationally when a random dip into the pot of a genome, or the determination of a new protein structure, is far more likely to turn up something already known, rather than to uncover something new. Nature is, after all, a system of unlimited possibilities but finite choices.

Applications will become more feasible, and mature ever more quickly from 'blue-sky' research to standard industrial and clinical practice. Some of the higher levels of biological information transfer—such as the programmes of genetic development during the lifetime of individuals, and the activities of the human mind—will come to be included in the processes we can describe quantitatively and analyse at the level of molecules and their interactions.

In Michaelangelo's frescos on the ceiling of the Sistine Chapel, the serpent offering Eve the fruit of the tree of knowledge is represented with its legs coiled around the tree in the form of a double helix. We can hope that our new temptation to knowledge embodied in another double helix will have more fortunate consequences.

Answers to Exercises

Answers are provided to all exercises except those requiring annotation or drawing of diagrams.

1.1 (a) 5000 (b) 5×10^6 (c) Taking the population of the U.S.A. as 3×10^8, the storage requirements for all DNA sequences of all inhabitants of the U.S.A exceeds the storage requirements for EOS/DIS by a factor of 60.

1.2 (a) 20833 (b) 4.6 (c) 1

1.3 neurodegenerative disease, polyglutamine tract, huntingtin, anticipation, counselling

1.4 (a) Leu-Ala-His-Lys-Tyr

(b) . . . cta . . .

(c) . . . ccg . . . (Leu → Pro . . .)

(d) . . . aag → tag . . .

(e) . . . taa → tac

1.7 Choose (a) $E < 10^{-200}$ (b) E ~ 0.003

1.8 at least 15 bases long

1.9 (a) about 100000 human generations. (b) 100000 bacterial generations = 3.8 years

1.10 (a) isoleucine (b) glutamic acid (c) serine (d) arginine (e) lysine (f) phenylalanine

1.11 (a) down (b) down

1.12 two times

1.14 m

1.16 75%

1.17 change `([A-Z][a-z]+ [a-z]+)` to `([A-Z]([a-z]+|\.) [a-z]+)`

1.18 change `$species{$1} = 1;` to `$species{$1}++;`

1.19 CGCAAAAAAGCG or GCGTTTTTTCGC

2.1 almost 300000

2.2 If another enzyme took over the function of the product of the gene knocked out, there might be no effect on phenotype

2.3 all of them

2.4 moon landing

2.5 autosomal dominant

2.6 phenotype correlates with allele of retinoblastoma gene (RB1), not with allele of esterase D gene.

2.7 different expression patterns in different tissues

2.8 7.5×10^5 bp, 6 genes

2.9 risk no higher than if father normal (but see Schwartz, M. & Vissing, J. (2002). Paternal inheritance of mitochondrial DNA, New Eng. J. Med. 347, 576–80.)

2.10 search human genome sequence

2.11 exceptions: trpE and trpD, and trpB and trpA, are two pairs of genes contributing to one enzyme. trpD encodes [parts of] two enzymes.

2.12 inversion has occurred in the region around the centromere

2.13 The fluorescent regions would not disappear but appear elsewhere

2.14 average transfer rate: 5.24×10^{-5} ORFs/year, 5×10^{-5} kbp per year

2.15 (a) both, (b) living genome, (c) both, (d) both, (e) both, (f) computer databases.

3.1 bicycle, tricycle: human propulsion.
bicycle, motorcycle: number of wheels = 2.
bicycle, car: (number of wheels = 2 AND human propulsion) OR (number of wheels = 4 AND engine propulsion).
tricycle, motorcycle: (number of wheels = 3 AND human propulsion) OR (number of wheels = 2 AND engine propulsion).
tricycle, car: number of wheels > 2; motorcycle, car: engine propulsion

3.3
```
while(<>)}
  if (/\^DEFINITION/) \{s/^DEFINITION\s*//;print \"$_\";\}
  elsif (/^\s*\d+\s*(\[a-z +])\s*$/) {$seq = $1; $seq =~ s/ //g;
  print \"$1\n\";}
}
```

4.1 4

4.2 6

4.3 agtcc → cgtcc → cgtca → cgctca

4.4 (a) Off main diagonal, expect run of matches on word time and waste, parallel to main diagonal. (b) use PERL program on p. 162, use window = 4, thresh = 2.

4.5 window 2 threshold 2

4.6 PAM250: H↔R more probable, BLOSUM62: W↔F more probable.

4.7
```
THE.RETORT.COURTEOUS-
THE.REPLY-.CHURLI--SH
```

4.8 Set weights of all routes into and out of Uppsala to very large negative values.

4.9 Do a dotplot of one sequence against the reverse complement of the other.

4.10 replace line

```
$_ =~ /^(.*)\n\s*(\d+)\s+(\d+)\s*\n(.*)\n([A-Za-z\n]*)
    \*\s*\n(.*)\     n([A-Za-z\n]*)\*/;
```

with

```
$_ =~ /^(.*)\n\s*(\d+)\s+(\d+)\s*\n(^\s*\>.*)\
    n([A-Za-z\n]*)\n(^\s*\>.*)\ n([A-Za-z\n]*)\*/;
```

4.11 0.16

4.12 (a) more similar (b) more similar (c) *just* more similar (d) less similar

4.15 4000000

4.16 65 = Min(40 + 25, 45 + 20, 50 + 25) where 40 + 25 corresponds to the vertical move, 45 + 20 to the diagonal move (mismatch) and 50 + 25 to the horizontal move. The two arrows appear because two possible predecessors give the same value.

4.18 30, 60, 70, 84

4.19 (a) VDFSAT (b) score of VDFSAT = −1533 score of VDFSAE = −1503

Scores computed by adding up six terms, the first of which is:
inventory score of V $\times \sum_{i=1}^{20}$ *BLOSUM62*(V, amino acid *i*)

4.20 (a)

Residue									number of											
number	**A**	**C**	**D**	**E**	**F**	**G**	**H**	**I**	**K**	**L**	**M**	**N**	**P**	**Q**	**R**	**S**	**T**	**V**	**W**	**Y**
90					8			2	2	1								3		
91	1			1				2						1		7		4		
92						16														
93	15	1																		
94			2						2	2		6				1		3		
95				4					5				3			2		1		

(b) 32 (c) 49

4.21 (a) yes, (b) no

4.22 9: a(bc), a(bc), a(cb), . . .

4.23 The reduced distance matrix after combining ATCC and ATGC is shown in the text. The next step is to combine TTCG and TCGG. In the further-reduced distance matrix following that combination, the distance from {ATCC, ATGC} to {TTCG, TCGG} = 3. Therefore the distances from the root to {ATCC, ATGC} or to {TTCG, TCGG} = $\frac{1}{2} \times 3 = 1.5$.

4.24 distance between ATCC and ATGC = 1 in both original and as sum of edges along shortest path through tree.

distance between ATCC and TTCG = 2 in original and 4.5 as sum of edges along shortest path through tree.

distance between ATCC and TCGG = 4 in original and 4.5 as sum of edges along shortest path through tree.
distance between ATGC and TTCG = 3 in original and 4.5 as sum of edges along shortest path through tree.
distance between ATGC and TCGG = 3 in original and 4.5 as sum of edges along shortest path through tree.
distance between TTCG and TCGG = 2 in both original and as sum of edges along shortest path through tree.

4.25 three nodes, all pairs connected

4.26 sheep

4.28 (a) as n^2 (b) as n.

5.1 25.5 kJ/hydrogen bond

5.2 (a) orthologue (b) paralogue (c) paralogue (d) paralogue (e) paralogue (f) neither orthologue nor paralogue

5.5 b

5.6 (a) The third through the sixth strands from the right in the template, and the helices that connect them, have the same topology as the second through the fifth strands from the right in the target. The rightmost strand in the target corresponds to a helix and the second from rightmost strand in the template (which points in the opposite direction from the rightmost strand in the target). The rightmost strand in the template does not correspond to a strand in the target. An extra $\beta - \alpha - \beta$ unit appears in the target, at the left of the sheet. (b) The six strands in Murzin's prediction have the correct topology as the target; the leftmost (N-terminal) strand in the target is missing from the prediction. (c) The connectivity of the rightmost strand are different; the prediction is like the target structure and the parent is not. The direction of the sixth strand from the right is different between the template and the target.

5.8 You would search for homologous proteins that had different (and ideally separable) modes of determining specificity. Even better, you would look for proteins carrying out functions essential to the pathogen that had NO human homologue.

5.9 24

5.10 It selects points above and to the right of the line $x + 2y = 2$. (This line intersects the x-axis at $x = 2$ and intersects the y-axis at $y = 1$.)

5.11 Sketch looks like top left diagram on p. 245. Geometric interpretation: selects points below and to the right of the line y = x.

6.1 (a) yes (b) yes (c) no

6.3 (a) RVVWLL (b) y5

6.4 (a) 22 (b) A↔C (c) yes (d) 23 (e) A↔C (f) yes

6.5 about 400 distinct sequences if recombination occurred with equal probability everywhere. 0.75%

6.6 (a) phage display (b) surface plasmon resonance

6.7 (a) V_1–V_3

(b) V_1–V_3

(c) V_5–V_6

(d) V_1–V_4

(e) V_4–V_3

6.8 (a) phylogenetic trees, parts of metabolic pathways, citation patterns, the World Wide Web. (b) metabolic pathways, chemical bonding patterns (if single/double/triple bonds are distinguished).

(c, d)

Graph	Nodes	Edges
Sets of people who have met each other	people	relationships
Electricity distribution systems	power stations	cables
Phylogenetic trees	species (or taxa)	lines of descent
Metabolic pathways	metabolites	enzyme-catalyzed reactions
Chemical bonding patterns in molecules	atoms	bonds
Citation patterns in the scientific literature	articles	references
The World Wide Web	sites	references

6.9 (a) 10 (b) 17 (c) 178.5 kJ

6.10 1/21

6.11 (a) Moorgate → Bank → Waterloo → Embankment

(b) King's Cross → Russell Square → Holborn → Tottenham Court Road → Oxford Circus → Euston Square → King's Cross

(c) 2/15

6.12 (a) Amersham - Upminster

(b) Stations between Upney and Upminster, stations between West Brompton and Wimbledon, Chiswick Park (allowing access by National Rail on Richmond spur).

(c) Stations between Stanmore and Baker Street (except for Wembley Park, West Hampstead - accessible by National Rail - and Finchley Road, Southwark, Bermondsey.

6.15 (a) closed (b) β-sheet

6.16 0.201

Glossary

B-factor a measure of the precision to which an atom position is determined in X-ray crystallography

binomial identifier of a taxonomic unit in terms of the genus and species

bootstrapping a protocol for validating a method in which data sets are sampled randomly from a corpus of data, allowing multiple copies of the same data set

character string a linear array of text, each item chosen from a fixed alphabet

connected graph a graph in which there is at least one path between every two nodes

contig map a chart of the order of appearance in a full DNA molecule of small sequenced fragments

controlled vocabulary fixed and carefully-defined set of terms, to which feature tables in database annotations are to be restricted

convergent evolution appearance of similar features not shared by a common parent, arising independently rather than by descent from a common ancestor

differential genomics and proteomics comparisons of sequences or expression patterns between individuals; for instance, healthy and diseased humans or animals, or drug-susceptible and resistant infectious microorganisms

docking prediction of protein (or nucleic acid)-ligand interactions

domain a compact structural subunit of a protein

drug target a protein, the function of which would be useful to modify to alleviate a symptom of disease

ecosystem a coherent localized grouping of living things interacting with one another and with their environment

edge length a number assigned to each edge of a graph signifying in some sense the distance between the nodes connected by the edge

Electrospray Ionization (ESI) a non-destructive method for vaporizing and ionizing proteins in which the sample is sprayed through a small capillary with an electric field at its tip

eukaryote an organism whose cells contain a nucleus

exon an expressed region of a split eukaryotic gene

expression chip a microarray containing oligonucleotides derived from cDNAs

fluorescent *in situ* hybridization identification of the position of a sequence on a chromosome by tagging a complementary oligonucleotide with a fluorescent marker

fold recognition attempt to decide which (if any) of a given library of folding patterns is adopted by a polypeptide sequence

genetic drift change in allelic frequencies within a population by mechanisms other than natural selection

gene duplication a type of mutation in which two or more copies of a gene appear corresponding to a single copy in the parent

gene flow redistribution of allele frequencies within or between populations

genomic hybridization microarray technique to look for gains or losses of genes or of copy number

genotype the hereditary information contained in the nucleotide sequence of the genetic material in the cells of an organism

graph an abstract structure containing nodes (points) and edges (lines connecting points)

Hamming distance a measure of dissimilarity between two character strings of the same length, equal to the number of mismatches in characters at corresponding positions

homologous descended from a common ancestor

homology modelling prediction of the three-dimensional structure adopted by a polypeptide chain on the basis of the known structure adopted by a similar sequence

hydrophobicity profile graph of hydrophobicity of successive residues in a protein, giving clues to positions of turns, or buried/exposed regions

hydrophobic effect unfavourable interaction of non-polar solutes with water accounting for low solubility

intron an intervening, non-expressed, region of a split eukaryotic gene

jackknifing a protocol for validating a method in which data sets are sampled randomly from a corpus of data

Levenstein distance a measure of dissimilarity between two character strings, equal to the minimum number of edit operations (substitution, insertion, deletion) required to convert one sequence into the other

LINE long interspersed nuclear element

macroevolution large-scale changes in genomes within populations

MATRIX-ASSISTED Laser Desorption Ionization (MALDI) a non-destructive method for vaporizing proteins in which the sample is embedded in surroundings (the matrix) that initially absorbs the energy from a laser pulse

microevolution small-scale gene changes within a population, some not even affecting phenotype

modular proteins proteins composed of a linear concatenation of domains

multiple sequence alignment assignment of mutual residue correspondences in three or more related sequences

mutation inheritable change in the nucleotide sequence of genetic material

mutation microarray analysis microarray set up to search for single-nucleotide polymorphisms (SNPs)

natural selection mechanism for changing the distribution of characteristics of a population, based on differential reproduction according to fitness

pairwise sequence alignment assignment of residue correspondences in two related sequences

PAM (percent accepted mutation) a measure of dissimilary between two sequences

path a consecutive set of edges in a graph

path length the sum of the lengths of the edges in a path

Quantitative Structure-Activity Relationships (QSAR) correlations between structural features of molecules and their reactivities or pharmacological effects

quorum sensing intercellular communication in a population of bacteria in which a secreted messenger molecule induces a physiological switch when its concentration exceeds a threshold value

pharmacophore substructure common to different molecules that share a pharmacological activity

phenotype the features of an organism, including macroscopic and biochemical, exclusive of the nucleotide sequence of the genetic material in the cells of the organism

population a group of interacting organisms from the same species

post-translational modification chemical change in a polypeptide chain after synthesis on the ribosome

primary structure the set of chemical bonds in a protein or nucleic acid

prokaryote an organism the cells of which do not contain a nucleus

quaternary structure the composition and assembly of subunits

recombination a rearrangement of genetic material by exchange of portions of chromosomes

reduction of dimensionality simplification of complex data set by extraction of a relatively small number of quantities that account for all or a large part of the variation

regular expression pattern within text, including specific characters or classes of characters, searchable for by computer programs

restriction enzyme an enzyme that cleaves DNA at a specific base sequence

restriction map a chart of the positions of cleavage sites for one or more restriction enzymes

reverse genetics identification of the gene sequence corresponding to a phenotypic character

RFLP (restriction fragment length polymorphism) distribution of sizes of fragments produced by cleaving DNA with a restriction enzyme

secondary structure interresidue hydrogen-bonded standard structures in a protein or nucleic acid

secondary structure prediction an attempt to compute from the amino acid sequence which regions of a polypeptide chain form α-helices and strands of β-sheet, without knowing the three-dimensional structure; or an analogous prediction of base-pairing regions of single-stranded RNA

sequence tagged site short sequenced region of DNA, typically 200–600 base pairs long, appearing in a unique and mapped position in a genome

similarity resemblance not necessarily derived from homology

SINE short interspersed nuclear element

SNP (single nucleotide polymorphism) one or more possible bases appearing at a given position of a genome

somatic cell hybrids rodent cells containing one or a few human chromosomes or fragments of human chromosomes

STRP (short tandem repeat polymorphism) or microsatellite short repeated unit of only 2–5 base pairs but appearing as many consecutive copies

substitution matrix a table specifying the relative expected probability of replacement of any amino acid (or nucleotide) by another

supersecondary structure combinations of secondary structure elements that are consecutive in the sequence and interacting in space

tertiary structure the spatial arrangment of the polymer chain of a protein or nucleic acid

tree a connected graph with exactly one path between every two points

turns regions in which a polypeptide (or polynucleotide) chain changes direction in space

twilight zone region from 18%–25% sequence identity, in which it is difficult to decide on the basis of sequence alone whether two proteins are related or not

VNTR (variable number tandem repeat) or minisatellite short repeated units about 10–100 base pairs long

window the size of each successive consecutive region in a sequence considered at one time

Z-score a measure of the unexpectedness of a result in a distribution of possible results; Z-score = (score – mean of distribution)/standard deviation of distribution

Index

A

accessible surface area 323
algorithm 16
alignment 26, *plate III*
alignment *plate IV*
alleles 5
analgesic drugs 277, 281
analysis of algorithms 16
anthropology 103
apoptosis 320
archaea 22
artificial chromosome 77
aspirin 283

B

B-factor 133
bacteria 22
bacterial artificial chromosome 77
banding patterns of chromosomes 72
Berardinelli-Seip syndrome 79
biological classification 21
BioMagResBank 131
BLOCKS database 173
BLOSUM matrices 173
bootstrapping 209
BRCA1 152, 299, *plate VIII*

C

Cambridge Crystallographic Data Centre 131
CATH (Class/Architecture/Topology/Homologous superfamily) 135
cattle domestication 104
chaos 319
chemical cross-linking 325
chemoinformatics 274
chloroplast DNA 88
chromatin immunoprecipitation 325
chromosome banding pattern 89, 114
cladistic methods 206
clinical trials 270
CLUSTAL-W 62, 149
clustering 199, 205
  hierarchical 199
coexpression pattern 326
coimmunoprecipitation 325
collision-induced dissociation (CID) 304
comparative genomics 110
conformational energy calculations 255
contiguous clone map (contig) 78
controlled vocabulary 119
convergent evolution 21
core of homologous proteins 237
Critical Assessment of Fully-Automatic Structure Prediction (CAFASP) 242
Critical Assessment of Structure Prediction (CASP) 52, 242, 287
cystic fibrosis 75

D

DALI (Distance-matrix ALIgnment) 135, 235
data structures 16
databanks 9
database annotation 13
database indexing 118
database searching 11
differential genomics 57, 273
divergence 320
divergence of function 266
divergence of sequence 237
divergence of structure 237
DNA methylation 70
DNA sequence databases 122
DNA structure 6, 8, *plate I*
docking 276
domain 42, 111, *plate VI*
domain recombination network 326
dotplot 160
  and alignment 165, 210
drug discovery 269
dynamic programming 176

E

E-value 183
edge (of a graph) 204
edge length 204
elastase *plate IV*
electron microscopy 338
electronic publication 15
Electrospray ionization (ESI) 303
Ensembl 151
ENTREZ 124, 141
Enzyme Commission 51
ENZYME DB 124

equilibrium 319
eukarya 22
EVA 249
'Eve', mitochondrial 60
evolution 4
Expert Protein Analysis System (ExPASy) 150
expressed sequence tag (EST) 136, 301
expression profile 136, 293, 309, *plates VII, VIII, IX*
eyeless gene, in Drosophila 34

F

FASTA format 25
feature table 123
FISH (Fluorescent *in situ* hybridization) 77, 113, *plate II*
flavodoxin *plate V*
fluorescence resonance energy transfer 326
fold recognition 51, 242, 252
fork 335

G

gap weighting 174, 181
GenBank 1
gene duplication 5, 91, 111
gene expression table 295
gene flow 5
gene maps 73
gene therapy 57
genetic code 7
genetic drift 5
genome 68
  databases 124
  evolution 105
  sizes 67, 68
genotype 4
global alignment 26, 175
graph 204, 315

H

haemoglobin gene cluster 91
Hamming distance 171, 210
haplotype 307
helical wheel 231
helix 42, 129, 224
heteroplasmy 104
heterozygote 5
Hidden Markov Model (HMM) 188, 193
HIV protease database 135
homology 21, 29, 197
homology modelling 51, 242, 250
homoplasmy 104
homozygote 5
horizontal gene transfer 108, 113
HOX genes 218
Huntington disease 56
hydrophobic effect 226
hydrophobicity profile 229

I

information retrieval 16
interaction domain 336
interaction network 212, *plate X*
International Human Genome Sequencing Consortium 97
International Immunogenetics database (IGMT) 136
isoniazid 309, *plate IX*

J

jackknifing 209

K

knockout 268
Kyoto Encyclopedia of Genes and Genomes (KEGG) 138

L

lead compound 271, 274
leukaemia *plate II*
Levenshtein distance 171, 210
LINE (long interspersed nuclear element) 30
linkage maps 72
LINUS 259
local alignment 26, 175
  Smith-Waterman algorithm 175
London Underground 314

M

Macromolecular Structure Database (MSD) 50
mass spectrometry 54, 71, 301
  nucleic acid sequencing 306
  protein sequencing 304
Matrix-assisted laser desorption ionization-Time of flight (MALDI-TOF), mass spectrometry 303
maximum likelihood 207
maximum parsimony 206
MEDLINE 139
metabolic pathway database 138
metazoa, phylogenetic tree of 23
Michaelis constant 322
microarray 53, 293
minimal genome 107
mitochondrial DNA 88
mitochondrial 'Eve' 60
MODBASE 241
modular protein 44, 111
molecular dynamics 255
molecular graphics 141, 222
Molecular Modelling Database (MMDB) 50
molecular modelling, in drug discovery 275
Monte Carlo algorithm 261
multiple sequence alignment 186
mutation 5

N

native state of proteins 8
natural selection 5
networks 313
  small-world 317
neural networks 246
nomenclature 21
nonsynonymous substitutions (Ka) 105
Nuclear Magnetic Resonance (NMR) 134, 324
Nucleic Acid Database (NDB) 131
nucleic acid sequence databases 122

O

OCA 133
'one-two punch' 335
Online Mendelian Inheritance in Man
    (OMIM[TM]) 142, 147
open reading frame (ORF) 80
organelle 88
orthologues 266

P

P-score 183
PAM250 matrix 173
paralogues 266
path in a graph 204
path length 204
PAX-6 genes 33
percent accepted mutation (PAM) 172
periodic variation 319
PERL 16
personal identification 104
phage dispay 325
pharmacogenomics 56
pharmacophore 235, 275
phenotype 4
Phobius 196
phylogenetic distribution pattern 326
phylogenetic relationship 24, 31
phylogenetic tree 203
phylogeny 198
polyacrylamide gel electrophoresis (PAGE) 71
polyglutamine repeats 56
population 5
position-specific scoring matrix 191
positional cloning 76
post-translational modification 72
prenatal diagnosis 308
primary structure of protein 41
primosome 327
principal component analysis (PCA) 298
profiles 188
  3D 252
PROSITE 124
protein complexes 320
Protein Data Bank (PDB) 1, 50, 128
  worldwide (wwPDB) 128
protein denaturation 225
protein engineering 51
protein folding 8, 221, 228
protein folding pattern 41
protein function prediction 265
Protein Identification Resource (PIR) 124, 149
protein interaction network 324
protein sequence databases 124
protein stability 225
protein structure 40
  classification 44, 135, 238
  evolution 236
  prediction 51, 240
protein-protein interface 323
proteome 52, 54, 68, 71, 136
pseudogene 90, 100
PSI-BLAST 32, 33, 185, 188, 191
PubMed 139
pyruvate decarboxylase, *plate VI*

Q

quantitative structure-activity
    relationships (QSAR) 272
quaternary structure of protein 41
quorum sensing 313

R

radiation hybrids 77
Ramachandran plot (see Sasisekharan-
    Ramakrishnan-Ramachandran plot) 223
recombination 5
reduction of dimensionality 297
regulatory networks 327
resolution, of X-ray structure determination 134
retinoblastoma 112
reverse genetics 74, 76
root, of a tree 201
root-mean-square (r.m.s.) deviation 233
Rosetta 259
runaway 320

S

Sasisekharan-Ramakrishnan-
    Ramachandran plot 223
scatter 335
SCOP (Structural Classification of Proteins)
    50, 135, 239, 289, *plate V*
scoring schemes, in sequence alignment 171
secondary structure 41, 129
  prediction 51, 242, 244, 287
selectivity and sensitivity contrasted 32
sequence alignment 158, 26
  multiple 159
  pairwise 159
  scoring schemes in 171
sequence databanks 1, 9
sequence logo *plate III*

Sequence Retrieval System (SRS) 148
sequence searching 32
sequence tagged site (STS) 80
sheet 42, 129
short tandem repeat polymorphism (STRP) 78
sidechains 225
similarity 29
simulated annealing 262
SINE (short interspersed nuclear element) 30
single-nucleotide polymorphism (SNP) 101
singular value decomposition 298
small-world networks 317
software engineering 16
somatic cell hybrids 77
steady state 319
strand 224
structural alignment 233
structural genomics 241
structural superposition 233
structure databanks 1
Structure Explorer 132
substitution matrix 172
superposition 233
supersecondary structure of protein 42
surface plasmon resonance 325
SWISS-MODEL 251
SWISS-PROT 124
synonymous substitutions (Ks) 105
systems biology 9, 54, 311

T

target selection 273
TATA box 81
taxonomy 21
tertiary structure of protein 41
thioredoxin *plate III*
threading (method of fold recognition) 254
time scale of Earth history 199
transition 172
transmembrane helix prediction 193, 263
transversion 172
tree 204
tree of life 22
trp operon 82
tuberculosis 308
tumour suppressor gene *plate II*
twilight zone 185
two-hybrid screening 324
typhoid Mary 317

U

UniProt 124
unweighted pair group method with
arithmetic mean (UPGMA) 206

V

variable number tandem repeat (VNTR) 78

W

wheat 88
world-wide-web 14
worldwide Protein Data Bank (wwPDB) 128

X

X-ray structure determination 324

Y

yeast artificial chromosome 77

Z

Z-score 183

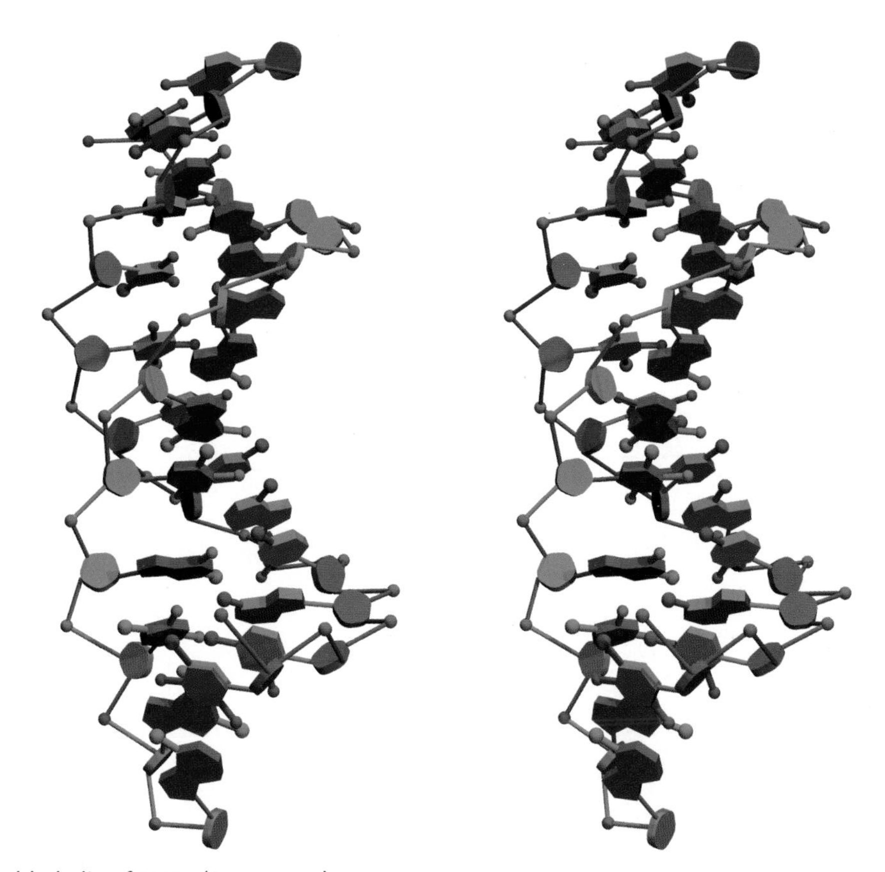

Plate I Double-helix of DNA. (See page 6.)

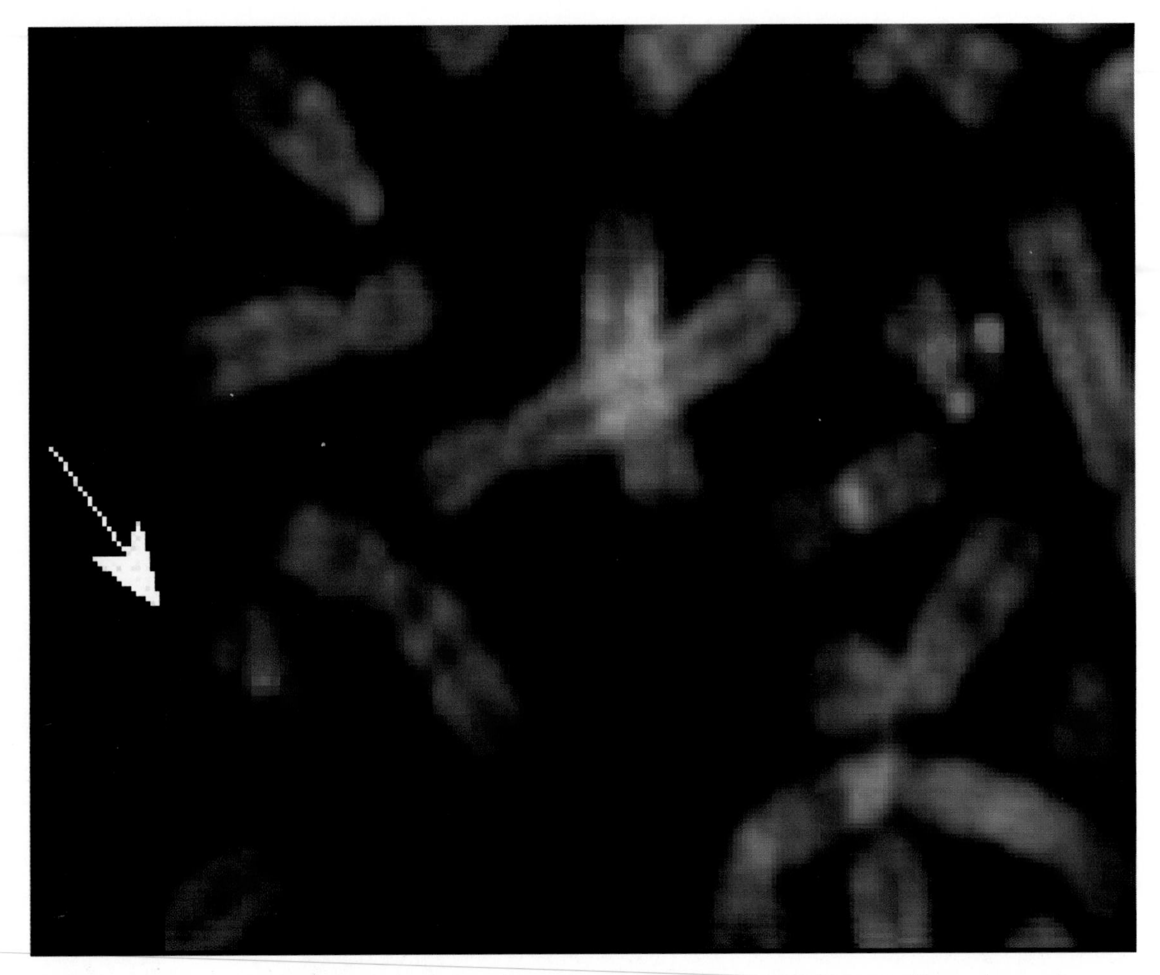

Plate II FISH (Fluorescent *in situ* hybridization) can detect the presence of locus-specific probes and visualize their chromosomal positions. Shown in red is a probe for the centromeric region of chromosome 20, to identify the two homologous copies of the chromosome that appear at metaphase. Shown in green is a probe for D20S108, within the region 20q11.2–13.1, which is present in one copy of chromosome 20 and deleted from the other (see arrow). This cell came from a patient suffering from polycythaemia rubra vera, (an abnormal increase in blood cells, primarily erythrocytes, arising from abnormality in bone marrow). The region deleted in the long arm of chromosome 20 is believed to contain tumour suppressor gene(s), the loss of which contributes to the development of leukaemias. (See page 75.) (Courtesy of Dr E. Nacheva, Department of Academic Haematology, Royal Free & University College London School of Medicine, London.)

Mammalian elastases

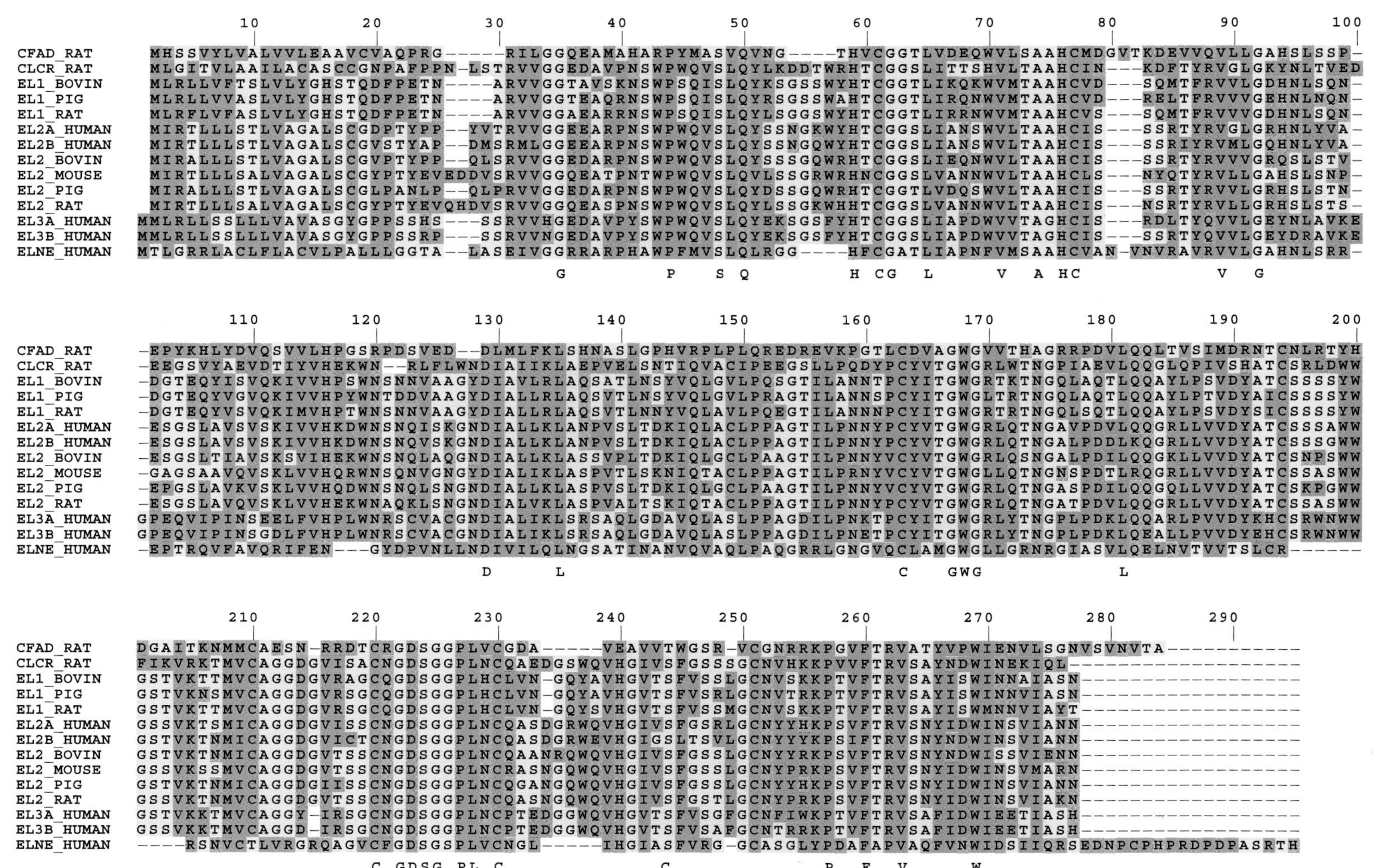

Plate III Alignment of amino-acid sequences of mammalian elastases. (See pages 148–149.)

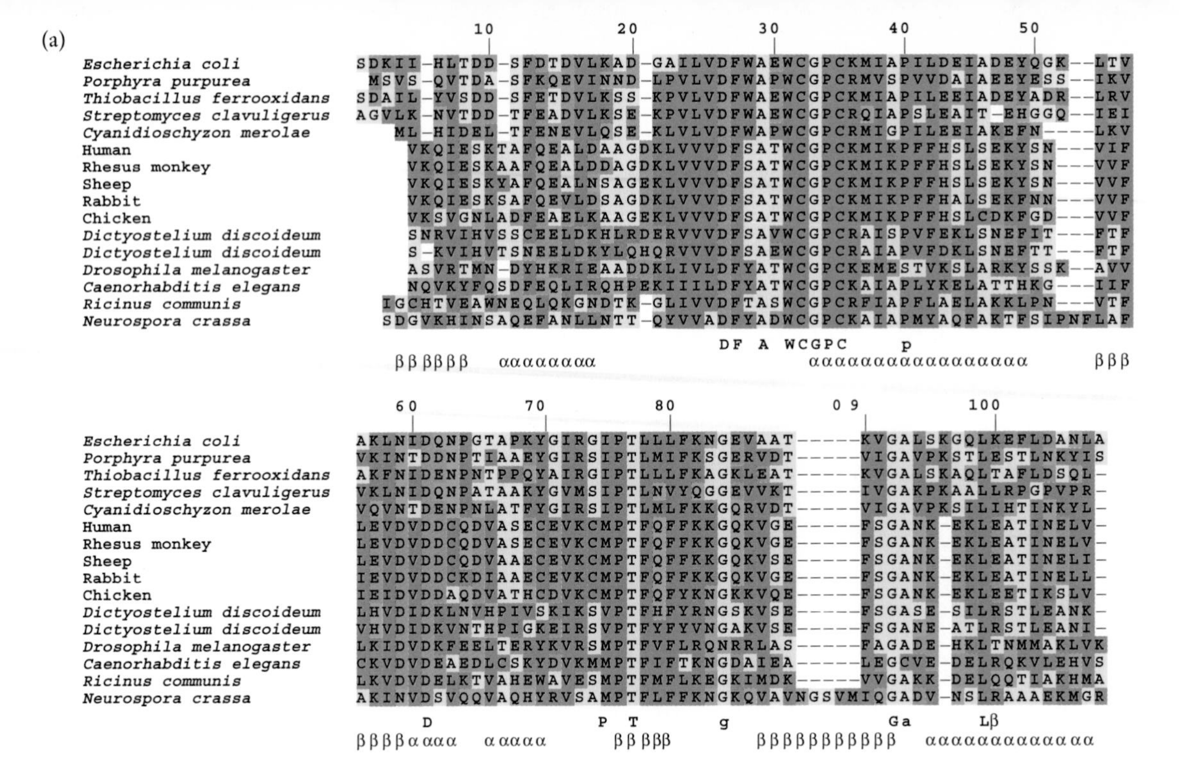

(b) **Thioredoxin**

(c)

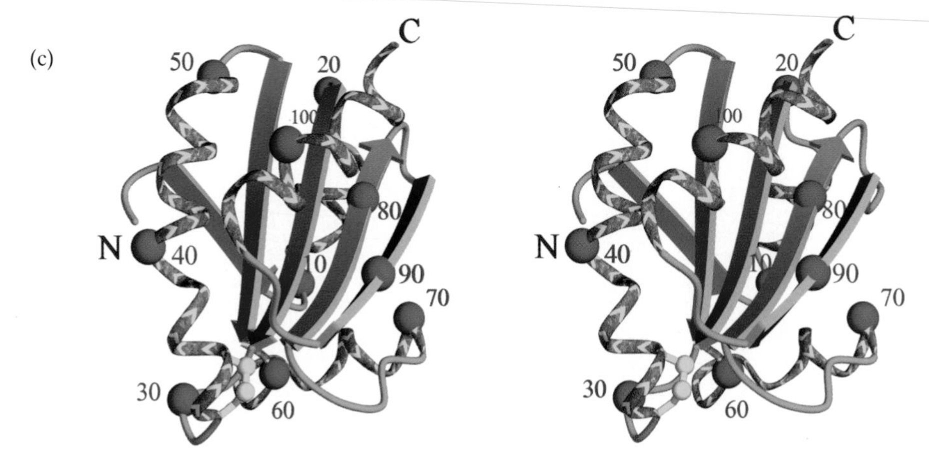

Plate IV (a) Alignment of amino acid sequences of *E. coli* thioredoxin and homologues. Some of the sequences have been trimmed at their termini. Residue numbers in this table correspond to positions in the *E. coli* sequence (top line). α-helix and β-strand assignments for *E. coli* thioredoxin are from PDB entry 2TRX. (b) Sequence logo derived from this multiple sequence alignment. (c) The structure of *E. coli* thioredoxin [2TRX] contains a central five-stranded β-sheet flanked on either side by α-helices. Residue numbers correspond to those in the multiple sequence alignment table. The N- and C-termini are also marked. Spheres indicate positions of the Cα atoms of every tenth residue. The reactive disulphide bridge between Cys32 and Cys35 appears in yellow. (See page 187.)

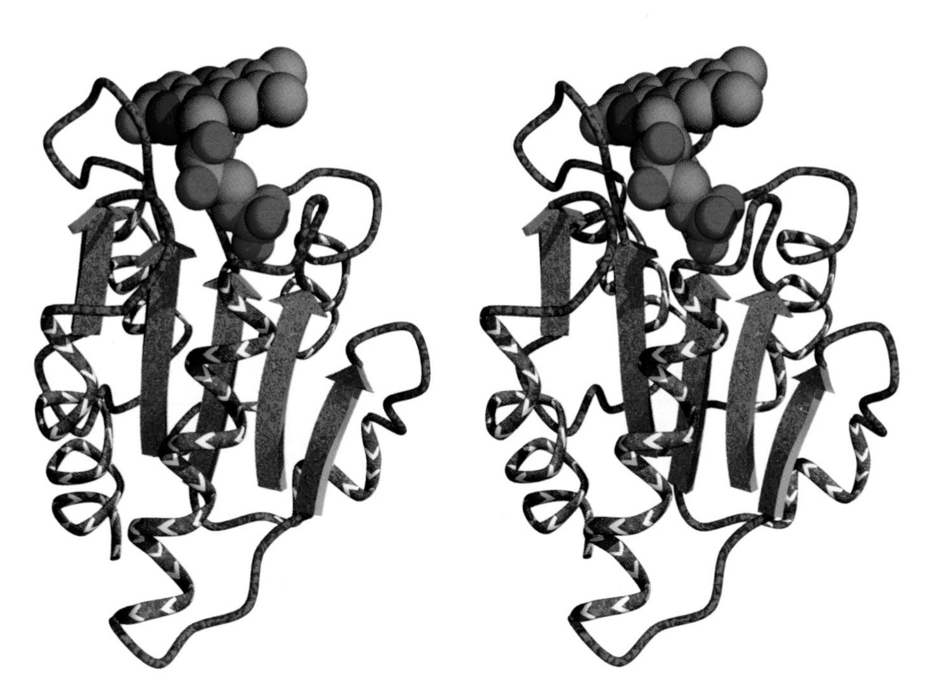

Plate V Flavodoxin from *Clostridium beijerinckii*, binding cofactor FMN [5NLL]. Large arrows represent strands of sheet. Placement of this structure in a hierarchical classification of protein structures according to the SCOP database is described on page 239.

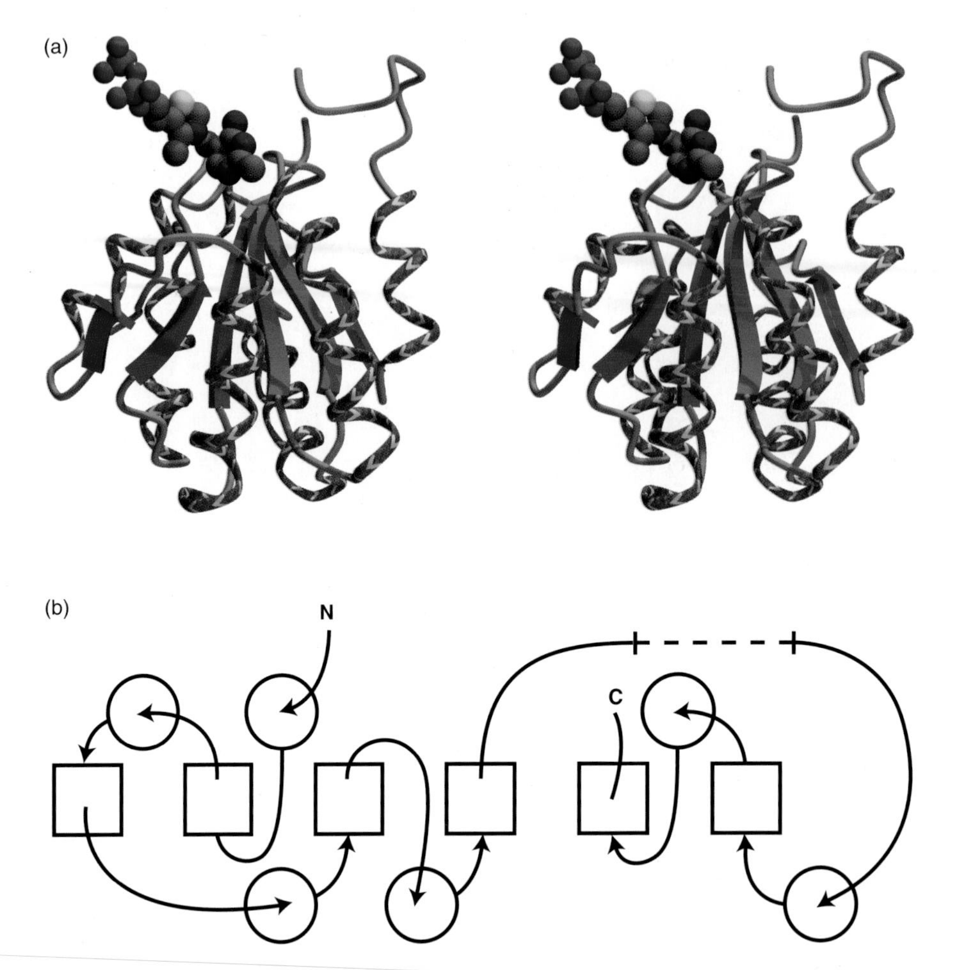

Plate VI The thiamin-binding domain from Yeast pyruvate decarboxylase. Thiamin-binding domains, identified by M. Gerstein as one of the five most common folding patterns, have been found in archaea, bacteria and eukarya. (a) Three-dimensional structure. (b) Schematic topology diagram. (See page 265.)

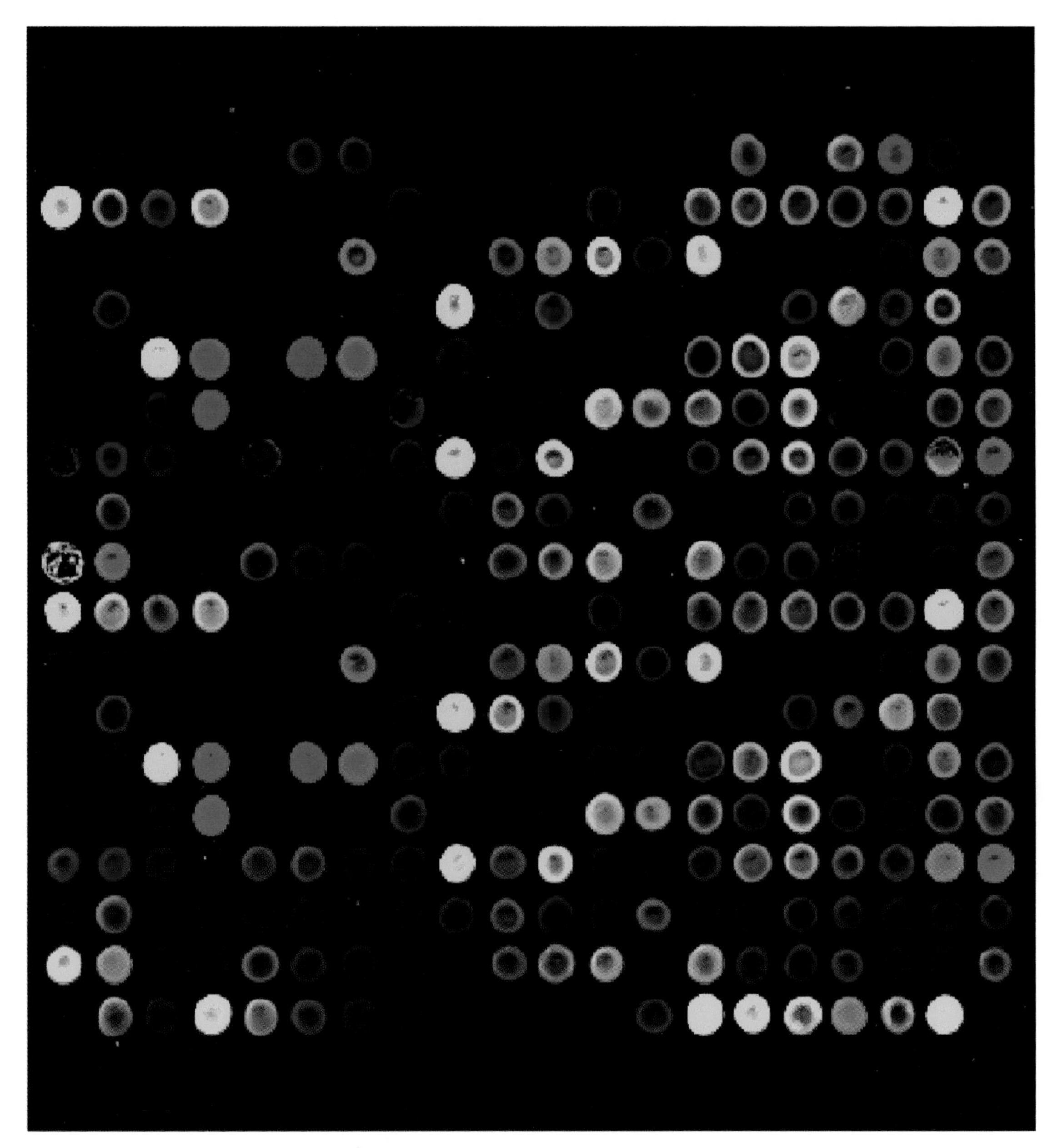

Plate VII Comparison of gene expression patterns in liver (red) and brain (green). The liver RNA is tagged with a red fluorophore, the brain RNA with a green one, then both are exposed to the array. Red spots correspond to genes active in the liver but not in the brain. Green spots correspond to genes active in the brain but not in the liver. Yellow spots correspond to genes active in both brain and liver. (See page 294.) (Courtesy of Dr. P. A. Lyons.)

Plate VIII Clustering of gene expression data in cells expressing high and low levels of BRCA1. BRCA1^{-} experiments, with low expression levels of BRCA1, appear in the left-hand six columns. BRCA1^{+} experiments, with high expression levels of BRCA1, appear in the right-hand six columns. The intensity of the colour reflects the ratio of the expression to that of a control. Red reflects genes with higher expression levels in response to BRCA1. Green reflects genes with lower expression levels in response to BRCA1. From: Welcsh, P. L., Lee, M. K., Gonzales-Hernandez, R. M., Black, D. J., Mahadevappa, M., Swisher, E. M., Warrington, J. A. & King, M.-C. (2002). BRCA1 transcriptionally regulates genes involved in breast tumourigenesis. *Proc. Natl. Acad. Sci. USA*, **99**, 7560–7566. Reproduced by permission. (See page 300.)

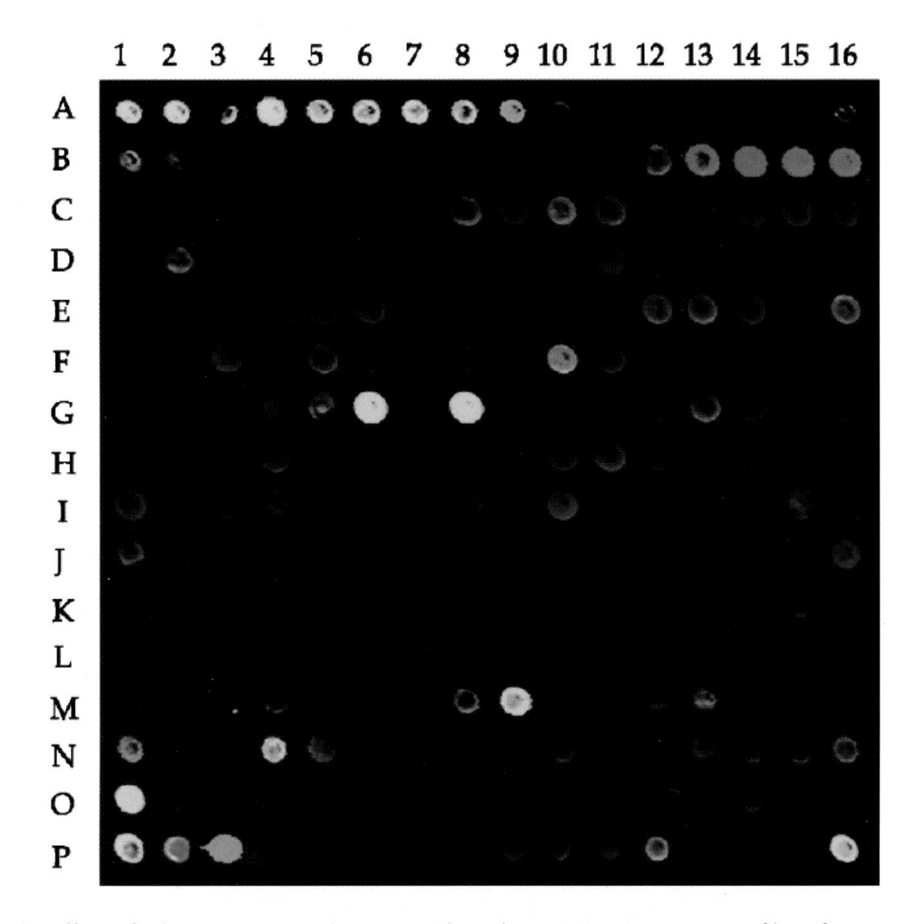

Plate IX The effect of 4-hour treatment by isoniazid on the mRNA expression profiles of 203 ORFs from *M. tuberculosis*. Red = expressed in cells treated with isoniazid, Green = expressed in untreated cells, Yellow = expressed in both treated and untreated cells. The row of red spots at the upper right corresponds to genes of the FAS-II gene cluster. (From: Wilson M., DeRisi, J., Kristensen, H. H., Imboden, P., Rane, S., Brown, P. O. & Shoolnik, G. K., Exploring drug-induced alterations in gene expression in *Mycobacterium tuberculosis* by microarray hybridization, *Proc. Natl. Acad. Sci. USA*, **96**, 12833–12838. Reproduced by permission.) (See pages 308–309.)

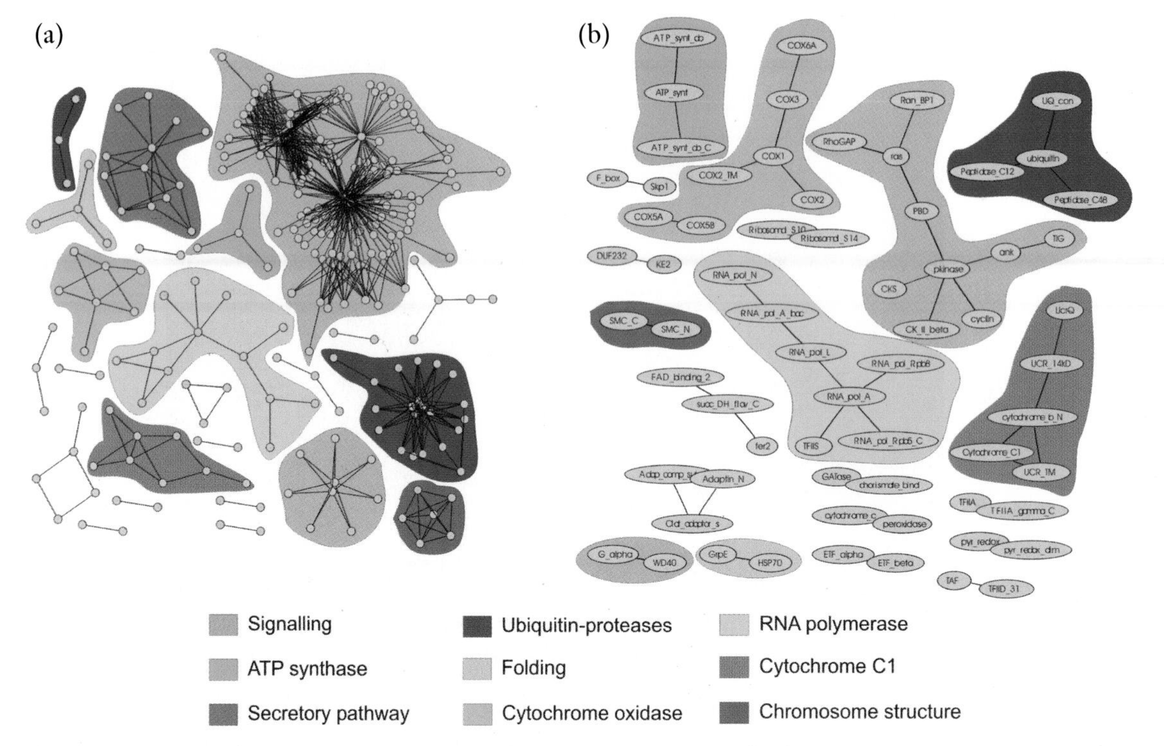

Plate X A portion of an interaction network of yeast proteins. Part A (left) describes the interactions of individual proteins, and part B (right) shows the interactions within a subnetwork based on representations of different protein families, in different functional catgories, linked in part A. This figure is based on structural data and modelling. Each relationship implies a physical interaction between the proteins. Some of the interactions involve stable complexes (for instance, RNA polymerase II); others involve transient complexes. (Picture courtesy of P. Aloy & R. B. Russell.) (See page 327.)